Polylactic Acid-Based Nanocellulose and Cellulose Composites

Polylactic Acid-Based Nanocellulose and Cellulose Composites

Edited by

Jyotishkumar Parameswaranpillai, Suchart Siengchin,
Nisa V. Salim, Jinu Jacob George,
and Aiswarya Poulose

CRC Press
Taylor & Francis Group
Boca Raton London New York

CRC Press is an imprint of the
Taylor & Francis Group, an **informa** business

First edition published 2022
by CRC Press
6000 Broken Sound Parkway NW, Suite 300, Boca Raton, FL 33487-2742

and by CRC Press
4 Park Square, Milton Park, Abingdon, Oxon, OX14 4RN

Library of Congress Cataloging-in-Publication Data

Names: Parameswaranpillai, Jyotishkumar, editor.
Title: Polylactic acid-based nanocellulose and cellulose composites / edited by
    Jyotishkumar Parameswaranpillai, Suchart Siengchin, Nisa V. Salim, Jinu Jacob
    George, and Aiswarya Poulose.
Description: First edition. | Boca Raton, FL : CRC Press, 2022. | Includes bibliographical references
    and index. | Summary: "Polylactic Acid-Based Nanocellulose and Cellulose Composites offers a
    comprehensive account of methods for the synthesis, characterization, processing, and
    applications of these advanced materials. It fills a gap in the literature as the only available
    book on this topic. Describes procedures for the extraction of cellulose materials from different
    sources and characterization methods adopted in analyzing their properties. Covers properties,
    processing, and applications of PLA biocomposites made using the extracted cellulose, Discusses
    the effect of reinforcement of cellulose in the biopolymer matrix and enhancement of properties.
    Examines current status, challenges, and future outlook. The book serves as a reference for
    researchers, scientists, and advanced students in polymer science and engineering and materials
    science interested in cellulose polymer composites and their applications"— Provided by publisher.
Identifiers: LCCN 2021053913 (print) | LCCN 2021053914 (ebook) | ISBN 9780367749521 (hbk) | ISBN
    9780367749538 (pbk) | ISBN 9781003160458 (ebk)
Subjects: LCSH: Polymeric composites—Materials. | Polylactic acid. | Cellulose.
Classification: LCC TA418.9.C6 P6355 2022 (print) | LCC TA418.9.C6 (ebook) | DDC 620.1/18—dc23/
    eng/20211227
LC record available at https://lccn.loc.gov/2021053913
LC ebook record available at https://lccn.loc.gov/2021053914

ISBN: 978-0-367-74952-1 (hbk)
ISBN: 978-0-367-74953-8 (pbk)
ISBN: 978-1-003-16045-8 (ebk)

DOI: 10.1201/9781003160458

Typeset in Times
by KnowledgeWorks Global Ltd.

Contents

Preface

Biopolymers are in increasingly high demand due to their biological degradation and environmental friendliness. Polylactic acid is one of the important biopolymers that can be synthesized either by fermentation of biomass or by chemical methods using petroleum products. Polylactic acid (PLA) has good thermomechanical properties and is an excellent replacement for polystyrene and polypropylene in many applications. Nanocellulose is isolated mainly from plants and bacteria. Most of the studies are focused on the isolation of cellulose from plants using mechanical and chemical methods. The properties of PLA can be improved with the incorporation of nanocellulose. The book comprises 15 chapters covering an introduction to PLA composites; synthesis and production of PLA; manufacturing methods of PLA composites; chemical modification of cellulose and PLA/cellulose composites; PLA hybrid composites; the morphology of PLA/cellulose composites; thermo-mechanical properties of PLA/cellulose composites; water barrier properties of PLA/cellulose composites; aging studies of PLA composites; biocompatibility, biodegradability, and environmental safety aspects of PLA/cellulose composites; recycling and reuse of PLA/cellulose composites; recent technologies such as electrospinning and electrospraying of PLA/cellulose composites; and finally the potential applications of PLA/cellulose composites. We hope that the priceless information present in the book will be useful to students, faculty, scientists, environmentalists, and government policymakers. The editors thank the authors for their excellent contributions to the text.

Dr. Jyotishkumar Parameswaranpillai (India)
Dr. Suchart Siengchin (Thailand)
Dr. Nisa V. Salim (Australia)
Dr. Jinu Jacob George (India)
Aiswarya Poulose (India)

Contributors

N. Abdullah
Universiti Pertahanan Nasional
  Malaysia (UPNM)
Kuala Lumpur, Malaysia

S. Ahmad
Universiti Teknologi MARA (UiTM)
Shah Alam, Malaysia

Ana Gabrielle Pires Alvarenga
Federal University of Rio Grande
Rio Grande, Brazil

A. Arbelaiz
University of the Basque Country UPV/
  EHU
Donostia-San Sebastian, Spain

M. R. M. Asyraf
Universiti Putra Malaysia (UPM)
Serdang, Malaysia

Amritha Bemplassery
National Institute of Technology
Calicut, India

Satinder Kaur Brar
York University
Toronto, Canada

Heidy Burrola-Núñez
Universidad Estatal de Sonora
Hermosillo, Mexico

Sagle Chan
Chiao Fu Enterprise
Taiwan

Poushali Das
Bar-Ilan University
Ramat-Gan, Israel

Michele Greque de Morais
Federal University of Rio Grande
Rio Grande, Brazil

Brian G. Falzon
RMIT University
Melbourne, Australia
and
Queen's University Belfast
Belfast, UK

Sayan Ganguly
Bar-Ilan University
Ramat-Gan, Israel

Georg Graninger
Queen's University Belfast
Belfast, UK

Warren J. Grigsby
Scion
Rotorua, New Zealand
Present employer: Henkel NZ, Ltd

Cheng-Han Hsieh
Industrial Technology Research
  Institute
Hsinchu, Taiwan

Martin A. Hubbe
North Carolina State University
Raleigh, NC

Guang-Way Bill Jang
Industrial Technology Research Institute
Hsinchu, Taiwan

Aswathy Jayakumar
King Mongkut's University of
  Technology North Bangkok
Bangkok, Thailand

R. A. Ilyas
Universiti Teknologi Malaysia
Skudai, Malaysia

S. H. Kamarudin
Universiti Teknologi MARA (UiTM)
Shah Alam, Malaysia

Jasila Karayil
Government Women's Polytechnic College
Calicut, India

Guneet Kaur
York University
Toronto, Canada

Premkumar Anil Kothavade
CSIR-National Chemical Laboratory
Pune, India
and
Academy of Scientific and Innovative
    Research
Ghaziabad, India

Sandeep Kumar
University of Warwick
Coventry, UK

Suelen Goettems Kuntzler
Federal University of Rio Grande
Rio Grande, Brazil

M. Kuzmin
Ogarev Mordovia State University
Saransk, Russia

Allen Lai
Chiao Fu Enterprise
Taiwan

Loong-Tak Lim
University of Guelph
Guelph, Canada

Tomás Jesús Madera-Santana
Centro de Investigación en
    Alimentación y Desarrollo
Hermosillo, Mexico

Shiji Mathew
Mahatma Gandhi University
Kottyam, India

Juliana Botelho Moreira
Federal University of Rio Grande
Rio Grande, Brazil

S. U. F. S. Najmuddin
Universiti Putra Malaysia (UPM)
Serdang, Malaysia

J. Naveen
Vellore institute of Technology
Vellore, India

M. N. F. Norrrahim
Universiti Pertahanan Nasional
    Malaysia (UPNM)
Kuala Lumpur, Malaysia

N. M. Nurazzi
Universiti Pertahanan Nasional
    Malaysia (UPNM)
Kuala Lumpur, Malaysia

A. Orue
University of the Basque Country
    UPV/EHU
Donostia-San Sebastian, Spain

Jyotishkumar Parameswaranpillai
Alliance University
Bengaluru, India

Sabarish Radoor
King Mongkut's University of
    Technology North Bangkok
Bangkok, Thailand

Kiana Rafiee
York University
Toronto, Canada

M. Rayung
Universiti Putra Malaysia (UPM)
Serdang, Malaysia

Jesús Rubén Rodríguez-Núñez
Universidad de Guanajuato
Celaya, Mexico

Kadhiravan Shanmuganathan
CSIR-National Chemical Laboratory
Pune, India
and
Academy of Scientific and Innovative
    Research
Ghaziabad, India

S. S. Shazleen
Universiti Putra Malaysia (UPM)
Serdang, Malaysia

Jyothi Mannekote Shivanna
Dayananda Sagar College of Engineering
Bengaluru, India

Suchart Siengchin
King Mongkut's University of
    Technology North Bangkok
Bangkok, Thailand

U. Txueka
University of the Basque Country
    UPV/EHU
Donostia-San Sebastian, Spain

Luis Ángel Val-Félix
Centro de Investigación en
    Alimentación y Desarrollo
Hermosillo, Mexico

Jorge Alberto Vieira Costa
Federal University of Rio Grande
Rio Grande, Brazil

T. A. T. Yasim-Anuar
Nextgreen Pulp & Paper Sdn. Bhd.
Pahang, Malaysia

About the Editors

Jyotishkumar Parameswaranpilla is currently an Associate Professor at Alliance University, Bangalore. He received his Ph.D. in Polymer Science and Technology (Chemistry) from Mahatma Gandhi University, Kottayam, India in the year 2012. He has published more than 120 papers in high-quality international peer-reviewed journals on polymer nanocomposites, polymer blends, biopolymers, and food packaging; has published around 50 book chapters; and is the editor of 22 books. He has received numerous awards and recognitions, including the prestigious KMUTNB Best Researcher Award 2019, the Kerala State Award for the Best Young Scientist 2016, and the INSPIRE Faculty Award 2011.

Suchart Siengchin is the President of King Mongkut's University of Technology North Bangkok (KMUTNB), Thailand. He received his Dipl.-Ing. in Mechanical Engineering from the University of Applied Sciences Giessen/Friedberg, Hessen, Germany in 1999; M.Sc. in Polymer Technology from the University of Applied Sciences Aalen, Baden-Wuerttemberg, Germany in 2002; M.Sc. in Material Science at the Erlangen-Nürnberg University, Bayern, Germany in 2004; Doctor of Philosophy in Engineering (Dr.-Ing.) from the Institute for Composite Materials, University of Kaiserslautern, Rheinland-Pfalz, Germany in 2008; and Postdoctoral Research from Kaiserslautern University and School of Materials Engineering, Purdue University, West Lafayette, IN. In 2016 he received the residency at the Chemnitz University in Sachen, Germany. He worked as a Lecturer for the Production and Material Engineering Department at The Sirindhorn International Thai-German Graduate School of Engineering (TGGS), KMUTNB. He has been a full Professor at KMUTNB and became the President of KMUTNB. He won the Outstanding Researcher Award in 2010, 2012, and 2013 at KMUTNB. His research interests include polymer processing and composite material. He is the Editor-in-Chief: *KMUTNB International Journal of Applied Science and Technology* and the author of 200 peer-reviewed journal articles. He has participated in presentations in more than 39 international and national conferences with respect to materials science and engineering topics.

Nisa V. Salim is currently a Research Fellow at the VC Initiative and received her Ph.D. from Deakin University in 2013 in materials engineering. Her research is mainly focused on advanced carbon materials and functional fibers. She has published over 50 high-impact journal papers, 1 book, and 3 book chapters, even with a career interruption of almost 5 years. She has won many awards in her research career, including the AINSE Gold Medal for Outstanding Ph.D. and Smart Geelong Early Researcher Award, the Victoria Fellowship, the Endeavour Fellowship, the Alfred Deakin Fellowship, and many more. She has held visiting appointments at the University of Southern Mississippi, the University of Kentucky, CNRS Montpellier, and the Indian Institute of Technology Madras. Her vision is to develop multifunctional materials that are enablers for digitalization and the internet of things – living materials that sense, actuate, and harvest energy.

Jinu Jacob George received his Ph.D. degree from the Indian Institute of Technology Kharagpur, India, in 2009. He completed his B.Tech. from Mahatma Gandhi University, Kottayam, and M.Tech. from Cochin University of Science and Technology, Kochi, India. After pursuing the Ph.D., he moved to Leibniz Institute for Polymer Research (IPF), Dresden, Germany, for his Post-Doctoral Research. In 2011, he joined the Rubber Research Institute of India, Kottayam, as a Scientist. In 2015 he joined the Department of Polymer Science and Rubber Technology, CUSAT, as a faculty member. At present, he is actively involved in research in the various advanced fields of Polymer Science and Rubber Technology and has co-authored more than 25 peer-reviewed journal articles. He has a general research interest in the fields of polymer micro-/nanocomposites, thermoplastic elastomers, functional additives, and shape memory polymers.

Aiswarya Poulose is a senior research fellow at the Department of Polymer Science and Rubber Technology, Cochin University of Science and Technology, India. Her research work mainly focuses on the preparation and characterization of biopolymeric materials.

1 Introduction to Polylactic Acid (PLA) Composites

State of the Art, Opportunities, New Challenges, and Future Outlook

Georg Graninger
Queen's University Belfast
Belfast, UK

Brian G. Falzon
RMIT University
Melbourne, Australia

and

Queen's University Belfast
Belfast, UK

Sandeep Kumar
University of Warwick
Coventry, UK

CONTENTS

1.1 INTRODUCTION: BACKGROUND AND MOTIVATION

In the 20th century, developments in the field of composite materials were driven by the need for robust and highly reliable systems that could provide consistent performance. As a result, composites made from petroleum-based resins, reinforced with engineered

DOI: 10.1201/9781003160458-1

fibers, such as glass and carbon, have dominated the industry due to their superior specific strength and stiffness compared to metallics (Akampumuza, 2016). Ceramics and metals which have been used in specific medical applications, such as orthopedic tissue replacement, suffer from being non-biodegradable and have strong limitations in terms of processability (Lopes, 2012). Reinforced polymer composites provide a high degree of structural tailoring with good processing control. However, polymer/synthetic fiber composites are also not biodegradable (Peter, 1998; Chen, 2002; Nair, 2007; Choudhury, 2021). Traditional management, i.e., incineration and land filling, of this kind of waste results in environmental pollution. Incineration generates greenhouse gases (GHGs) hazardous to human health, while landfilling contaminates soil and water, threatening the environment. Among synthetic polymers, thermoplastics offer the possibility of recycling, while thermosets are not easily recycled. Though recyclability is an advantageous property, the recycling potential is usually not fully exploited due to regional availability, contamination and deteriorating properties. Packaging materials, designed for single use, significantly contribute to disposal-related concerns (Davis, 2006; Akampumuza, 2016).

In automotive applications, 25–35% of components can neither be safely disposed of, nor recycled (Ghomi, 2021). This bears increasingly critical relevance as the kerb weight of cars has steadily increased over the years to accommodate customers' requests and stricter safety standards (Akampumuza, 2016). An automobile's mass accounts for 75% of its energy consumption, meaning fuel efficiency has also been negatively affected (Mohanty, 2002).

In order to tackle growing environmental concerns, governments around the globe, in cooperation with environmental bodies, have launched policies and set carbon emission targets. This has sparked an enormous research interest to find renewable material alternatives and provide solutions for companies seeking to meet environmental requirements (Mooney, 2009).

A new class of composite materials, so-called "green composites", is investigated as an alternative to petroleum-based composite systems, owing to their renewability and sustainability. Green composites are created when two or more renewable source-based materials are combined. This can include the combination of biopolymers with natural fibers (Ben, 2007; Gejo, 2010).

Ideally, a biopolymer originates from a renewable biological source and provides end-of-life conversion into simpler compounds. This conversion characteristic is referred to as bio-degradability (Kargarzadeh, 2018). The resulting elements include nitrogen, sulfur and carbon, which can then be redistributed (Choudhury, 2021). Natural polymers such as cellulose, chitosan and starch share these characteristics and can be extracted from biomass directly. They are sometimes referred to as "agro-polymers". "Biopolyesters" such as polylactic acid (PLA) and polycaprolactone (PCL) are synthetically obtained from biomass but possess biodegradability (Kargarzadeh, 2018; Choudhury, 2021).

Among emerging biopolymers such as polyamide 11 (PA11) and polyhydroxybutyrate (PHB), PLA has shown great commercial success with a world production of 240 kt/a; a figure which is said to double by 2023 (Birat, 2015; Getme, 2020).

1.2   PLA PRODUCTION AND CHARACTERISTICS

Lactic acid (2-hydroxypropionic acid), $CH_3–CHOHCOOH$, is a hydroxycarboxylic acid and the monomer to PLA. Due to its wide occurrence, it is applied in a multitude of pharmaceutical, food, chemical, textile and leather products (Vickroy, 1985; John, 2009b).

Due to the chirality of the lactic acid molecule (two molecule sets existing as non-superimposable mirror images), optically active (plane-polarized light is rotated when passing through the chiral molecules) L- (levorotatory, counterclockwise rotation) and D- (dextrorotatory, clockwise rotation)-enantiomers (mirror images) can be found. As a result, PLA appears as four different stereoisomers: isotactic poly-L-lactic acid (PLLA), isotactic poly-D-lactic acid (PDLA), atactic poly-D, L-lactic acid (PDLLA) and syndiotactic PDLLA (Lopes, 2012; Zhou, 2021). Each type of PLA exhibits different mechanical, thermal, degradable and barrier properties. These physical properties can be tailored to various different applications via the adaptation of processing parameters (temperature, shear, etc.), route of synthesis and the lactic acid source (Liu, 2014; Zhou, 2021). While lactic acid can be manufactured via chemical synthesis (hydrolysis of lactonitrile), 90% of total lactic acid production is fermentative (Adsul, 2007; Gupta, 2007). In order to produce lactic acid via microbial fermentation, bacteria are provided with either pure sugar (lactose, sucrose, glucose) or sugar-containing materials (sugarcane bagasse, molasses, whey) as a carbon source. In 2008, Brazil, the world's largest producer, supplied 130 million tons of bagasse generated from 650 million tons of sugarcane (Lopes, 2012). Fermentation also has the added advantage of low energy consumption, low production temperature and low cost of substrates in comparison to chemical synthesis with strong acids. This process also generates optically pure L- or D-lactic acid, which can produce PLLA and PDLA (Adsul, 2007; John, 2007; Lopes, 2012). Both PLA types are crystallizable, meaning they can show improved mechanical and thermal properties. PLLA, in particular, shows great biocompatibility and high crystallinity, resulting in excellent mechanical and thermal properties (Lopes, 2012; Zhou, 2021). Due to its high crystallinity, the degradation time of PLLA is enhanced compared to the optically inactive PDLLA, which is amorphous. In addition, inflammatory reactions can be triggered by high-crystalline fragments produced during the degradation process of PLLA in the body. In order to avoid the formation of potentially harmful crystalline fragments during degradation, D, L-lactic acid monomers are added to L-lactic acid monomers during the PLA synthesis (Lopes, 2012).

In general, PLA is an aliphatic biopolyester usually produced from lactic acid (via condensation) or lactide (via ring-opening polymerization [ROP]) (Kargarzadeh, 2012; Lopes, 2012; Choudhury, 2021). Polycondensation of lactic acid involves the production of PLA oligomers. Those molecular structures are limited to low average molecular weights (tens of thousands). This is due to the high polymerization-related viscosity of the solution, inhibiting the effective removal of water in order to shift the chemical equilibrium towards the formation of PLA (Morteza, 2014). During the polymerization process, side reactions including transesterification cause the formation of side products such as lactides (ring structures), which, in turn, negatively affect the properties of the oligomeric lactic acid (Auras, 2010; Lopes, 2012; Zhou, 2021). Low molecular weight from direct polycondensation is used for applications in medicine requiring high biodegradability (Hamad, 2018).

If polymerization conditions are controlled, functionalizing additives and catalysts are added accordingly and water is removed using a decompression method, high amounts of lactide can be generated. The lactide ring structures open subsequently, polymerizing into long molecules chains of PLA (Mehta, 2005; Cheng, 2009). This process is called ROP. As it involves additional processing steps and is harder to manipulate, manufacturing costs are increased compared to direct

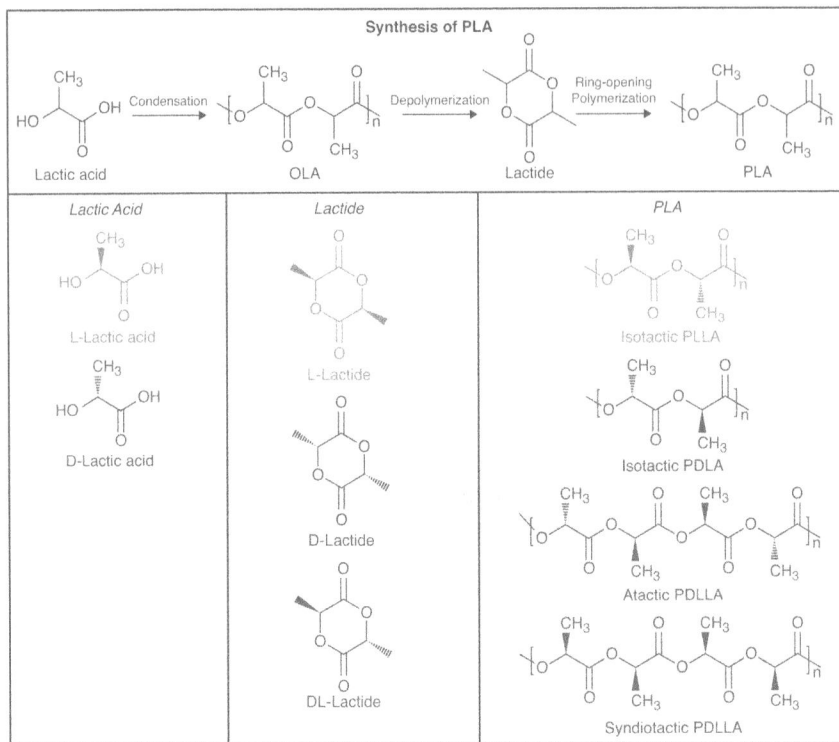

FIGURE 1.1   Synthesis of PLA stereoforms via ring-opening polymerization. (With permission from Koh, 2018.)

polycondensation. However, high-molecular-weight PLA can be obtained with this process enabling its use in packaging applications (Lopes, 2012; Hamad, 2018). Stannous octoate catalyst and tin (II) chloride are popular catalysts for the ROP as they initiate reactions with high conversion and reaction rates, resulting in PLA with high molecular weight (Witzke, 1997; Morteza, 2014). Figure 1.1 illustrates the ROP process and shows the L- and D-enantiomers of PLA as well as the stereoisomers of lactides and PLA. Other potentially viable polymerization methods for the production of high-molecular-weight PLA include azeotropic dehydration condensation and enzymatic polymerization. In the case of azeotropic dehydration, this method suffers from catalysts remaining after the process (Cheng, 2009; Sreekumar, 2021).

In addition to its biological properties (biocompatibility, biodegradability), PLA offers tunable mechanical characteristics as well as thermoplastic processability, making this material one of the most promising biodegradable polymeric alternatives (Gupta, 2007). The stereochemistry of the lactic acid and the processing-related thermal history are the main influencing factors in PLA production and application (Nampoothiri, 2010). The resulting crystallinity and molecular weight determine the polymer's physical property profile including mechanical and rheological properties, density and transition temperatures (Henton, 2005).

Three different crystal types (defined by helix confirmation and cell symmetry) can be found in PLA: α, β, and γ. As a thermoplastic polymer, PLA liquefies upon reaching its melting point during heating. Once melted, it can be shaped to replicate various structures depending on the process, e.g., injection molding. The parts can be removed from the processing machine after the polymer has cooled down. PLA can later be reheated without significant degradation to be recycled for further use (Sreekumar, 2021). During heating (cold crystallization) and/or cooling (melt crystallization), the growth of α-crystals with a melting temperature of 185°C can occur (Lim, 2008). B-crystals, with a lower melting temperature of 175°C, are formed when α-crystals are mechanically stretched. Γ-crystals grow as spherulites when PLA crystallizes on top of a substrate, e.g., hexamethylbenzene (Di Lorenzo, 2005). Among the different PLA crystal forms, α-crystals are comparatively more stable than the others. Copolymer composition directly affects the crystallization rate, while molecular weight affects it inversely (Sreekumar, 2021). As such, the presence of more than 90–93% L-lactic acid in PLLA results in a semi-crystalline polymer with a maximum crystallinity of up to 40% (heat of fusion ΔH_m: 93 J/g [up to 148 J/g]). For values below this threshold, i.e., 50–93% L-lactic acid, PLA is entirely amorphous (Lopes, 2012; Sangeetha, 2016). In addition to crystallinity, enantiomer structure and molecular weight, PLA properties are further influenced by plasticizers, catalysts, copolymerization of lactide and blending of PLA with other materials (Cheng, 2009). As a result, the mechanical behavior of PLA can range within certain limits, generally exhibiting properties of a stiff and brittle polymer, see Table 1.1 (Ghomi, 2021; Sreekumar, 2021). When comparing the mechanical performance of PLA to other commodity plastics such as polypropylene (PP), high-density polyethylene (HDPF), polystyrene (PS) and polyamide 6 (PA6), PLA shows an increased tensile modulus. At the same time, the elongation-at-break is comparatively lower, resulting in brittleness and low impact resistance (Sreekumar, 2021).

The density of solid crystalline polylactide (L-lactide) was measured to be 1.36 g/cm³, while 1.25 g/cm³ was reported for amorphous PLA (Auras, 2004).

Physical properties such as modulus, strength, elongation-at-break and density are strongly influenced by changes in the polymer chain mobility, occurring at the

TABLE 1.1
PLA Property (Mechanical, Thermal) Spectrum

Properties	PLA
Polymer density (g/cm³)	1.21–1.36
Tensile modulus (GPa)	2.1–16
Tensile strength (MPa)	15.5–150
Ultimate strain (%)	2–10
Specific tensile strength (Nm/g)	16.8–66.8
Specific tensile modulus (kNm/g)	0.28–3.85
Glass transition temperature (°C)	58–65
Melting temperature (°C)	130–180

Sources: With permission from Ghomi (2021) (Witzke, 1997; Chen, 2003; Auras, 2004; Mathew, 2005; Ochi, 2008; Yang, 2008; Suryanegara, 2009; Lim, 2013; Scaffaro, 2017; Sedničková, 2018; Sreekumar, 2021).

glass transition temperature, T_g (for amorphous and semi-crystalline PLA), and at the melting temperature, T_m (for semi-crystalline PLA) (Lopes, 2012). As highlighted in Table 1.1, T_g of PLA can range between 58°C and 65°C. Once the polymer temperature drops below T_g, PLA transitions from the rubbery to the glassy state, inhibiting chain mobility. Yet, creep behavior can be observed during cooling until the β-transition temperature is reached (45–60°C) (Ghomi, 2021; Sreekumar, 2021). T_g itself is influenced by optical purity (higher T_g in the same percentage PLLA compared to PDLA) and molecular weight (Sreekumar, 2021). The relation between molecular weight and T_g is described by the Flory-Fox equation (Stoddart, 2012):

$$T_g = T_{g\infty} - \frac{K}{M_n} \tag{1.1}$$

M_n is the number average molecular weight. Using T_g values for known molecular weights, Equation 1.1 shows a linear relationship between T_g and $1/M_n$. The excess free volume of polymer chain end groups, K, represents the gradient of this straight line and is a constant. Extrapolating the line to intercept the y-axis gives $T_{g\infty}$, the predicted T_g at infinite molecular weight. High-molecular-weight PLA exhibits a higher T_g, requiring higher amounts of thermal energy to enable chain mobility (Stoddart, 2012; Sreekumar, 2021). The melt temperature ranges from 130°C to 180°C for amorphous and semi-crystalline PLA polymers; see Table 1.1 (Rajeshkumar, 2021).

One of the most attractive features of PLA is its biodegradability. At high temperatures and high humidity, PLA degrades in two stages. In the first stage, macromolecules are hydrolyzed via random chain scission (main and side chains), leading to low-molecular-weight lactic acid oligomers. The second stage begins when a value of around 10,000 Da is reached in average molecular weight. Within the compost atmosphere, micro-organisms metabolize the low-molecular-weight PLA to produce water and carbon dioxide. The molecular weight of PLA prior to degradation determines the duration of the process itself, which can be desirable, or not, depending on the application (Auras, 2004; Oyama, 2009; Lopes, 2012; Farah, 2016; Rajeshkumar, 2021).

The commercial success of the PLA polymer can be justified by its bio-behavior (biocompatibility, biodegradability), availability and processability as well as optical transparency (Khalil, 2016; Kargarzadeh, 2018). The production of a PLA biopolymer requires 27.2 MJ/kg and is more energy efficient than petrol-based high-density polyethylene (HDPE) and polypropylene (PP), requiring 73.7 MJ/kg and 85.9 MJ/kg, respectively (Khalil, 2016). In general, around 25% to 55% less fossil energy is needed to produce PLA compared to petrol-based plastics. This reduction can be up to 90% when wind energy and other renewable energy sources are used to power different stages of PLA production (Vink, 2003).

When PLA degrades, around 1600 kg/metric ton CO_2 is emitted by the time biodegradation is complete. This is noteworthy in two ways. Firstly, the CO_2 emission rate of PLA is comparatively lower than the rate of traditional polymers such as polypropylene (1850 kg/t), polystyrene (2740 kg/t), polyethylene terephthalate (4140 kg/t) and polyamide (7150 kg/t). Secondly, when the agricultural feedstocks for the production of PLA are grown, CO_2 is taken from the atmosphere. This intake balances the release of CO_2 during PLA degradation, resulting in low GHG emissions.

According to Mohanty et al. (2005), if sourcing is extended towards agricultural residues or biomass, the production of PLA will actually reduce the release of GHGs into the atmosphere (Rajeshkumar, 2021).

Many factors influence the properties of PLA, enabling a tailored use of the bio-polymer (especially due to the recent availability of high-molecular-weight PLA) in a variety of applications. This potential was reflected in 2020 in the global PLA market with an estimated worth of 525.47 million USD. During the coming years, between 2021 and 2028, a consecutive annual growth rate (CAGR) of the market of 18.1% is expected (Balla, 2021). At the same time, inherent drawbacks inhibit PLA from substituting traditional polymers. These challenges include a hydrophobic nature, comparatively low transition temperatures, low toughness and barrier properties (oxygen and moisture) as well as a slow rate of degradation (at room temperature and 50% humidity). The glass transition of PLA of around 60°C causes poor thermal resistance. During melt processing, poor shape stability is found in PLA due to its low melting strength. Here, the low melting temperature of PLA is of advantage as it reduces the consumption of processing-related energy (Okamoto, 2005; Getme, 2020; Choudhury, 2021). Consequently, the widespread application of PLA, e.g., in bio-polymer-based packaging, has been limited by its poor mechanical and barrier properties (Khalil, 2016).

1.3 CELLULOSE AS THE IDEAL REINFORCEMENT IN PLA COMPOSITES

In order to improve PLA properties for successful practical use, chemical modification of PLA, gamma irradiation and the addition of plasticizer, among other methods, have been proposed (Khalil, 2016). One approach that preserves the biodegradability of PLA is blending PLA with other biopolymers (Kargarzadeh, 2018; Zhu, 2019; Zhou, 2021). Here, PLA acts as a polymer matrix with natural fibers being incorporated as reinforcements, yielding the desired property and performance improvements (Kargarzadeh, 2018).

Three distinct groups can generally be defined when talking about natural fibers, as depicted in Figure 1.2. Animal fibers, mineral fibers and cellulose fibers all share biodegradability and renewability, low cost and easy availability. They are lightweight with good acoustic and electrical insulation properties (Rajeshkumar, 2017; Sanjay, 2019; Rajeshkumar, 2020).

Natural fibers show environmental superiority compared to synthetic fibers throughout the whole life cycle, illustrated by the energy requirements for production, emissions and crude oil consumption. Measured in British Thermal Units (BTU), the energy needed to produce one kilogram of glass fibers (47,000 BTUs) far exceeds the amount required for a pound of Kenaf fibers (13,000 BTUs) (Mohanty, 2002). Traditional fibers such as glass, aramid and carbon fibers cause the release of huge quantities of harmful gases including CO_2, NOx and SOx, as well as dust into the atmosphere during their production and disposal (Marsh, 2003; Mohanty, 2005). In comparison, the carbon balance of natural fibers theoretically remains favorable after incineration and composting. This is because the amount of CO_2 emitted into the atmosphere during thermal and microbial degradation is negligible compared to the amount sequestered by the plant during its growth (De Bruijn, 2000).

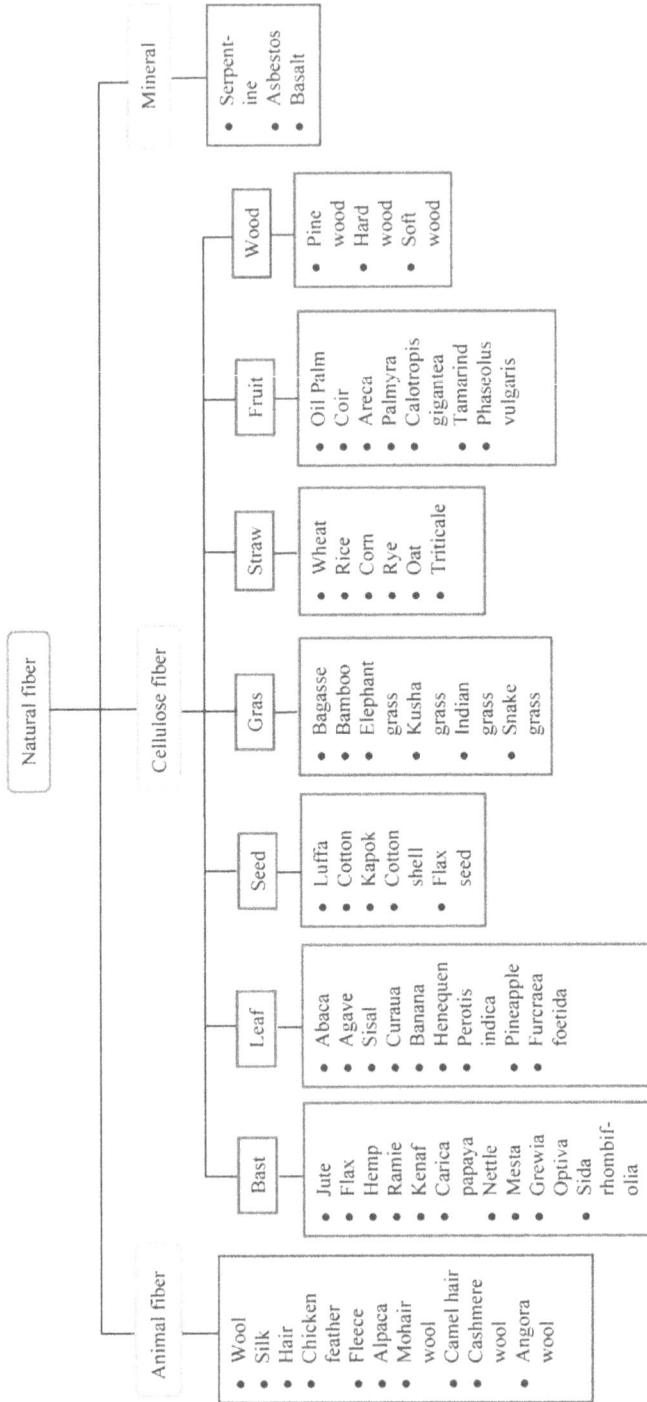

FIGURE 1.2   Classification system for natural fibers. (With permission from Rajeshkumar, 2021.)

According to Pervaiz and Sain in a publication from 2003, 1.19 million m^3 of crude oil consumption and 3.07 million tons of CO_2 emissions were saved when natural fiber composites had substituted 50% of glass fiber-reinforced composites in North American automobiles (Pervaiz, 2003). High specific strength and stiffness of natural fibers combined with their low density (1.5 g/cm^3) causes a weight reduction that results in a higher fuel efficiency and reduced air pollution (Akampumuza, 2016). Biocomposites reinforced with natural fibers have seen increasing interest over the last few decades. Industry and academia are on a quest to find a valuable alternative to composites consisting of conventional fibers and depletable petrol-based polymer products, alleviating GHGs emissions (Patel, 2005; Rajeshkumar, 2021).

Cellulose, next to hemicellulose and lignin, is one of the three main chemical components of plant fibers (Nagarajan, 2021). Repeating D-anhydroglucose units ($C_6H_{11}O_5$) make up the linear 1,4-β-glucan cellulose polymer (Rajeshkumar, 2021). Intra- and inter-molecular hydrogen bonding occurs between the cellulose's hydroxyl groups, providing the plant fiber with stiffness and strength (Ramamoorthy, 2015). The mechanical properties determined by cellulose depend on a variety of parameters. These include the location of cellulose extraction (seeds, stem, leaves, etc.), the plant species and age of the fibers, the type of soil and the climatic conditions (e.g., the availability of water) as well as storage and transportation conditions (Dittenber, 2012; Thakur 2014; Rajeshkumar, 2016). Hemicellulose determines the behavior of the plant in terms of moisture absorption and thermal and biodegradation (Saheb, 1999). Lignin fulfils the role of a structural support material. It is a phenolic compound with a hydrophobic nature. When in solid state, it possesses an amorphous structure (Mohanty, 2000; Hatakeyama, 2009, Rajeshkumar, 2021). As such, cellulose is the strongest and stiffest out of these three major plant fiber components, making it a material of choice for reinforcing polymer matrix composites (Ray, 2002; Dwivedi, 2009; John, 2009a; Scarponi, 2009; Hammajam, 2019; Rajeshkumar, 2021).

Different types of cellulose such as microcrystalline cellulose and nanocellulose interact with PLA on different levels (macromolecular, molecular and atomic), thus affecting the properties of the final bio-composite including mechanical and thermal properties, barrier and antimicrobial performance as well as crystallization and degradability (Khalil, 2016; Khosravi, 2020). Depending on the pre-treatment and preparation methods to isolate nanocellulose, two types are typically defined: Cellulose nanofibrils (CNFs) are long and flexible chains of cellulose polymer that are commonly found entangled, forming a network of microfibrils, connected via inter- and intramolecular bonds. CNFs contain both amorphous and crystalline regions, as shown in Figure 1.3 (Khalil, 2016).

In order to isolate individual CNFs movement and twisting of the amorphous regions needs to be enabled, disrupting the microfibril arrangement. Pre-treatment facilitates this process, with the two most common chemical methods being hydrolysis (enzymatic or acidic cleavage of chemical bonds) and 2,2,6,6-tetramethylpiperidinyloxyl radical (TEMPO) oxidation (use of TEMPO-CuCl as a catalyst to aerobically oxidize primary and secondary alcohols). Pre-treatment holds significant benefits resulting in the facilitated disintegration of CNFs, a more efficient nanofibrillation process, highly purified cellulose, removal of non-cellulosic constituents (hemicelluloses, lignin), a hydrophobic surface, and a lower energy consumption during mechanical CNF extraction, such as microgrinding, high-pressure homogenization and micro-fluidization.

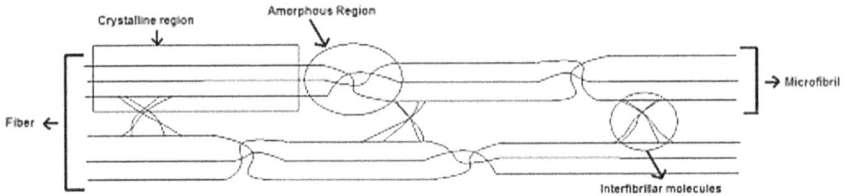

FIGURE 1.3 Schematic illustration of semi-crystalline cellulose fibers containing microfibrils and CNFs. (With permission from Khalil, 2016.)

Various pre-treatments and extraction methods for the CNF isolation are summarized in Figure 1.4 (Kalia, 2011; Zhang, 2011; Khalil, 2012; Benhamou, 2014; Lee, 2014). With lengths of a few micrometers and 3–50 nm diameters, CNFs boast high aspect ratios (length/diameter) (Petersen, 1999; Lavoine, 2014; Kargarzadeh, 2018).

Cellulose nanocrystals (CNCs) have a rod-like shape and are highly crystalline. High crystallinity is achieved via acid hydrolysis, in which acid – most commonly sulfuric acid – is used to remove most of the amorphous parts of the polymer, producing CNCs (sulfonated or with other surface groups depending on the acid) (Kargarzadeh, 2012). With a lower density of 1.5 g/cm^3 compared to the density of glass fibers of 2.5 g/cm^3, CNCs can serve as lightweight reinforcement (Ferreira, 2018). The range

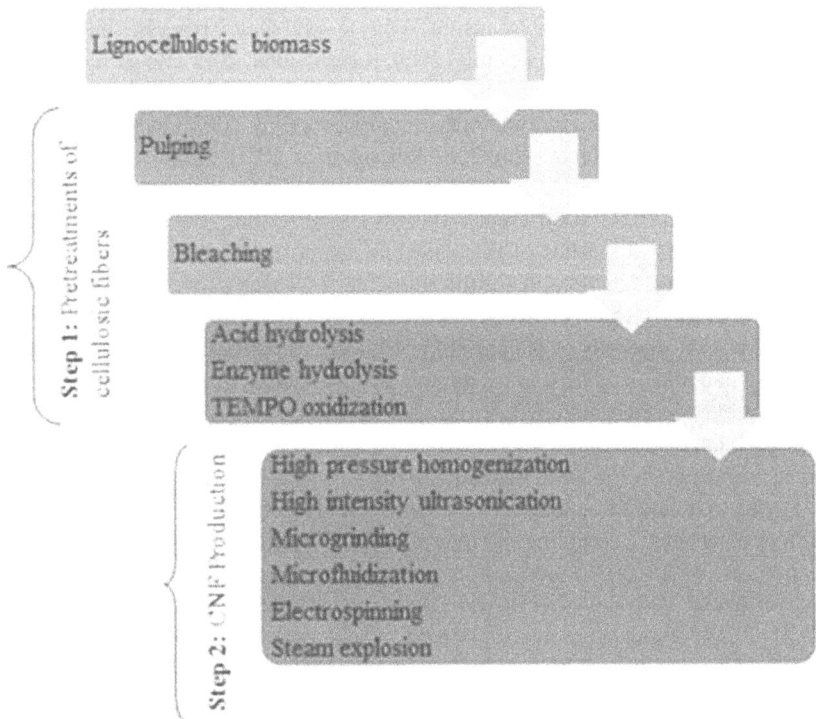

FIGURE 1.4 Pre-treatment and mechanical extraction methods to obtain cellulose. (With permission from Khalil, 2016.)

of polymer composites is limited by the hydrophilic nature of CNCs (due to the abundance of hydroxyl groups on the surface), causing agglomeration and insufficient compatibility with hydrophobic polymers such as PCL and PLA (Zhou, 2021).

Compared to CNFs, the less-amorphous CNCs possess smaller geometries with 10–500 nm of length and diameters of a few nanometers. The geometries and presented forms of CNFs and CNCs result in a significant difference in mechanical properties such as tensile modulus and strength, with CNCs showing much higher levels (CNFs: 23 GPa and 383.3 MPa vs. CNCs: 150 GPa and 7.5 Mpa) (Kargarzadeh, 2018; Zhou, 2021). Further properties of CNFs and CNCs as well as various applications in composites are illustrated in Table 1.2.

TABLE 1.2

Comparison of CNFs and CNCs: Properties and Applications

	CNFs	CNCs
Properties	Usually 3–50 nm in diameter	Few nm in diameter
	Few micrometers in length	From 10 to 500 nm
	Crystallinity below 50%	Crystallinity 60–90%
	Few-fold increase in tensile strength and modulus when incorporated in hydrophilic polymers	Few-fold increase in tensile strength and modulus when incorporated in hydrophilic polymers
	Thermally stable up to 260°C	Thermally stable up to 200–230°C
	Tensile modulus 23 GPa	Tensile modulus ~150 GPa
	Tensile strength 383.3 MPa	Tensile strength ~7.5–7.7 GPa
Applications	Support for antibacterial coating	Drug delivery vehicle
	Rheology modifier	Rheology modifier
	Support for catalysts and sensors	Support for catalysts and sensors
	Tissue engineering, scaffolds	Diaphragms in earphones
	Network structure	Tissue engineering, scaffolds
	Paper making as carrier and/or binders and coating	Biomimetic foams
	Food coating	Optical application
	Never dried membranes	Additive to drilling fluids
	Support for flexible light-emitting diodes	Additive to cement-based materials
	Biocompatible nanocomposites	Water pollutant remediation
	Dental applications	Toughened paper
		Flexible panels for flat-panel displays
	Polymer nanocomposites by solvent casting and melt mixing	Polymer nanocomposites for developing membranes, fibers, textiles, batteries, supercapacitors, electroactive polymers, sensors, and actuators
	Packaging products	
	Used in display devices	
	Water pollutant remediation	

Sources: With permission from Kargarzadeh (2018) (Kalia, 2011; Wie, 2014; George, 2015; Li, 2015; Kaushik, 2016; Moon, 2016; Hubbe, 2017; Kafy, 2017).

1.4   MODIFICATION STRATEGIES FOR IMPROVED CELLULOSE-PLA INTERACTION

Looking at bio-composites manufactured from CNFs and PLA, pristine CNFs can be introduced as direct reinforcement into the PLA matrix, causing an increase in strength and elastic modulus. The reinforcing effect is attributed to the long, flexible chains of CNF, offering high aspect ratios for enhanced stress transfer at the interface from PLA to the CNF reinforcement. While PLA is non-polar, CNFs and CNCs both have a polar nature. Consequently, only weak compatibility is enabled between the hydrophobic matrix and the hydrophilic fillers (Kargarzadeh, 2018). This, in turn, results in low loading levels with an optimal concentration of only about 0.5–2 wt.%. Wang et al. presented an approach combining microencapsulation-mixing and melt-compression to manufacture hydrophobic PLA containing high loading levels (8–32 wt.%). In addition to a large increase in modulus and strength of up to 58% and 210%, respectively, CNFs acting as nucleating agents cause positive changes in the crystallization behavior of the matrix (homogeneous and stable crystals, faster crystallization rate and lower cold crystallization temperature) (Song, 2013; Wang, 2012). A high loading of 17 wt.% was also achieved by Tingaut et al. when applying surface acetylation to modify CNFs. This promoted improved compatibility in the CNF-PLA composites, resulting in an increased glass transition temperature (T_g) (Tingaut, 2009). In comparison to CNFs, CNCs with their rod- or whisker-like shape are shorter and more rigid, causing agglomeration of CNCs in the PLA matrix and more brittle behavior. Therefore, physical (adsorption of a surfactant) and chemical (small molecule modification, grafting of PLA) surface modifications of CNC are required (Kargarzadeh, 2018; Zhou, 2021). The use of surfactant-modified CNCs in PLA has been reported to improve dispersion as well as increase thermal stability resulting in higher storage modulus at elevated temperatures in comparison to pristine CNC-PLA blends (Kvien, 2005; Petersson, 2007). Small molecule modification is performed to promote the hydrophobicity of cellulose. Three different chemical reaction types are most commonly used to apply this technique, during which hydrophilic hydroxyl groups on the cellulose surface are replaced by small molecules. Esterification, oxidation/amidation and silanization are three diverse chemical reactions that can be performed in facile reaction conditions. The advantage of this technique includes high grafting efficiency as well as the possibility to further modify the cellulose via optional functional groups. Fischer esterification (acylating agents: carboxylic acid (acid anhydride) under strong acid catalysis) and Steglich esterification (acylation catalyst: 4-dimethylaminopyridine [DMAP]) are two methods to esterify cellulose (acylation of hydroxyl groups). Depending on the small molecule substitute, three kinds of acid can be used for this procedure: mono-carboxylic acid (fatty/aromatic acid), lactic acid and multiple carboxylic acids (maleic/succinic anhydride) (Zhou, 2021).

Esterified cellulose decorated with hydrophobic aliphatic branches changes its hydrophilic to a hydrophobic nature, causing better dispersion compared to pristine cellulose in the PLA matrix. When esterification is performed with multiple carboxylic acid groups, a large quantity of carboxyl groups can be achieved at the end of the cellulose molecules. As a consequence, hydrogen bonding interaction between the carbonyl groups of PLA, and the hydroxyl and carboxylic groups of modified cellulose, are strengthened (Zhou, 2021). A dicarboxylic anhydride that has been studied in cellulose/PLA composites due to its low cost and low toxicity is maleic anhydride (MAH). Using dicumyl peroxide (DCP) as an initiator, MAH connects to both the PLA chains as well as to the hydroxyl groups of cellulose, causing the formation of a crosslinking structure with improved stress transfer and heightened mechanical properties due to enhanced interfacial interaction between the composite system constituents (Hwang, 2013; Johari, 2016; Zhou, 2018).

The introduction of functional groups onto the cellulose surface can also be performed via oxidation. The best-known oxidation reaction used on cellulose is the TEMPO oxidation (position-selective catalytic oxidation of C6 primary hydroxyls). TEMPO oxidation has become a common pre-treatment for nanocellulose, enabling further functionalization, e.g., in-situ crosslinking of TEMPO-oxidized cellulose with PLA during melt processing (Soman, 2017). Silanization of cellulose is a very simple reaction where hydroxyl-rich surfaces are equipped with different functional groups according to the polymer matrix using silanes with an alkoxysilane group. Despite its facile approach, silane modification suffers from low grafting degrees, resulting in limited interaction and, thus, mechanical property improvement. The small molecule modifications (esterification, oxidation/amidation and silanization) described above can all improve the dispersion of cellulose in PLA by providing a hydrophobic cellulose surface. It is important to optimize reaction conditions, grafting efficiency and the formation of crosslinked networks in order to provide sufficient interfacial interaction between cellulose and PLA to improve mechanical properties of cellulose/PLA composites (Zhou, 2021). Table 1.3 shows how different types of small-molecule-modified cellulose affect the tensile properties of PLA-based composites.

Next to small molecule modification, polymer grafting is another strategy that focuses on enhancing the interaction between cellulose and PLA via chemical and hydrogen bonding as well as physical entanglement. Two grafting procedures are commonly employed to graft PLA molecular chains onto cellulose and to avoid phase separation of the constituents. Dubbed a "grafting from" strategy, surface-initiated ring-opening polymerization (SI-ROP) uses the concept of traditional ROP to polymerize PLA on cellulose (using the abundant hydroxyl groups on the cellulose surface) (Zhou, 2021).

Here, freeze-dried nanocellulose is either directly dispersed into melted lactide or a toluene/lactide solution, respectively, or dispersed in toluene via solvent exchange with the lactide monomer being present. This is followed by the use of stannous octoate $Sn(Oct)_2$ as a catalyst at high temperatures to initiate the ROP reaction. While a relatively high degree of polymerization and grafting efficiency is promoted by this method, properties of the final cellulose composites are limited as hydrophilic nanocellulose subjected to freeze-drying and hydrophobic toluene/lactide melt is hard to redisperse during processing. In order to better harvest the synergistic effect of cellulose-g-PLA in the PLA matrix, various approaches have been investigated as possible solutions (Zhou, 2021). Using hydrophilic solvents to improve dispersion, stereocomplex crystallization, annealing treatment, orientation and PLA-PCL copolymerization all have the potential to enhance composite performance as they

TABLE 1.3

Property Enhancement in Small-Molecule-Modified Cellulose/PLA Composites

Cellulose Source	Filler (Content)	Modification	Processing	Tensile Properties (Pristine PLA: 100%)			Enhancement in Other Property	References
				Strength	Young's Modulus	Elongation		
MCC	CNC (10%)	Formic acid	Solution casting	310%	350%	60%	Crystallization Barrier	Yu (2017)
Ginger fibers	CNF (10%)	Formic acid	Solution casting	360%	450%	55%	Crystallization Barrier	Yu (2017)
MCC	CNF (1%)	Acetic anhydride	Solution casting	125%	–	150%	Hydrophilicity	Jamaluddin (2019)
Pulp	MFC (3%)	Acetic anhydride	Solution casting	125%	125%	55%	Crystallization	Bin (2017)
Unknown	CNF (1%)	Acetic anhydride	Solution casting	40.6%	88%	1493%	Toughness	Raisipour-Shirazi (2018)
Wood	CNC (3%)	Valeric acid	Extrusion	116%	145%	143%	Toughness	Shojaeiarani (2019)
Wood	CNC (3%)	Benzoic acid	Extrusion	110%	127%	100%	Toughness	Shojaeiarani (2018)
Pulp	CNF (8%)	Rosin	Solution casting	128%	250%	220%	Antimicrobial	Niu (2018)

Source: With permission from Zhou (2021).

increase crystalline density, facilitate fibrillation of the cellulose particles to create larger surfaces for interaction and restrict the free movement of PLA chains. This can result in improved mechanical properties, high transparency as well as good barrier properties (reduced water vapor and oxygen permeability) (Dong, 2017; Gupta, 2017).

The second grafting procedure is called a "grafting onto" method based on a free radical-initiated reaction different from ROP as crosslinking of cellulose and PLA happens in a single step during a reactive extrusion process. Here, DCP is used as a radical initiator triggering C–C bonding between the CNC methylene and PLA methine groups during the grafting reaction. As a result, CNCs are inserted as chain extenders, creating a branched structure involving the PLA matrix. Mechanical properties and biodegradability of CNC/PLA composites are improved compared to pristine PLA via this facile, non-toxic and commercially scalable technique (Zhou, 2021).

Both grafting procedures, "grafting from" and "grafting onto", can be used to graft various other polymers to the cellulose surface, such as poly(glutamic acid) (PGA) (Averianov, 2019), polyethylene glycol (PEG) (Li, 2019; Pal, 2019), poly(ethylene oxide) (PEO), poly(butylene succinate) (PBS) (Zhang, 2016) and polymethylmethacrylate (PMMA) (Rosli, 2019; Singh, 2020) (see Table 1.4).

Depending on the polymer used for grafting, unique functions can be introduced onto the cellulose surface, giving polymer grafting an edge over small molecule modification in terms of tailoring PLA properties. As large steric hindrances and limited reactive ends of polymers make it challenging to realize high grafting degrees, reversible addition-fragmentation chain transfer (RAFT) polymerization, atom transfer radical polymerization (ATRP) and click chemistry offer mild and efficient chemical reaction alternatives. Industrial manufacturing methods including extrusion, blow or injection molding will have to be combined with advanced processing technology to offer a scalable alternative for solution casting with possibilities to control how cellulose is dispersed, distributed and oriented in the PLA and how cellulose affects the matrix' crystallization behavior (Zhou, 2021).

1.5 CHALLENGES IN MANUFACTURING GREEN PLA COMPOSITES

Concerning the fabrication of green composite materials such as cellulose/PLA composites, it is imperative to acknowledge the differences to traditional polymer composite processing caused by thermal instability of the constituents, differing rheological behavior, morphological differences in natural fiber types and species and tendency to absorb water (Sain, 2008). A new processing window, informed by the influence of processing parameters on surface interaction, morphological changes and chemical crosslinking between constituents, needs to be developed (Gällstedt, 2004). Key challenges in the processing of bio-composites include reduced processability caused by a rise in viscosity with increasing natural fiber content, poor flow properties and high crystallinity of PLA and other biopolymers; a relatively low degradation onset temperature of around 190°C; reduced adhesion between fiber and matrix; dispersion difficulties due to natural fibers being hydrophilic; and morphology and chemical composition being determined by the plant source used to manufacture bio-polymers and natural fibers (Zampaloni, 2007; Lim, 2008; Sain, 2008; La Mantia, 2011; Thakur, 2011; Faruk, 2012). On the one hand, this causes a large

TABLE 1.4

Property Enhancement in Polymer-Modified (Grafted) Cellulose/PLA Composites

Matrix	Filler (Content)	Modification	Processing	Tensile Properties (Pristine Matrix: 100%)			References
				Strength	Young's Modulus	Elongation	
PLA/PBS (70/30)	CNC-g-PBS (1%)	In-site polymerization	Melt blending	112%	112%	102%	Zhang (2016)
PLA/NR (90/10)	Cellulose-g-PMMA (7.5%)	In-site polymerization	Melt blending	160%	115%	–	Rosli (2019)
PLA	CNC-g-PGA (5%)	Grafting onto	Solution casting	67%	94%	32%	Averianov (2019)
PLA	CNC-g-PEG (0.5%)	Grafting onto	Solution casting	120%	114%	113%	Li (2019)
PLA	CNC-g-PEG (5%)/rGO (0.5%)	Grafting onto	Solution casting	158%	160%	56%	Pal (2019)
PLA	CNF-g-PEO (30%)	Hydrogen bonding	Extrusion	82%	95%	140%	Singh (2020)

Source: With permission from Zhou (2021).

variation in the structure and properties of natural fiber-reinforced bio-composites, making the composite design process a balancing act between providing uniform structural and functional stability during use and storage and ensuring susceptibility to fast environmental and microbial degradation when disposed of (Mohanty, 2005; Kim, 2011). On the other hand, limitations such as low thermal stability, formation of aggregates and low moisture resistance deny the bio-composites meeting the required criteria of high thermomechanical performance, e.g., in automotive applications (Kim, 2011; Pilla, 2011). In addition, the low temperature required for natural fiber processing can render the final product less cost-effective as available processing technology and logistics need to be adapted to suit the new material system (Puglia, 2008).

PLA itself comes with a number of challenges and limitations for use in bio-composites, which has hindered industry take-up, which would entail considerable investment in new processing technologies to replace current practices (Akampumuza, 2016). Up until now, lactic acid has been primarily derived from soybeans and corn. While industrial scale-up is possible, they are primary food sources (Akampumuza, 2016; Balla, 2021). Alternate biomass sources need to be developed and applied to produce lactic acid (Balla, 2021). Another drawback of PLA is its degradation profile, which requires specific conditions difficult to find in nature. In nature, micro-organisms such as fungi and bacteria cause the bio-degradation of polyesters like PLA. During these natural biological processes, a combination of hydrolysis and enzymatic degradation using enzymes as biocatalysts is happening as part of a surface or bulk erosion procedure. As enzymes such as Proteinase K, which are strictly necessary for the hydrolysis of PLA, are rare to find in nature, bio-degradation of PLA can take up to 1000 years with a minimum of 80 years (Balla, 2021). Studies investigating PLA degradation found that when PLA is submerged in water, hydrolysis is hindered by low temperatures (below 30°C) and the absence of required enzymes, further prolonging the degradation onset of PLA (Chamas, 2020). While material disintegration takes a long time to start, mechanical properties deteriorate much earlier (Rudnik, 2011a,b). Ideal conditions for PLA degradation can (presently) only be achieved in industrial composters (Balla, 2021). The prevalent specific conditions including the presence of micro-organisms (thermophilic bacteria), high humidity (>60% moisture) and a rich oxygen environment with high temperatures (58–80°C) can deliver an almost complete degradation (>90%) within 30–150 days. Micro-organisms produce compost ingredients next to CO_2 and H_2O (Balla, 2021).

1.6 NECESSARY CONSIDERATIONS TO ENABLE FUTURE APPLICATIONS OF CELLULOSE-PLA COMPOSITES

The most common application sectors include packaging, automotive, medical, single-use plastics, textiles, agricultural, construction and electronics. The fastest-growing regional markets can be found in the United States and Canada (packaging), Mexico and Latin American countries (automotive, electronics), Asia Pacific (automotive), Middle East and Africa (textiles). Europe, together with the United Kingdom and Russia, has shown continuous market development in the transportation industry (Balla, 2021).

One of the main uses of PLA is in the packaging industry, with a revenue share of more than 36% in 2020. The increasing popularity of PLA in food containers, as well as fresh food packaging, is, on the one hand, related to new regulations aiming to benefit the environment, such as the ban of single-use plastics, as in the Directive (EU) 2019/904. On the other hand, PLA is favored as it possesses biocompatibility and non-toxicity, good thermoplastic processability as well as high strength. In order to make up for its lacking barrier properties, low flexibility and crystallization behavior, PLA needs to be blended with other polymers or additives, which provide the necessary gas barrier performance (Balla, 2021; Fredi, 2021). Here, biocomposites of PLA and natural fibers offer promising alternate solutions with low oxygen and water permeability for food packaging (Martínez-Sanz, 2012; Espino-Pérez, 2013). Determining the target market, availability, price and suitable technology is considered an essential prerequisite for incorporating nanocellulose into sustainable packaging (Khalil, 2016). Only then can an engineering design be successfully implemented with both optimized bio-based materials (appearance, quality, physical performance, and contamination of recycled materials) and maximum product quality (durability and physical properties being comparatively higher than existing commercially available packaging materials and being tailored to account for property changes caused by storage temperatures ranging from below 25°C to above 60°C) (Khalil, 2016; Ghomi, 2021). Adul Khalil et al. recommend performing lab-scale studies across fields of expertise (engineers, designers and scientists) as well as life cycle assessment (LCA) in order to produce nanocellulose-enhanced solutions for sustainable packaging (Khalil, 2016).

Ghomi et al. compared multiple LCAs using GHG emissions as a parameter to compare materials and energy use during the whole life cycle of PLA (Ghomi, 2021). According to this comparison, the most energy-intensive process during the life cycle is the conversion of bio-sources to lactic acid and lactic acid to PLA. There is huge potential to make PLA a more environment-friendly, low-carbon material by optimizing the conversion process, which accounts for more than 50% of all the CO_2 released during the PLA life cycle (2.8 kg CO_2/kg PLA). Another recommended option to remove end-of-life emissions from the LCA calculations is to develop recycling facilities to provide recycled PLA products with sufficient properties and quality (Ghomi, 2021). In industry, LCAs are not often performed due to the inherent complexity of the method and the absence of techniques to assess new products in terms of their environmental impact (Poole, 1997).

The international bio-plastics market is constantly growing, with an expected 500,000 tons of PLA being required by 2021–2022. According to Jem and Tan, the global market demands of PLA can be estimated to double every 3 to 4 years (Jem, 2020). This can be attributed to increasing interest from governments and international environmental agencies in bio-degradable plastics and globalization of the products, and more engagement of industries that want to future-proof their businesses against adverse government legislations, to current practices, which may be brought into law (Balla, 2021).

While the scale-up of PLA production facilities is actively pursued by many companies, manufacturing composites with nanocellulose on an industrial scale is difficult. In terms of CNF, it is the mechanical disintegration consuming high amounts of energy, which poses a challenge. It was recently shown that energy consumption can be decreased significantly upon combining certain pre-treatments with the mechanical treatment. As for CNC, the acid used during the production of CNCs is a large cost factor, especially as there is no effective option to recover the acid. Using centrifugation or filtration and treatment with activated charcoal helps to address this issue as trace contaminants can be removed; hence, the acid can be re-used. Another challenge to overcome is to commercialize PLA-cellulose composites, which exhibit both the stable formation of a percolating nanocellulose network as well as a high interfacial adhesion between PLA and nanocellulose (Kargarzadeh, 2018). To predict the enhancement effects based on compatibility and percolating network, modelling and simulation will become increasingly important in the future. As an example, using solid and shell composite models in ANSYS, simulation results close to experimental tensile data of bamboo-reinforced PLA have been successfully generated (Rao, 2021).

While solution casting and other wet processing methods can achieve reasonable dispersion of cellulose particles retaining their nano-size while promoting a strong percolating network, these methods suffer from high energy consumption and hard-to-recover solvent, causing environmental problems. The future go-to processing step to effectively determine morphology and properties of the nanocomposites is seen in melt processing as it is a quick method, requiring no solvents and promising reproducible fabrication. In addition, nanocomposites manufactured via melt processing possess improved thermal stability compared to solution casting. However, in order to produce a homogeneous nanocellulose dispersion in the PLA nanocomposites, issues of the structural integrity of nanocellulose and cellulose orientation need to be addressed. Mixing nanocellulose and PLA both in powder form prior to melt processing has been devised as one suitable strategy. While many challenges remain, nanocellulose-based bio-composites with their renewability and sustainability aspects as well as unique properties will continue to be investigated for use in a multitude of applications (Kargarzadeh, 2018). In automotive applications, the use of natural fibers could lead to a significant reduction in vehicle weight and a commensurate increased fuel efficiency as almost 75% of the energy consumption is related to car weight (Friedrich, 2013).

In order to offer nanocellulose-PLA composites as commercial products with stable performance, improved compatibility between the hydrophilic nanocellulose and the hydrophobic PLA, and low costs for target applications, future research is expected to continuously grow, focusing on environment-friendly production, modification and processing routes and methods (Kargarzadeh, 2018).

1.7 CONCLUSION

PLA from bio-renewable sources has become a popular alternative to traditional polymers in recent years, owing to high mechanical stiffness and strength, commercial availability via ROP and the possibility to degrade under industrial composting conditions. Cellulose being a natural fiber offers the possibility to reinforce PLA, creating fully bio-degradable composite materials. Insufficient interaction between the polar cellulose and the non-polar limits the performance of the resulting composites and their widespread use in industry and everyday life. Successful chemical modification strategies to improve the cellulose-PLA interface and, subsequently, the stress transfer from the PLA matrix into the cellulose fiber have been devised, including small molecule modification and grafting procedures ("grafting from"/"grafting onto"). Enhanced interaction offers the possibility to address PLA shortcomings, such as low barrier properties, enabling the use in applications where biodegradability is essential, e.g., food packaging. Building processing-related know-how is another important step towards the successful commercialization of PLA composites.

REFERENCES

Adsul, M. G., Varma, A. J., Gokhale, D. V., "Lactic acid production from waste sugarcane bagasse derived cellulose", *Green Chemistry* 2007, 9, 58–62.

Akampumuza, O., Wambua, P. M., Ahmed, A., Li, W., Qin, X.-H., "Review of the applications of biocomposites in the automotive industry", *Polymer Composites* 2016, 38 (11), 2553–2669.

Auras, R., Harte, B., Selke, S., "An overview of polylactides as packaging materials", *Macromolecular Bioscience* 2004, 4, 835–864.

Auras, R., Lim, L. T., Selke, S. E. M., Tsuji, H., *"Poly(Lactic Acid): Synthesis, Structures, Properties, Processing, and Application"*, John Wiley & Sons, Inc. 2010, ISBN: 978-0-470-29366-9.

Averianov, I. V., Stepanova, M. A., Gofman, I. V., Nikolaeva, A. L., Korzhikov-Vlakh, V. A., Karttunen, M., Korzhikova-Vlakh, E. G., "Chemical modification of nanocrystalline cellulose for improved interfacial compatibility with poly(lactic acid)", *Mendeleev Communications* 2019, 29, 220–222.

Balakrishnan, H., Hassan, A., Imran, M., Wahit, M. U., "Toughening of polylactic acid nanocomposites: a short review", *Polymer-Plastics Technology and Engineering* 2012, 51 (2), 175–192.

Balla, E., Daniilidis, V., Karlioti, G., Kalamas, T., Stefanidou, M., Bikiaris, N. D., Vlachopoulos, A., Koumentakou, I., Bikiaris, D. N., "Poly(lactic acid): a versatile bio-based polymer for the future with multifunctional properties – from monomer synthesis, polymerization techniques and molecular weight increase to PLA applications", *Polymers* 2021, 13, 1–50.

Ben, G., Kihara, Y., "Development and evaluation of mechanical properties for kenaf fibers/PLA composites", *Key Engineering Materials* 2007, 334–335, 489–492.

Benhamou, K., Dufresne, A., Magnin, A., Mortha, G., Kaddami, H., "Control of size and viscoelastic properties of nanofibrillated cellulose from palm tree by varying the TEMPO-mediated oxidation time", *Carbohydrate Polymers* 2014, 99, 74–83.

Bin, Y., Yang, B., Wang, H., "The effect of a small amount of modified microfibrillated cellulose and ethylene–glycidyl methacrylate copolymer on the crystallization behaviors and mechanical properties of polylactic acid", *Polymer Bulletin* 2017, 75, 3377–3394.

Birat, K. C., Pervaiz, M., Faruk, O., Tjong, J, Sain, M., "Green Composite Manufacturing via Compression Molding and Thermoforming", Chapter 3 in Sapuan, M. S., Jawaid, M., Yusoff, N. B., Hoque, M. E. (Editors), *"Manufacturing Process of Natural Fibre Reinforced Polymer Composites"*, Springer International Publishing 2015, ISBN: 978-3-319-07943-1, 45–63.

Chamas, A., Moon, H., Zheng, J., Qiu, Y., Tabassum, T., Jang, J. H., Abu-Omar, M., Scott, S. L., Suh, S., "Degradation rates of plastics in the environment", *ACS Sustainable Chemistry & Engineering* 2020, 8, 3494–3511.

Chen, C.-C., Chueh, J.-Y., Tseng, H., Huang, H.-M., Lee, S.-Y., "Preparation and characterization of biodegradable PLA polymeric blends", *Biomaterials* 2003, 24, 1167–1173.

Chen, G., Ushida, T., Tateishi, T., "Scaffold design for tissue engineering", *Macromolecular Bioscience* 2002, 2, 67–77.

Cheng, Y., Deng, S., Chen, P., Ruan R., "Polylactic acid (PLA) synthesis and modifications: a review", *Frontiers of Chemistry in China* 2009, 4, 259–264.

Choudhury, M. R., Debnath, K., "Green Composites: Introductory Overview", Chapter 1 in Thomas, S., Balakrishnan, P. (Editors), *"Green Composites"*, Springer Nature Singapore Pte Ltd 2021, ISBN: 978-981-15-9643-8, 1–20.

Davis, G., Song, J., "Biodegradable packaging based on raw materials from crops and their impact on waste management", *Industrial Crops and Products* 2006, 23 (2), 147–161.

De Bruijn, J. C. M., "Natural fibre mat thermoplastic products from a processor's point of view", *Applied Composite Materials* 2000, 7, 415–420.

Di Lorenzo M. L., "Crystallization behavior of poly(L-lactic acid)", *European Polymer Journal* 2005, 41 (3), 569–575.

Dittenber, D. B., GangaRao, H. V., "Critical review of recent publications on use of natural composites in infrastructure", *Composites: Part A* 2012, 43, 1419–1429.

Dong, J., Li, M., Zhou, L., Lee, S., Mei, C., Xu, X., Wu, Q., "The influence of grafted cellulose nanofibers and postextrusion annealing treatment on selected properties of poly(lactic acid) filaments for 3D printing", *Journal of Polymer Science Part B: Polymer Physics* 2017, 55, 847–855.

Dwivedi, U. K., Chand, N., "Influence of MA-g-PP on abrasive wear behaviour of chopped sisal fibre reinforced polypropylene composites", *Journal of Materials Processing Technology* 2009, 209, 5371–5375.

Espino-Pérez, E., Bras, J., Ducruet, V., Guinault, A., Dufresne, A., Domenek, S., "Influence of chemical surface modification of cellulose nanowhiskers on thermal, mechanical, and barrier properties of poly(lactide)-based bionanocomposites", *European Polymer Journal* 2013, 49, 3144–3154.

Farah, S., Anderson, D. G., Langer, R., "Physical and mechanical properties of PLA, and their functions in widespread applications – a comprehensive review", *Advanced Drug Delivery Reviews* 2016, 107, 367–392.

Faruk, O., Bledzki, A. K., Fink, H. P., Sain, M., "Biocomposites reinforced with natural fibers: 2000–2010", *Progress in Polymer Science* 2012, 37, 1552–1596.

Ferreira, F. V., Dufresne, A., Pinheiro, I. F., Souza, D. H. S., Gouveia, R. F., Mei, L. H. I., Lona, L. M. F., "How do cellulose nanocrystals affect the overall properties of biodegradable polymer nanocomposites: a comprehensive review", *European Polymer Journal* 2018, 108, 274–285.

Fredi, G., Rigotti, D., Bikiaris, D. N., Dorigato, A., "Tuning thermo-mechanical properties of poly(lactic acid) films through blending with bioderived poly(alkylene furanoate)s with different alkyl chain length for sustainable packaging", *Polymer* 2021, 218, 123527.

Friedrich, K., Almajid, A. A., "Manufacturing aspects of advanced polymer composites for automotive applications", *Applied Composite Materials* 2013, 20, 107–128.

Gällstedt, M., Mattozzi, A., Johansson, E., Hedenqvist, M. S., "Transport and tensile properties of compression-molded wheat gluten films", *Biomacromolecules* 2004, 5, 2020–2028.

Gejo, G., Kuruvilla, J., Boudenne, A., Sabu, T., "Recent advances in green composites", *Key Engineering Materials* 2010, 425, 107–166.

George, J., Sabapathi, S., "Cellulose nanocrystals: synthesis, functional properties, and applications", *Nanotechnology, Science and Applications* 2015, 8, 45–54.

Getme, A. S., Patel, B., "A review: bio-fibers as reinforcement in composites of polylactic acid (PLA)", *Materials Today* 2020, 26, 2116–2122.

Ghomi, E. R., Khosravi, F., Ardahaei, A. S., Dai, Y., Neisiany, R. E., Foroughi, F., Wu, M., Das, O., Ramakrishna, S., "The life cycle assessment for polylactic acid (PLA) to make it a low-carbon material", *Polymers* 2021, 13 (1854), 1–16.

Gupta, A., Katiyar, V., "Cellulose functionalized high molecular weight stereocomplex poly-lactic acid biocomposite films with improved gas barrier, thermomechanical proper-ties", *ACS Sustainable Chemistry & Engineering* 2017, 5, 6835–6844.

Gupta, B., Revagade, N., Hilborn, J., "Poly(lactic acid) fiber: an overview", *Progress in Polymer Science* 2007, 34, 455–482.

Hamad, K., Kaseem, M., Ayyoob, M., Joo, J., Deri, F., "Polylactic acid blends: the future of green, light and tough", *Progress in Polymer Science* 2018, 85, 83–127.

Hammajam, A. A., El-Jummah, A. M., Ismarrubie, Z. N., "The green composites: millet husk fiber (MHF) filled poly lactic acid (PLA) and degradability effects on environment", *Open Journal of Composite Materials* 2019, 9, 300.

Hatakeyama, H., Hatakeyama, T., "Lignin Structure, Properties, and Applications", Chapter 1 in Abe, A., Dusek, K., Kobayashi, S. (Editors), "*Biopolymers*", Springer 2009, ISBN: 978-364-21-3630-6, 1–63.

Henton, D. E., Gruber, P., Lunt, J., Randall, J., "Polylactic Acid Technology", Chapter 16 in Mohanty, A. K., Misra, M., Drzal, L. T. (Editors), "*Natural Fibers, Biopolymers and Biocomposites*", CRC Press 2005, ISBN: 978-0-849-31741-5.

Hubbe, M. A., Ferrer, A., Tyagi, P., Yin, Y., Salas, C., Pal, L., Rojas, O. J., "Nanocellulose in thin films, coatings, and plies for packaging applications: a review", *Bioresources* 2017, 12, 2143–2233.

Hwang, S. W., Shim, J. K., Selke, S., Soto-Valdez, H., Rubino, M., Auras, R., "Effect of maleic-anhydride grafting on the physical and mechanical properties of poly(L-lactic acid)/starch blends", *Macromolecular Materials and Engineering* 2013, 298, 624–633.

Jamaluddin, N., Kanno, T., Asoh, T.-A., Uyama, H., "Surface modification of cellulose nanofiber using acid anhydride for poly(lactic acid) reinforcement", *Materials Today Communications* 2019, 21, 100587.

Jem, K. J., Tan, B., "The development and challenges of poly (lactic acid) and poly (glycolic acid)", *Advanced Industrial and Engineering Polymer Research* 2020, 3, 60–70.

Johari, A. P., Mohanty, S., Kurmvanshi, S. K., Nayak, S. K., "Influence of different treated cellulose fibers on the mechanical and thermal properties of poly(lactic acid)", *ACS Sustainable Chemistry & Engineering* 2016, 4, 1619–1629.

John, M. J., Anandjiwala, R. D., "Chemical modification of flax reinforced polypropylene composites", *Composites: Part A* 2009a, 40, 442–448.

John, R. P., Nampoothiri, K. M., Pandey, A., "Fermentative production of lactic acid from biomass: an overview on process developments and future perspectives", *Applied Microbiology and Biotechnology* 2007, 74, 524–534.

John, R. P., Anisha, G. S., Nampoothiri, K. M., Pandey, A., "Direct lactic acid fermenta-tion: focus on simultaneous saccharification and lactic acid production", *Biotechnology Advances* 2009b, 27, 145–152.

Kafy, A., Kim, H. C., Zhai, L., Kim, J. W., Hai, L. V., Kang, T. J., Kim, J., "Cellulose long fibers fabricated from cellulose nanofibers and its strong and tough characteristics", *Scientific Reports* 2017, 7, 1–8.

Kalia, S., Dufresne, A., Cherian, B. M., Kaith, B., Avérous, L., Njuguna, J., Nassiopoulos, E., "Cellulose-based bio-and nanocomposites: a review", *International Journal of Polymer Science* 2011, 2011, 1–35.

Kargarzadeh, H., Ahmad, I., Abdullah, I., Dufresne, A., Zainudin, S. Y., Sheltami, R. M., "Effects of hydrolysis conditions on the morphology, crystallinity, and thermal sta-bility of cellulose nanocrystals extracted from kenaf bast fibers", *Cellulose* 2012, 19, 855–866.

Kargarzadeh, H., Huang, J., Ahmad, I., Mariano, M., Dufresne, A., Thomas, S., Galeski, A., "Recent developments in nanocellulose-based biodegradable polymers, thermo-plastic polymers, and porous nanocomposites", *Progress in Polymer Science* 2018, 87, 197–227.

Kalia, S., Dufresne, A., Cherian, B. M., Kaith, B. S., Averous, L., Njuguna, J., Nassiopoulos, E., "Cellulose-based bio- and nanocomposites: a review", *International Journal of Polymer Science* 2011, 2011, 1–35.

Kaushik, M., Moores, A., "Nanocelluloses as versatile supports for metal nanoparticles and their applications in catalysis", *Green Chemistry* 2016, 18, 622–637.

Kim, J. K., Pal, K. (Editors), *"Recent Advances in the Processing of Wood-Plastic Composites"*, Springer, Berlin 2011, ISBN 978-364-21-4877-4.

Khalil, H. P. S. A., Bhat, I. U. H., Jawaid, M., Zaidon, A., Hermawan, D., Hadi, Y. S., "Bamboo fibre reinforced biocomposites: a review", *Materials & Design* 2012, 42, 353–368.

Khalil, H. P. S., Davoudpour, Y., Saurabh, C. K., Hossain, M. S., Adnan, A. S., Dungani, R., Paridah, M. T., Sarker, M. Z. I., Fazita, M. R. N., Syakir, M. I., Haafiz, M. K. M., "A review on nanocellulosic fibres as new material for sustainable packaging: Process and applications", *Renewable and Sustainable Energy Reviews* 2016, 64, 823–836.

Khosravi, A., Fereidoon, A., Khorasani, M. M., Naderi, G., Ganjali, M. R., Zarrintaj, P., Saeb, M. R., Gutierrez, T. J., "Soft and hard sections from cellulose-reinforced poly (lactic acid)-based food packaging films: a critical review", *Food Packaging and Shelf Life* 2020, 23, 100429.

Koh, J. J., Zhang, X., He, C., "Fully biodegradable poly(lactic acid)/starch blends: a review of toughening strategies", *International Journal of Biological Macromolecules* 2018, 109, 99–113.

Kvien, I., Tanem, B. S., Oksman, K., "Characterization of cellulose whiskers and their nano-composites by atomic force and electron microscopy", *Biomacromolecules* 2005, 6, 3160–3165.

La Mantia, F. P., Morreale, M., "Green composites: a brief review", *Composites Part A: Applied Science and Manufacturing* 2011, 42, 579–588.

Lavoine, N., Givord, C., Tabary, N., Desloges, I., Martel, B., Bras, J., "Elaboration of a new antibacterial bio-nano-material for food-packaging by synergistic action of cyclodex-trin and microfibrillated cellulose", *Innovative Food Science & Emerging Technologies* 2014, 26, 330–340.

Lee, J.-A., Yoon, M.-J., Lee, E.-S., Lim, D.-Y., Kim, K.-Y., "Preparation and characterization of cellulose nanofibers (CNFs) from microcrystalline cellulose (MCC) and CNF/poly-amide 6 composites", *Macromolecular Research* 2014, 22, 738–745.

Li, F., Mascheroni, E., Piergiovanni, L., "The potential of nanocellulose in the packaging field: a review", *Packaging Technology and Science* 2015, 28, 475–508.

Li, L., Bao, R. Y., Gao, T., Liu, Z. Y., Xie, B. H., Yang, M. B., Yang, W., "Dopamine-induced functionalization of cellulose nanocrystals with polyethylene glycol towards poly (L-lactic acid) bionanocomposites for green packaging", *Carbohydrate Polymers* 2019, 203, 275–284.

Lim, J. S., Park, K.-I., Chung, G. S., Kim, J. H., "Effect of composition ratio on the thermal and physical properties of semicrystalline PLA/PHB-HHx composites", *Materials Science and Engineering* 2013, 33, 2131–2137.

Lim, L. T., Auras, R., Rubino, M., "Processing technologies for poly(lactic acid)", *Progress in Polymer Science* 2008, 33 (8), 820–852.

Liu, G., Zhang, X., Wang, D., "Tailoring crystallization: towards high-performance poly(lactic acid)", *Advanced Materials* 2014, 26, 6905–6911.

Lopes, M. S., Jardini, A. L., Filho, R., M., "Poly (lactic acid) production for tissue engineering applications", *Procedia Engineering* 2012, 42, 1402–1413.

Marsh, G., "Next step for automotive materials", *Materials Today* 2003, 6 (4), 36–43.

Martínez-Sanz, M., Lopez-Rubio, A., Lagaron, J. M., "Optimization of the dispersion of unmodified bacterial cellulose nanowhiskers into polylactidevia melt compounding to significantly enhance barrier and mechanical properties", *Biomacromolecules* 2012, 13, 3887–3899.

Mathew, A. P., Oksman, K., Sain, M., "Mechanical properties of biodegradable composites from poly lactic acid (PLA) and microcrystalline cellulose (MCC)", *Journal of Applied Polymer Science* 2005, 97, 2014–2025.

Mehta, R., Kumar, V., Bhunia, H., Upadhyay, S., "Synthesis of poly(lactic acid): a review", *Polymer Reviews* 2005, 45, 325–349.

Mohanty, A. K., Liu, W., Tummala, P., Drzal, L. T., Misra, M., Narayan, R., "Soy Protein-Based Plastics, Blends, and Composites", Chapter 22 in Mohanty, A. K., Misra, M, Drzal, L. T., *"Natural Fibers, Biopolymers, and Biocomposites"*, Taylor & Francis, CRC Press, Boca Raton, FL 2005, ISBN 978-084-93-1741-5.

Mohanty, A. K., Misra, M., Drzal, L. T., "Sustainable bio-composites from renewable resources: opportunities and challenges in the green materials world", *Journal of Polymers and the Environment* 2002, 10, 19–26.

Mohanty, A. K., Misra, M., Drzal, L. T., Selke, S. E., Harte, B. R., Hinrichsen, G., "Natural Fibers, Biopolymers, and Biocomposites: An Introduction", Chapter 1 in Mohanty, A. K., Mohanty, A. K., Misra, M. A., Hinrichsen, G. I., *"Biofibres, Biodegradable Polymers and Biocomposites: An Overview"*, *Macromolecular Materials and Engineering* 2000, 276, 1–24.

Misra, M, Drzal, L. T., *"Natural Fibers, Biopolymers, and Biocomposites"*, Taylor & Francis, CRC Press 2005b, ISBN 978-084-93-1741-5, 1–35.

Moon, R. J., Schueneman, G. T., Simonsen, J., "Overview of cellulose nanomaterials, their capabilities and applications", *The Journal of The Minerals, Metals & Materials Society* 2016, 68, 2383–2394.

Mooney, B. P., "The second green revolution? Production of plant-based biodegradable plastics", *Biochemical Journal* 2009, 418 (2), 219–232.

Morteza, E., Khosrow, K., Mohammad A., "Lactide synthesis optimization: investigation of the temperature, catalyst and pressure effects", *e-Polymers* 2014, 14 (5), 353–361.

Nagarajan, K. J., Ramanujam, N. R., Sanjay, M. R., Siengchin, S., Surya Rajan, B., Sathick Basha, K., Madhu, P., Raghav, G. R., "A comprehensive review on cellulose nanocrystals and cellulose nanofibers: pretreatment, preparation, and characterization", *Polymer Composites* 2021, 42 (4), 1588–1630.

Nair, L. S., Laurencin, C. T., "Biodegradable polymers as biomaterials", *Progress in Polymer Science* 2007, 32, 762–798.

Nampoothiri, K. M., Nair, N. R., John, R. P., "An overview of the recent developments in polylactide (PLA) research", *Bioresource Technology* 2010, 101, 8493–8501.

Niu, X., Liu, Y., Song, Y., Han, J., Pan, H., "Rosin modified cellulose nanofiber as a reinforcing and co-antimicrobial agents in polylactic acid/chitosan composite film for food packaging", *Carbohydrate Polymers* 2018, 183, 102–109.

Ochi, S., "Mechanical properties of kenaf fibers and kenaf/PLA composites", *Mechanics of Materials* 2008, 40, 446–452.

Okamoto, M., "Biodegradable Polymer/Layered Silicate Nanocomposites: A Review", Chapter 8 in Mallapragada, S., Narasimhan, B. (Editors), *"Handbook of Biodegradable Polymeric Materials and Their Applications"*, American Scientific Publishers 2005, ISBN: 158-88-3053-5,1–45.

Oyama, H. T., Tanaka, Y., Kadosaka, A., "Rapid controlled hydrolytic degradation of poly(l-lactic acid) by blending with poly(aspartic acid-co-l-lactide)", *Polymer Degradation and Stability* 2009, 94, 1419–1426.

Pal, N., Banerjee, S., Roy, P., Pal, K., "Reduced graphene oxide and PEG-grafted TEMPO-oxidized cellulose nanocrystal reinforced poly-lactic acid nanocomposite film for biomedical application", *Materials Science and Engineering C: Materials for Biological Applications* 2019, 104, 109956.

Patel, M., Narayan, R., "How Sustainable Are Biopolymers and Biobased Products? The Hope, the Doubts, and the Reality", Chapter 27 in Mohanty, A. K., Misra, M, Drzal, L. T., *Natural Fibers, Biopolymers, and Biocomposites*", Taylor & Francis, CRC Press 2005, ISBN 978-084-93-1741-5.

Pervaiz, M., Sain, M. M., "Carbon storage potential in natural fiber composites", *Resources, Conservation and Recycling* 2003, 39, 325–340.

Peter, S. J., Miller, M. J., Yasko, A. W., Yaszemski, M. J., Mikos, A. G., "Polymers concepts in tissue engineering", *Journal of Biomedical Materials Research* 1998, 43, 422–427.

Petersen, K., Nielsen, P. V., Bertelsen, G., Lawther, M., Olsen, M. B., Nilsson, N. H., Mortensen, G., "Potential of biobased materials for food packaging", *Trends in Food Science & Technology* 1999, 10 (2), 52–68.

Petersson, L., Kvien, I., Oksman, K., "Structure and thermal properties of poly(lactic acid)/cellulose whiskers nanocomposite materials", *Composites Science and Technology* 2007, 67, 2535–2544.

Pilla, S. (Editor), *"Handbook of Bioplastics and Biocomposites Engineering Applications"*, Wiley-Scrivener 2011, ISBN: 978-047-06-2607-8.

Poole S., Simon, M., "Technological trends, product design and the environment", *Design Studies* 1997, 18 (3), 237–248.

Puglia, D., Biagiotti, J., Kenny, J. M., "A review on natural fibre-based composites – Part II", *Journal of Natural Fibres* 2008, 1 (3), 23–65.

Raisipour-Shirazi, A., Ahmadi, Z., Garmabi, H., "Polylactic acid nanocomposites toughened with nanofibrillated cellulose: microstructure, thermal, and mechanical properties", *Iranian Polymer Journal* (Engl. Ed.) 2018, 27, 785–794.

Rajeshkumar, G., "A new study on tribological performance of phoenix Sp. Fiber reinforced epoxy composites", *Journal of Natural Fibres* 2020, 17, 1–12.

Rajeshkumar, G., Hariharan, V., Sathishkumar, T. P., "Characterization of Phoenix sp. natural fiber as potential reinforcement of polymer composites", *Journal of Industrial Textiles* 2016, 46, 667–683.

Rajeshkumar, G., Hariharan, V., Sathishkumar, T. P., Fiore, V., Scalici, T., "Synergistic effect of fiber content and length on mechanical and water absorption behaviors of Phoenix sp. fiber-reinforced epoxy composites", *Journal of Industrial Textiles* 2017, 47, 211–232.

Rajeshkumar, G., Seshadri, S. A., Devnani, G. L., Sanjay, M. R., Siengchin, S., Maran, J. P., Al-Dhabi, N. A., Karuppiah, P., Mariadhas, V. A., Sivarajasekar, N., Anuf, A. R., "Environment friendly, renewable and sustainable poly lactic acid (PLA) based natural fiber reinforced composites – a comprehensive review", *Journal of Cleaner Production* 2021, 310, 1–26.

Ramamoorthy, S. K., Skrifvars, M., Persson, A., "A review of natural fibers used in biocomposites: plant, animal and regenerated cellulose fibers", *Polymer Reviews* 2015, 55, 107–162.

Rao, G. S., Debnath, K., Mahapatra, R. N., "Development and Characterization of PLA-Based Green Composites: Experimental and Simulation Studies", Chapter 7 in Thomas, S., Balakrishnan, P. (Editors), *"Green Composites"*, Springer Nature Singapore Pte Ltd 2021, ISBN: 978-981-15-9643-8, 209–224.

Ray, D., Sarkar, B. K., Das, S., Rana, A. K., "Dynamic mechanical and thermal analysis of vinylester-resin-matrix composites reinforced with untreated and alkali treated jute fibres", *Composites Science and Technology* 2002, 62, 911–917.

Rosli, N. A., Ahmad, I., Anuar, F. H., Abdullah, I., "Application of polymethylmethacrylate-grafted cellulose as reinforcement for compatibilised polylactic acid/natural rubber blends", *Carbohydrate Polymers* 2019, 213, 50–58.

Rudnik, E., Briassoulis, D., "Degradation behaviour of poly(lactic acid) films and fibres in soil under Mediterranean field conditions and laboratory simulations testing", *Industrial Crops and Products* 2011a, 33, 648–658.

Rudnik, E., Briassoulis, D., "Comparative biodegradation in soil behaviour of two biodegradable polymers based on renewable resources", *Journal of Polymers and the Environment* 2011b, 19, 18–39.

Saheb, D. N., Jog, J. P., "Natural fiber polymer composites: a review", *Advances in Polymer Technology* 1999, 18, 351–363.

Sain, M., Pervaiz, M., "Mechanical Properties of Wood-Polymers Composites", in Niska, K. O., Sain, M. (Editors), *"Wood-Polymer Composites"*, Woodhead Publishing 2008, ISBN: 978-184-56-9272-8, 101–116.

Sangeetha, V. H., Deka, H., Varghese, T.O., Nayak, S. K., "State of the art and future prospectives of poly(lactic acid) based blends and composites", *Polymer Composites* 2016, 10. DOI:10.1002/pc.23906.

Sanjay, M. R., Siengchin, S., Parameswaranpillai, J., Jawaid, M., Pruncu, C. I., Khan, A., "A comprehensive review of techniques for natural fibers as reinforcement in composites: preparation, processing and characterization", *Carbohydrate Polymers* 2019, 207, 108–121.

Scaffaro, R., Botta, L., Maio, A., Gallo, G., "PLA graphene nanoplatelets nanocomposites: physical properties and release kinetics of an antimicrobial agent", *Composites Part B: Engineering* 2017, 109, 138–146.

Scarponi, C., Pizzinelli, C. S., Sanchez-Saez, S., Barbero, E., "Impact load behavior of resin transfer molding (RTM) hemp fibre composite laminates", *Journal of Biobased Materials and Bioenergy* 2009, 3, 298–310.

Sedničková, M., Pekařová, S., Kucharczyk, P., Bočkaj, J., Janigová, I., "Changes of physical properties of PLA-based blends during early stage of biodegradation in compost", *International Journal of Biological Macromolecules* 2018, 113, 434–442.

Shojaeiarani, J., Bajwa, D. S., Hartman, K., "Esterified cellulose nanocrystals as reinforcement in poly(lactic acid) nanocomposites", *Cellulose* 2019, 26, 2349–2362.

Shojaeiarani, J., Bajwa, D. S., Stark, N. M., "Green esterification: a new approach to improve thermal and mechanical properties of poly(lactic acid) composites reinforced by cellulose nanocrystals", *Journal of Applied Polymer Science* 2018, 135, 46468.

Singh, A. A., Genovese, M. E., Mancini, G., Marini, L., Athanassiou, A., "Green processing route for polylactic acid–cellulose fiber biocomposites", *ACS Sustainable Chemistry & Engineering* 2020, 8, 4128–4136.

Soman, S., Chacko, A. S., Prasad, V. S., "Semi-interpenetrating network composites of poly(lactic acid) with cis-9-octadecenylamine modified cellulose-nanofibers from Areca catechu husk", *Composites Science and Technology* 2017, 141, 65–73.

Song, Y., Tashiro, K., Xu, D., Liu, J., Bin, Y., "Crystallization behavior of poly(lactic acid)/microfibrillated cellulose composite", *Polymer* 2013, 54, 3417–3425.

Sreekumar, K., Bindhu, B., Veluraja, K., "Perspectives of polylactic acid from structure to applications", *Polymers from Renewable Resources* 2021, 12 (1–2), 60–74.

Stoddart, A., Feast, W. J., Rannard, S. P., "Synthesis and thermal studies of aliphatic polyurethane dendrimers: a geometric approach to the Flory–Fox equation for dendrimer glass transition temperature", *Soft Matter* 2012, 8, 1096–1108.

Suryanegara, L., Nakagaito, A. N., Yano, H., "The effect of crystallization of PLA on the thermal and mechanical properties of microfibrillated cellulose-reinforced PLA composites", *Composites Science and Technology* 2009, 69, 1187–1192.

Thakur, V. K., Singha, A. S., Thakur, M. K., "Graft copolymerization of methyl acrylate onto cellulosic biofibers: synthesis, characterization and applications", *Journal of Polymers and the Environment* 2011, 20, 164–174.

Thakur, V. K., Thakur, M. K., "Processing and characterization of natural cellulose fibers/thermoset polymer composites", *Carbohydrate Polymers* 2014, 109, 102–117.

Tingaut, P., Zimmermann, T., Lopez-Suevos, F., "Synthesis and characterization of bionanocomposites with tunable properties from poly(lactic acid) and acetylated microfibrillated cellulose", *Biomacromolecules* 2009, 11, 454–464.

Vickroy, T. B., "Lactic Acid", in Moo-Young, M. (Editor), *"Comprehensive Biotechnology"*, Pergamon Press 1985, ISBN: 978-044-46-4047-5, 761–776.

Vink, E. T., Rabago, K. R., Glassner, D. A., Gruber, P. R., "Applications of life cycle assessment to NatureWorks™ polylactide (PLA) production", *Polymer Degradation and Stability* 2003, 80, 403–419.

Wang, T., Drzal, L. T., "Cellulose-nanofiber-reinforced poly (lactic acid) composites prepared by a water-based approach", *ACS Applied Materials & Interfaces* 2012, 4, 5079–5085.

Wie, H., Rodriguez, K., Renneckar, S., Vikesland, P. J., "Environmental science and engineering applications of nanocellulose-based nanocomposites", *Environmental Science: Nano journal* 2014, 1, 302–316.

Witzke, D. R., Narayan, R., Kolstad, J. J., "Reversible kinetics and thermodynamics of the homopolymerization of L-lactide with 2-ethylhexanoic acid tin(II) salt", *Macromolecules* 1997, 30, 7075–7085.

Yang, S.-L., Wu, Z.-H., Yang, W., Yang, M.-B., "Thermal and mechanical properties of chemical crosslinked polylactide (PLA)", *Polymer Testing* 2008, 27, 957–963.

Yu, H. Y., Zhang, H., Song, M. L., Zhou, Y., Yao, J., Ni, Q. Q., "From cellulose nanospheres, nanorods to nanofibers: various aspect ratio induced nucleation/reinforcing effects on polylactic acid for robust-barrier food packaging", *ACS Applied Materials & Interfaces* 2017, 9, 43920–43938.

Zampaloni, M., Pourboghrat, F., Yankovich, S. A., Rodgers, B. N., Moore, J., Drzal, L. T., Mohanty, A. K., Misra, M., "Kenaf natural fiber reinforced polypropylene composites: a discussion on manufacturing problems and solutions", *Composites Part A: Applied Science and Manufacturing* 2007, 38, 1569–1580.

Zhang, X., Tu, M., Paice, M. G., "Routes to potential bioproducts from lignocellulosic biomass lignin and hemicelluloses", *BioEnergy Research* 2011, 4 (4), 246–257.

Zhang, X., Zhang, Y., "Reinforcement effect of poly(butylene succinate) (PBS)-grafted cellulose nanocrystal on toughened PBS/polylactic acid blends", *Carbohydrate Polymers* 2016, 140, 374–382.

Zhou, L., He, H., Li, M.-C., Huang, S., Mei, C., Wu, Q., "Enhancing mechanical properties of poly(lactic acid) through its in-situ crosslinking with maleic anhydride-modified cellulose nanocrystals from cottonseed hulls", *Industrial Crops and Products* 2018, 112, 449–459.

Zhou, L., Ke, K., Yang, M.-B., Yang, W., "Recent progress on chemical modification of cellulose for high mechanical performance poly(lactic acid)/cellulose composite: a review", *Composites Communication* 2021, 23, 1–11.

Zhu, L., Qiu, J., Liu, W., Sakai, E., "Mechanical and thermal properties of rice Straw/PLA modified by nano attapulgite/PLA interfacial layer", *Composites Communications* 2019, 13, 18–21.

2 Synthesis and Production of Polylactic Acid (PLA)

Sayan Ganguly and Poushali Das
Bar-Ilan University
Ramat-Gan, Israel

CONTENTS

2.1   INTRODUCTION

Consumer plastics produced by several multinational companies throughout the world are concentrated on petroleum-based polymers [1, 2]. The use of naturally obtained polymers and biopolymers is somehow restricted in engineering and commodity polymer composite sections. Petroleum-based polymers are non-biodegradable, and most of the biodegradable polymers are not giving high strength for external use [3]. Synthetic biodegradable polymers are comparatively new than other commodity plastics [4]. The commonly used synthetic biodegradable polymers are polylactic acid (PLA), polyglycolic acid (PGA) and polycaprolactone (PCL) [5]. These all polymers belong to the polylactone family. Besides their biodegradability, these synthetic polymers are also non-toxic and less expensive compared to the naturally abundant polymers [6]. Among the polylactone-family, PGA was the first reported biodegradable synthesized polymer. As per history, Théophile-Jules Pelouze reported for the first time the synthesis of PLA adopting polycondensation pathway in the year 1845 [7]. But it was commercialized when coronavirus and coworkers developed an inexpensive method for polymerizing lactide to PLA [8]. In 1954 when du Pont covered the patent of PLA synthesis the commercialization started [9]. Initially, PLA was assumed to be a potential biomaterial due to its non-toxicity and excellent compatibility with human cells [10]. PLA has been used in several therapeutic and pharmaceutical industries such as drug carriers, encapsulation

DOI: 10.1201/9781003160458-2

29

TABLE 2.1

Properties of Commercially Available Plastic Films in Comparison to PLA

Parameters	PLA	PET	PP	Cellophane	Nylon
Density (g/cc)	1.25	1.4	0.9	1.45	1.2
Tensile strength (psi)	15,950	29,725	27,550	13,050	36,250
Tensile modulus (psi)	478,500	551,000	348,000	594,500	264,625
Elongation at break (%)	160	140	110	23	125
Tear (g/mil)	15	18	4–6	4	13
Haze (%)	2.1	2–5	1–4	1–2	2–3

Abbreviations: PET, poly(ethylene terephthalate); PLA, poly(lactic acid); PP, polypropylene.

controlled release microspheres and hydrogels [11]. For a few years, PLA is also used in tissue engineering applications by developing high strength compatible scaffolds [12]. Moreover, PLA is also used in biocompatible sutures and prostheses for in vivo applications. In 1974, 'vicryl' suture was marketed, which consists of PGA and PLA at a 90:10 blend ratio [13]. But the main drawback of PLA was their restriction in producing high-molecular weight polymers. Later on, high-molecular weight lactic acid was synthesized by adopting the ring-opening polymerization (ROP) route. The PLA of high molecular weight has excellent mechanical properties that are almost comparable to petroleum-based commodity plastics. In Table 2.1, we are trying to depict some basic features of PLA compared to other commercially available plastics. PLA is mostly composed of in contents and specific stereoisomers that enhance the melting point of the product and improve mechanical properties.

PLA and its derivatives are easily degraded by simple hydrolysis of the ester bonds present in the backbone of the polymer. This hydrolysis process is not only dependent on industrial hydrolysis systems but also propagated through enzyme interaction. The suture made of PLA is easily miscible to the human body after surgery. Besides the biomedical applications, PLA is also used in several packaging applications candy wraps, shrink films and high optically transparent films. PLA resin is used for thermoforming of cups containers and disposable bottles. Because of its thermoplastic nature, the polymer products are re-utilized and recycled several times that is highly advantageous for low polymer waste management. PLA synthesis demands several reaction conditions like temperature, pressure and pH. The choice of catalyst for preparing the polymer is also has a major significance in their mechanical properties. In this chapter, the synthetic routes to prepare PLA from the monomer will be discussed.

2.2 MONOMER (L-LACTIC ACID) SYNTHESIS

Lactic acid (2-hydroxypropanoic acid) is the monomer of PLA. The monomer lactic acid is optically active, and it has two enantiomers. The enantiomers are named as D and L. The configuration and chemical structures are shown in Figure 2.1. The monomer is considered as an efficient molecule for polycondensation reactions.

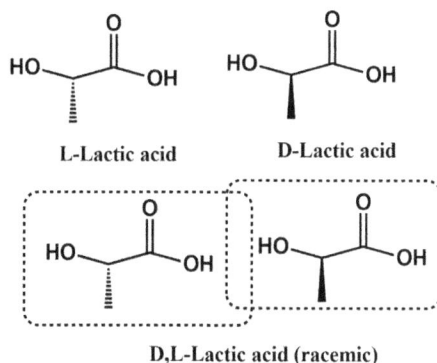

FIGURE 2.1 Possible isomers of lactic acid.

When the D and L isomers are present in equimolar proportion then it is called a racemic mixture which is optically inactive.

The fermentation of lactic acid is one of the common bacterial reactions that have taken place since the early days of mankind. The lactic acid fermentation bacteria are classified into several genres according to their cell morphology, like lactobacillus, streptococcus, pediococcus, leuconostoc, aerococcus and coryne species. These bacteria produce in lactic acid with the presence of the very small amount of D and DL lactic acid. Lactobacillus normally produces either D variety or L variety of lactic acid preferentially. Most of the bacteria turn saccharide molecules into lactic acid via fermentation. But the major problem is the purification process because in the case of bacteria as a sister mission process, the monomer is not pristine. After fermentation of the monomer, various adulteration like unreacted saccharides media amino acids, carboxylic acids proteins and several inorganic salts. Thus the purification and isolation have immense significance in lactic acid monomer production. The lactic acid monomer obtained by the bacterial fermentation process is initially neutralized through in situ method with calcium oxide or ammonia. When the monomer mixture is treated with alkalis like calcium oxide, calcium lactate is formed and precipitated out from the reaction mixture. Salt is very easy to handle and isolated by filtration followed by several times washing with water. After complete washing, the obtained monomer is acidified with sulfuric acid to produce free lactic acid without any calcium lactate. For ammonia treated monomer systems, the product is ammonium lactate which is directly converted into butyl lactate through an esterification reaction. Butyl lactate is a stable product that can be converted into lactic acid via simple hydrolysis and distillation process. This lactic acid is considered as a high purity monomer; that's why the calcium salt precipitation method is the most commercialized pathway to prepare lactic acid. The above-mentioned processes give almost 100,000 tons production of lactic acids in a year.

Besides bacterial fermentation, lactic acid monomers are also prepared by the chemical synthesis method. Lactic acid can be prepared from acetaldehyde through a cyanohydrin reaction, as shown in Figure 2.2. The chemical

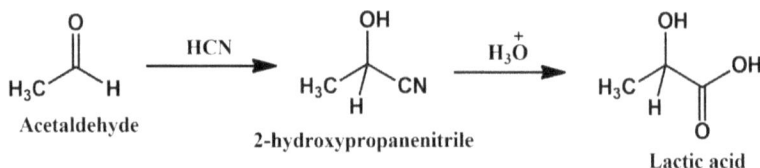

FIGURE 2.2 Synthesis of lactic acid by cyanohydrin process.

synthesis follows two consecutive steps; first step is the formation of cyano-hydrin and the second step is acidic hydrolysis. Industrially this process is favorable and is called the Sohio process. This process has only one demerit of adulteration by lactonitrile.

Lactides also consist of different stereoisomers like lactic acid, as shown in Figure 2.3. Lactides are also categorized into D, L, meso and DL-mixed varieties. The racemic lactide is a mixture of D and L pairs in an equimolar ratio. Both the D and L lactides are comparable in terms of their individual melting points. For meso lactide, the melting point is comparatively lower, and racemic lactide possesses the highest melting point among the lactide family.

The lactides mentioned here are normally synthesized after depolymerization of oligo lactic acid. The formation of lactic acid is abided by the condensation reaction. Figure 2.4 depicts the preparation of oligo lactic acid and the mecha-nism of the polymerization from oligo lactic acid. The formation of lactate from this oligomer is unzipping depolymerization. The backbiting phenomenon is also

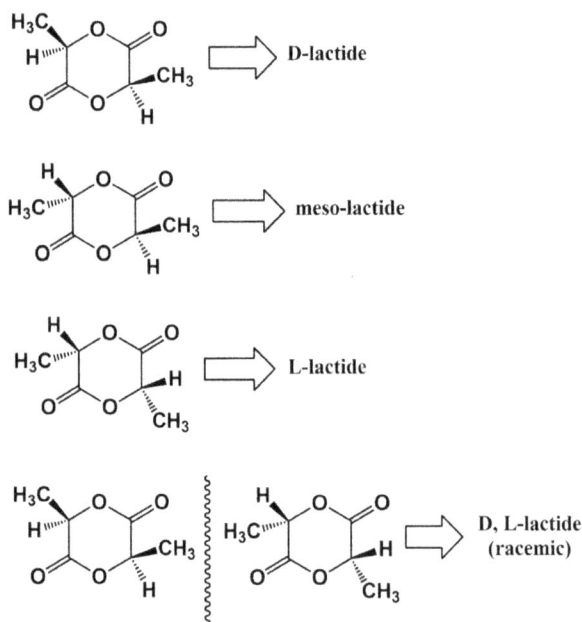

FIGURE 2.3 Structures of lactides.

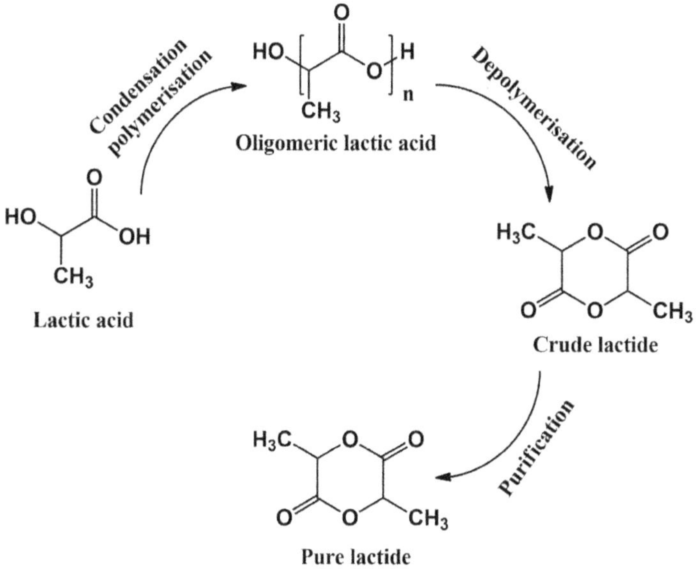

FIGURE 2.4 Synthesis of lactide from lactic acid adopting oligolactide route.

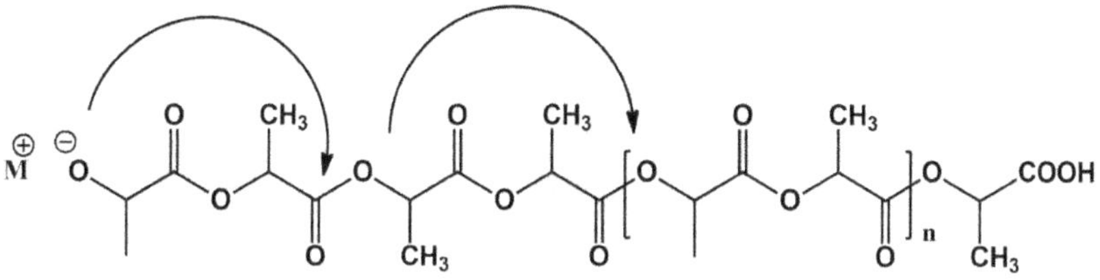

FIGURE 2.5 Plausible reaction of lactide showing backbiting process.

prominent in such depolymerization method, as shown in Figure 2.5. The catalysts used for these reactions are Sn, Al, Zn and Sb ions. The lactides are purified by the melt crystallization method or less expensive recrystallization from the crude solution.

2.3 POLYMERIZATION FROM LACTIC ACID

Lactic acid is the main monomer for preparing PLA and its derivative copolymers. For preparing PLA from lactic acid, two different polymerization techniques are adopted. The first one is the polycondensation reaction which is the first discovered method to prepare PLA. The second one is the ROP that can produce comparatively higher molecular weight PLA than polycondensation reactions. Other than these methods, dehydration and enzymatic polymerization are also taking place but in a very limited fashion. Figure 2.6 shows the basic adopted production techniques applied for the synthesis of PLA.

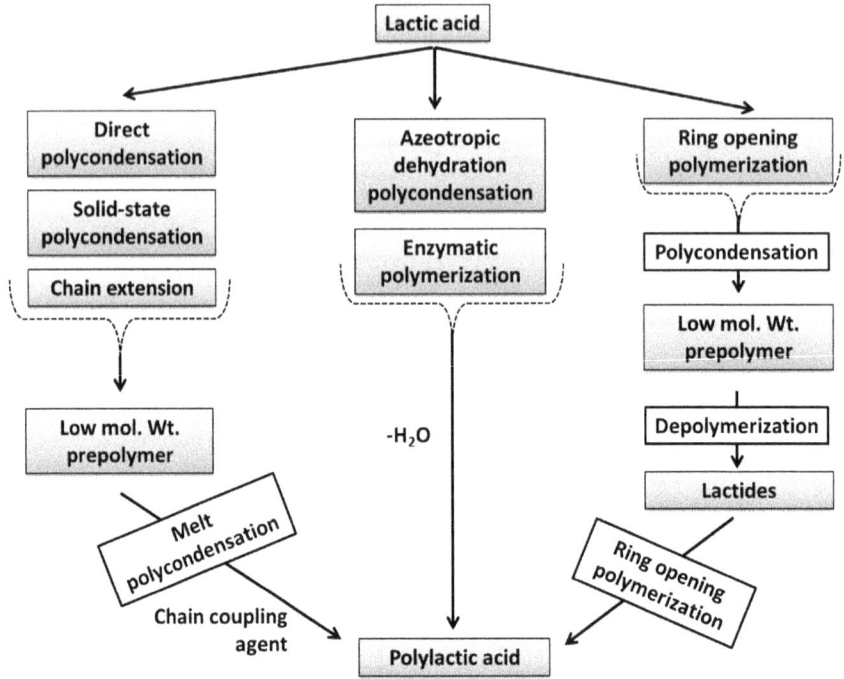

FIGURE 2.6   Different synthetic approaches for the preparation of poly(lactic acid).

2.4   PRODUCTION OF PLA BY POLYCONDENSATION

Condensation polymerization of lactic acid normally takes place in a solvent with high vacuum and high temperature. This polycondensation can be performed in a bulk system by using a distillation column. This process is not solely dependent on the catalyst that means polycondensation could take place with or without a catalyst system provided or the reaction conditions should be rigorous. But in this process, high-molecular weight polymer preparation is little difficult. The intermediate oligomer of PLA is sometimes used in polyurethane synthesis, which is controlled by a different catalyst system. This aforementioned approach was used by Carothers, and later that was adopted by Mitsui Toatsu Chemicals Inc. for manufacturing of low molecular weight for lactic acid, the catalyst system is very much significant in polymer synthesis. The most common catalyst systems used for this purpose are boric acid or sulfuric acid, which are well-known for their acceleration type of behavior in the esterification and transesterification process. But the high-temperature reaction always affects side reactions. Due to these side reactions, the resultant PLA causes the molecular weight of 3000 for racemic mixtures and 6500 for pristine L-lactide. The yield could be improved by using non-acidic transesterification catalysts like lead monoxide. Low-molecular weight PLA is used for biodegradable adhesive applications like glues and lacquers. Due to the highly abundant hydroxyl and carboxylic groups at their terminal of the polymer change, these are very much suitable for cross-linking with inorganic and organic multivalent moieties.

Polycondensation also produces a moderate molecular weight of 10,000 by using precise control in the solvent ratio or by using azeotropic solvents. This type of polymer is used independently as a thermoplastic matrix and as a coupling agent for isocyanates oxides and peroxides for better mechanical properties. Though polycondensation has several merits, for industrialization, this process has limited applicability.

2.5 PRODUCTION OF PLA BY RING-OPENING POLYMERIZATION (ROP)

Lactide is a cyclic dimer of lactic acid which is produced under water-less mild conditions. This monomer is verified under the vacuum distillation process without using any solvent. The molecular weight of the end product is highly dependent on the quality of the lactide dimer used in the polymerization process. Moreover, the purification of lactide dimer is also responsible for the preparation of a wide range of molecular weight distribution of the PLA.

PLA production has been started for the last ten years, which means it is a newcomer in the thermoplastic polymer industry compared to the other commodity thermoplastics. The currently used catalyst for this purpose is stannous octoate which is an industrially approved catalyst system. Besides, stannous octoate zinc salts are commonly used in France. ROP takes place in bulk, melt or solution processes where the precursor material is the lactide. In terms of the mechanic stick point of hue ROP of lactides also takes place in cationic anionic and coordination type modes. There are several synthetic parameters which are playing significant roles in the molecular weight obtained from ROP. The specific synthetic parameters are initiator systems, change control agent, catalyst concentration, co-initiator system, monomer to initiator ratio, reaction temperature and the duration of the polymerization process. These synthetic parameters affect the degree of polymerization and crystallinity of the end product. Low initiator concentration and high duration of polymerization provide a high-molecular weight polymer with high mechanical properties. The racemic mixture of the monomer imposes enantiomeric purity and distinguishable chain microstructure of the end product. In recent days, stannous octoate is the only used catalyst for preparing PLA from lactides. The high-temperature synthesis method is accepted by the US Food and Drug Administration because of low racemization and very low toxicity of the end product. Table 2.2 provides the information of PLA molecular weights against the reaction parameters.

Some other works revealed the use of alkoxy-amino-bis (phenolate)/ Group 3 metal complexes [14], dimeric aluminum chloride-based complexes of N-alkoxyalkyl-b-ketoimines (activated with propylene oxide) [15], heterobimetallic iron(+2) alkoxide/aryloxides [16], aluminum complexes having tetradentate bis(aminophenoxide) ligands [17], zinc alkoxide complex [18], stannous octoate and diethanolamine [19]. B-diiminate attached magnesium and zinc amides [20], iron alkoxide [21], titanium alkozide [22], 2,6-dimethyl aryloxide [23], calcium coordination complexes [24], complexes of Cu, Zn, Co and Ni Schiff base derived from salicylidene and L-aspartic acid [25], stannous octoate associated with oligomers of L-lactide and racemic lactide [26], dizinc monoalkoxide complex supported by

TABLE 2.2
Effect of Synthesis Parameters in Production of PLA

Polymer	Catalyst	Reaction Temp. (°C)	Time	Mol. Wt.	References
D,L-PLA	Al-isopropoxide	70–100	100 h	>90,000	[40]
L-PLA	Sn octoate	130	6 h	$DP_n \sim 43–178$	[41]
D,L-PLA	Sn octoate	200	60–75 min	<350,000	[42]
D,L-PLA	Sn octoate	130	2–72 h	<250,000	[43]
L-PLA	Sn octoate	130	72 h	20,000–680,000 (M_v)	[44]
L-PLA	Sn octoate and triphenylamine	180–185	7 min	91,000	[45]
L-PLA	Sn octoate and Ti/Zr compounds	180–235	15–180 min	40,000–1,000,000	[46]
D-PLA	Sn trifluoromethane sulfonate, Sc(+3) trifluoromethane sulfonate	40–65	50–100 h	$DP_n \sim 15–30$	[47]
L-PLA	Sn-mesoporous silica molecular sieve	130	72 h	<36,000	[48]
L-PLA	Mg/Al/Zn/Ti-alkoxides	100	–	$DP_n < 400$	[49]
L-PLA	Yttrium tris(2,6-di-tert butylphenolate)	22	2–5 min	<25,000	[50]
D,L-PLA	Butyl lithium, butyl magnesium	–	30 min	<45,000	[51]
D,L-PLA	Zn lactate	140	96 h	212,000	[34]
D,L-PLA	Butyl magnesium, Grignard reagent	0–25	4–8 days	<3,000,000	[34]
L-PLA	Potassium naphthalenide	40	48–120 h	<16,000	[52]
L-PLA	Fe, acetic, butyric, siobutyric and dichloroacetic acids	17–210	0.5–25 h	150,000 (M_w)	[32]
D,L-PLA	(Trimethyl triazacyclohexane) praseodymium triflate	120–200	18 h	10,000–20,000 (M_v)	[53]
D,L-PLA	La-isopropoxide	21	30 min	5300–21,900	[54]
PLA	PbO, Pb-stearate, basic lead carbonate, antimony trioxide, zinc oxide, zinc borate, cadmium oxide, titanyl stearate, magnesium oxide and calcium formate	140–180	10–111 h	IV (benzene) = 1.21	[19]

a dinucleating ligand [27], tertiary amines, phosphines and N-heterocyclic carbenes [28], alkyl aluminum [29] and aluminum–achiral ligand complexes [30]. In earlier stages, Baker and Smith reported unique strategies of PLA synthesis from racemic materials [31].

In some of the reported works, the lactide monomer was used without purification, whereas for some special cases, it was purified by recrystallization from solvents like acetone, chloroform, ethyl, etc. [32]. It has been revealed that recrystallized L-lactide has higher intrinsic viscosities compared to the single-recrystallized lactide monomer [33]. Moreover, it also has been noticed that dichlorodimethylsilane (DCMS, silanizer) treated reaction vessel/kettle gives a higher molecular weight product. The monomer to initiator ratio, a significant synthetic parameter, also has an effect on the molecular weight. The ratio varies from 50 to 50,000 in various studies as reported in literatures. Other functioning circumstances also have been altered substantially. In some reports, the stirring is also added parameters for mixing that are performed in the sealed reaction vessel [34]. The reaction environments are generally air [33], inert gas [35] or vacuum [34]. The reaction product is solubilized in a suitable solvent (dichloromethane, chloroform or acetone) and precipitated out by using diethyl ether/methanol as non-solvents followed by filtration and drying. The polymer characterization is performed by ^{1}H NMR, ^{13}C NMR and FTIR spectroscopic techniques. Characterization of several PLA by NMR [36] and by FTIR [37] has been studied. The molecular weights of the prepared PLA are found out by several fashions including calculation of intrinsic viscosity, calculation of number average molecular weight (M_n), and by using Mark-Houwink equation [38] in chloroform or benzene. Gel permeation chromatography (GPC) and ^{13}C NMR are the most powerful tools to confirm its structure and bonding [39].

2.6 SINGLE-STEP ECO-COMMERCIAL SYNTHESIS

Single-step production method is always welcomed due to commercial feasibility. A single-step process is less time-consuming and a continuous flow process. For the manufacturing of PLA single-step process has been used where the continuous reactive extrusion method is the primary strategy. In this process, bulk polymerization has taken place. Bulk polymerization was chosen for such a reactive extrusion method because of their very low cycle time, which is in the range of 5–7 minutes. The short time of polymerization is desirable for PLA because of their low thermal stability at high temperature and high shear force. As molecular weight is also an important parameter for thermoplastic PLAs, the reactive extrusion method is also superior in that case. In the bulk polymerization method, catalyst couple is used. Like the melt processing technique, reactive extrusion is also affected by the chain branching of PLA backbiting and intermolecular transesterification reaction. For the reactive extrusion method, Lewis base couple has been used where triphenylphosphine and stannous octoate are associated with forming a catalyst couple. Commercial scale polymerization of PLA is not restricted not only to the commodity plastic section, but also it is promoted as a good biomaterial in several biopharmaceutical industries. That's why catalyst choice is an important area of research where the material scientist has to ensure the compatibility of the end product to the human body. Stannous octoate is the most commonly used catalyst system that is compatible with biological systems, and till now, it is the best-known catalyst for PLA synthesis. But there were some problems with the stannous octoate because of its large molecular size that affects the renal threshold. Similarly, triphenylphosphine is also showing some adverse effects on human tissues.

2.7 CONTINUOUS PROCESS FOR PLA SYNTHESIS

The continuous process for preparing PLA is on demand due to its first production rate and tunable optical clarity of the end product. Figure 2.7 shows a schematic diagram of the continuous process applied for the manufacturing of PLA from its precursor material. A continuous process involves two consecutive steps where a pre-polymer is prepared initially, followed by polymerization to prepare PLA. Initially, lactide has been formed from lactic acid monomer and the prepared lactide is puri-fied by distillation. Crude lactic acid consists of 15% pure monomer by weight, and the rest of the part is water. The heat is normally purified by evaporation technique power by the addition of alcohol as solvents. Alcohol is used as the career for lactic acid. In this process, almost 99% of the lactic acid is concentrated and used for

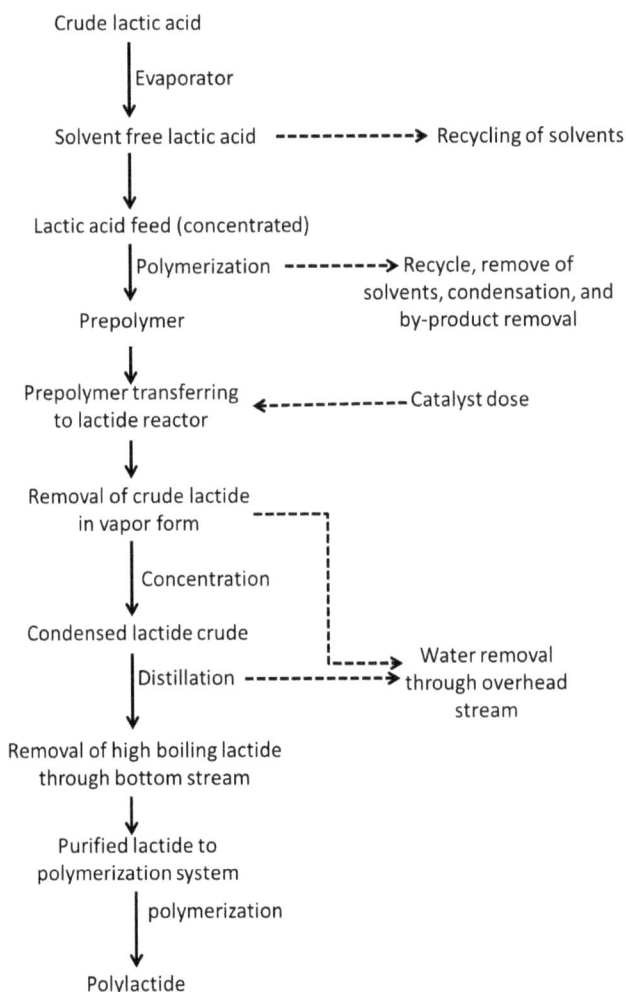

FIGURE 2.7   Schematic flow diagram of PLA preparation in a continuous process.

or the next step of prepolymer formation. In the prepolymer process, a very low-molecular weight polymer is prepared. The prepared polymer molecular weight is laid in between 4000 and 5000. This polymer mixture is transferred to the lactide reactor followed by catalyst charging. Various types of process aids and other stabilizers are added in this step for preparing masterbatches. During the polymerization of lactides, it is kept in mind that the monomer should be stored inside the reactor as less as possible. For removal of excess monomers from the reaction, vaporized forms of the monomers are transferred to a separate distillation unit for further purification. When the excess monomer is kept inside the reactor for a prolonged time, there would be no chance of the polymerization of the polylactides. This process affects the very low yield of the end product compared to the other conventional processes. In this process, both L and D forms are used as monomers or sometimes as a combination of them. During the process, sometimes racemization and inversion both are carried out that affect the optical transparency of the end product. The racemization is also an outcome of the reaction parameters like temperature, pressure and time. Moreover, catalyst purity is also an important parameter for the optical clarity of the polymer product. The relative concentrations of the lactide monomers (D and L) are also playing a good role in optical properties as well as crystal unity of that polymer product. In general, amorphous polymer films are the most transparent and less crystalline in nature. In this continuous process, PLA forms are very much amorphous and semi crystalline in nature.

2.8 MECHANISM OF PLA FORMATION

PLA synthesis involves serious roles of catalyst systems, especially for the ROP. Several catalyst systems provide different molecular weight and their distribution in the end products. The most common commercially available polymer which has been synthesized by ROP is lactone-based polymers. According to several scientists, it is already well known that electron polymerization occurs via an insertion mechanism [55]. The simple catalyst systems used for this insertion mechanism are tin alkoxides and aluminum alkoxides. Initially, halogen lights are used for the catalyst, but it is proved later that the halogenides are transferred to alkoxides during the polymerization process [56]. Kricheldorf et al. reported that teen alkoxides are liable for or oxygen Bond cleavage of the electron families [49]. This process involves the formation of alkyl ester groups which are normally persist at the terminal of the polymers. This process also has been supported by a coordination insertion mechanism where aluminum isopropoxide is the catalyst system35. Coordination insertion method age of the oxygen bond is also carried out in the presence of stannous octoate (Figure 2.8). The actual intermediate catalyst system is stannous alkoxide which is a product of stannous octoate and the alcohol media. Thus, alkoxide is the actual catalyst system for the coordination insertion method. Alcohols are also affecting significantly in the initiated formation chain termination and transesterification. At the same time, carboxylic acid has a negative impact on the rate of polymerization. In the PLA formation, alcohol improves the rate of polymerization, whereas carboxylic acid groups inhibit the propagation reaction. But after the complete synthesis of PLA, the product is not sensitive to the carboxylic acid concentration.

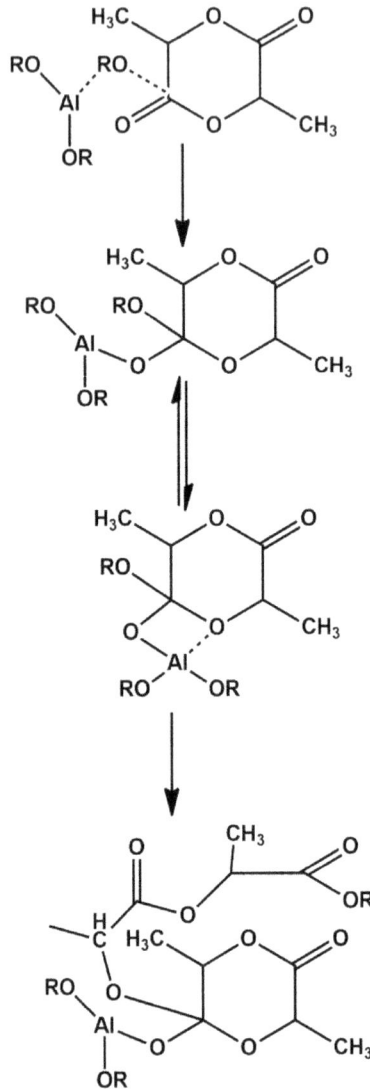

FIGURE 2.8 Mechanism of PLA synthesis by using aluminum isopropoxide catalyst system.

Stolt and Sodergard hypothesized the ROP of L-lactide applying several organic monocarboxylic iron complexes [32]. The acetate anions are prone to attach to the iron by coordination complexation. The complex encompasses the polymer chains followed by chain propagation. This mechanism is also called anionic type coordination polymerization. The mechanism has been depicted in Figure 2.9.

Schwach and his coworkers carried out ROP of PLA with the help of stannous octoate [57]. They used end group analysis by NMR spectra to examine the mechanism behind it. The monomer initiator ratio plays an important role in the quality

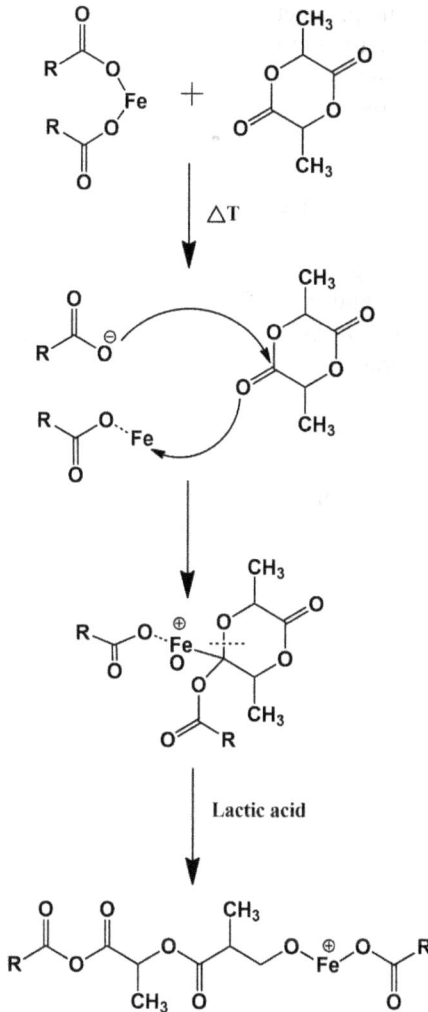

FIGURE 2.9 Schematic illustration of anionic type coordination insertion reaction for the monocarboxylic-Fe complex.

of the polymer product. For the low value of the monomer/initiator ratio, the lactyl octoate-terminated short chains are formed when the precursor is the racemic monomer. The cationic synthesis method is also reported by such coordination polymerization approaches for PLA. The cationic synthesis involves octanoic acid as an intermediate material.

The polymerization was also studied by using stannous octoate and elemental zinc as catalyst system, which was performed in bulk method [35]. Optimization study of the synthesis was evaluated by a second-order factorial model, which shows intermolecular transesterification dominance in PLA polymerization. The factorial study also showed the gradual dependency of the synthetic parameters. The dominance of the synthetic parameters is as follows: polymerization, temperature, monomer to initiator ratio, time of polymerization, initiator concentration and the required time for degassing of the monomer. For confirming the mechanism C^{13} NMR spectra have been studied elemental zinc induced polymerization. This reaction is very much sensitive in the presence of water, and as a result, a very small amount of elemental zinc has been activated [58]. During the polymerization process, a small amount of zinc lactate is formed, which is assumed to be the exact initiator for polymerization.

Kricheldorf et al. reported the synthesis of PLA when the precursor monomer systems are racemic and meso-DL-lactides [59]. The polymerization took place in xylem solution or in bulk method and the temperature was kept constant at 120°C throughout the reaction process. In this process, lead, zinc, antimony and bismuth salts were used as initiators. The main advantage of this process is that its yield is more than 90% is when the initiator system is Pb and Bi-based salts. When the temperature is increased, industrial regularity is changed and randomness appears in the macromolecular chain of PLA. In another work (Figure 2.10), acidic materials are used as a catalyst system for

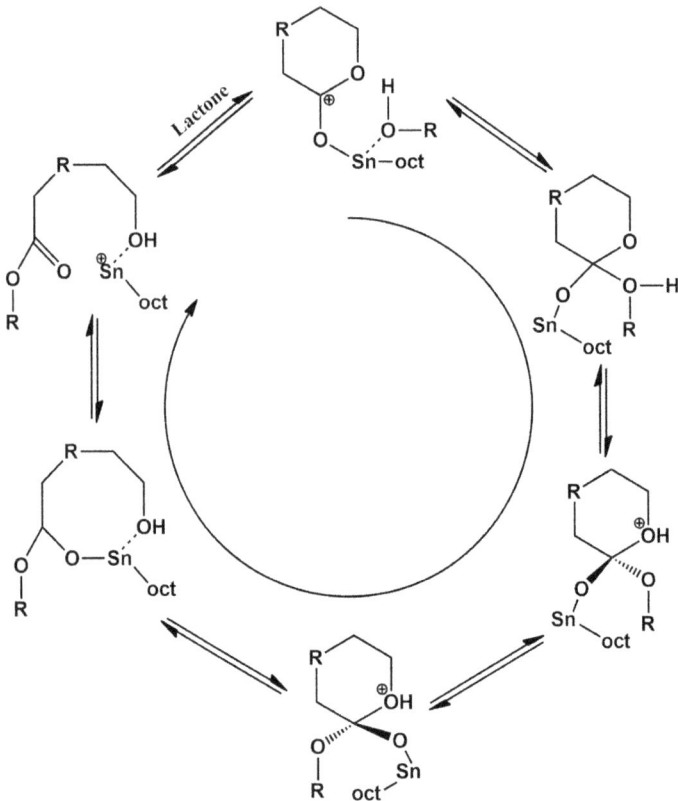

FIGURE 2.10 Schematic representation of ring-opening polymerization in the presence of stannous octoate.

the polymerization of L-lactide [60]. It has been proved that trifluoromethane-sulfonic acid and methyl triplet both are acting good initiator systems for PLA formation. Moreover, the effect of solvent and the reaction temperature could not be ruled out.

Stannous trifluoromethanesulfonic and scandium trichloromethane sulfonate; both are acting as good catalysts for PLA formation [47]. This catalyst system is desirable because of the tunable molecular weight and tailor maid polydispersity index (PDI) for the prepared polymer. It is also proved that a very small amount of base would be added as a solvent that improves the rate of polymerization with tiny loss to the polymerization control.

Another group of workers reported PLA synthesis from L-lactide monomers where the reaction parameters influence the rate of polymerization. This study was based on the reaction parameters of temperature time and catalyst concentration (here the catalyst used is stannous octoate) [61]. They also reported that the intrinsic viscosity of the prepared PLA was 10^6 with a very small amount of catalyst dose that is 0.015 wt.%. The reaction temperature was kept constant at 100°C. The PDI for this process was calculated at around 2–3; the results of this work support a non-ionic insertion method in the polymerization mechanism. Similar kind of catalyst system was also used by another group of researchers where they obtained very high-molecular weight PLA at a catalyst dose of 0.05% at 130°C reaction temperature. In this study, they also showed that polymerization at higher temperature and the prolonged residence time inside the polymerization reactor deteriorate the intrinsic viscosity of the end product. This drop in intrinsic viscosity confirms the depolymerization effect during the high temperature reaction process. Similar kind of drop was also noticed in number average molecular weight, which was also reported when aluminum isopropoxide was used at high temperature and high vomit to initiator ratio. This kind of drop was inferred as intra- and intermolecular transesterification reactions.

Another work synthesis of PLA was carried out in the presence of stannous octoate and zinc bis(2,2-dimethyl-3,5-heptanedionate-O,Ó) [62]. This catalyst system provided more than 80% conversion of the monomer. The catalyst system showed some anomalous behavior. The conversion rate up to 80% was accelerated in the presence of only zinc catalyst, but beyond 80%, the acceleration in the rate of polymerization was noticed due to the presence of zinc and tin containing catalyst systems. Monomer to catalyst dosage is also affecting the crystallinity of the end product. The impurities present in monomers as well as initiated systems also have a significant role in the rate of polymerization. The impurities actually accelerated the propagation step with a high yield. In the presence of Lewis acid, acceleration is more dominant because of the transesterification reaction between the activated lactone group and a hydroxyl group.

2.9 POLYMERIZATION KINETICS AND MODELING

Polymerization from monomer to polymer involves three distinct consecutive steps. The steps are initiation, propagation and chain termination [63]. The reaction kinetics is the time-dependent study of concentrations of the feed materials as well as the prepared polymer material. There are several articles that already depict the reaction Kinetics of polymerization. In the case of ROP, the reaction kinetics is almost similar to the change propagation polymerization reaction. The propagation steps involve the consecutive addition of monomer units resulting in the formation of long chains. But the molecules bigger than the monomer unit are unable to react with the propagating chains. There are several resemblances and characteristics of both the chain and step-growth polymerizations observed in ROP.

Initiation of ROP starts with the formation of an ionic species. The monomer units are ionized in this reaction at the first step of polymerization. The ionic monomer species propagates why chain reaction or coordination insertion to grow a large strand of polymer. The propagated chain would be living or terminated depending on the mode of reaction and the process kinetics. The initiation of the polymerization reaction is assumed below:

$$R - Z + C \rightarrow M*$$

Z is the functional group attached to the monomer and C is the ionic initiator moiety.

$$M* + nR - Z \rightarrow M - (R - Z)_n *$$

Dubois et al. reported that the reaction was 1st order compared to monomer concentration as well the initiator (aluminum isopropoxide) concentration [40]. The rate equation could be given as follows:

$$-\frac{dM}{dt} = k[M][I]$$

Here, [M] and [I] stand for the monomer and initiator concentrations, respectively. They calculated the value of rate constant, k, as 0.6 L mol^{-1}min^{-1}.

In another work, the rate constant was reported as 0.045 L/mol, where zinc lactate was used as catalyst 46. Moreover, it was also evidenced that the rate constant was higher for stannous octoate compared to zinc lactate which was around eight times higher [64]. But, the main disadvantage is that this rate constant is not depending on the reaction vessel and its internal effects. The experimental difficulties were also affecting the rate constants and also discussed elsewhere [65]. Under various steady state assumptions (rate of initiation and rate of termination are comparable), the ratio of rate constants (propagation rate constant/termination rate constant) can be found if the fluctuation of the degree of polymerization (number average molecular weight; M_n) with monomer concentration is documented. The finding of individual rate constants is a difficult job because it only is calculated during the propagation steps of the polymer synthesis. This could only be happened when the propagating intermediate species are precisely monitored during propagation. These steps are also not compared to the formation of initiator fragments generating and an actual number of reacting monomer species developed during polymerization. The common methods for the experimental evaluation of these concentrations are two basic quick methods. The first one is the end group analysis that is only done after stopping a propagating species in the middle of the reaction. During polymerization propagation step, a small amount of intermediate is taken and characterized to find out the terminal groups. The second one is the UV-Vis spectrophotometric assays of the materials [66]. The spectra also give the proof of bond formation and the intensity provided by the spectra is also significant to calculate the actual amount of the prepared polymer [67].

The rate constants at different steps (initiation, propagation and chain transfer/termination) are also calculated by using some molecular simulations and kinetics modeling approaches. For simplifying the mechanism of ROP, the successive steps have been given as follows:

Initiation:
$I + M \rightarrow P_1$; rate constant is k_1

Chain propagation:
$P_n + M \rightarrow P_{n+1}$; rate constant is k_p

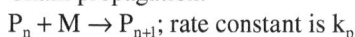

Chain termination/transfer:
$P_n + M \rightarrow M_n + P_1$; rate constant is k_t

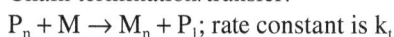

Here, P_n is the growing polymer chain having 'n' repeating units and M is the monomer.

One assumption has been made here, which is the spontaneity of the ROP. M_n is the deactivated macrochain with 'n' units. These segments of macro-chains are not taking part in any reactions. The number average molecular weight (M_n) can be calculated from the following equation of the respective polymer:

$$M_n = m \frac{\sum_{n=1}^{x} n\left([P_n] + [M_n]\right)}{\sum_{n=1}^{x} \left([P_n] + [M_n]\right)}$$

The weight average molecular weight (M_w) for the polymer will be

$$M_n = m \frac{\sum_{n=1}^{x} n^2 \left([P_n] + [M_n]\right)}{\sum_{n=1}^{x} n\left([P_n] + [M_n]\right)}$$

Here, 'm' stands for the molecular weight of the repeating unit of the prepared polymer.

If the reaction is carried out in batch reactor, then the reaction rate equations will be as follows:

$$\frac{d[M]}{dt} = -[M]\left\{ k_i[I] + \sum_{n=1}^{x} k_{pn}[P_n] + \sum_{n=1}^{x} k_{t,Mn}[P_n] \right\}$$

$$\frac{d[I]}{dt} = -k_i[I][M]$$

$$\frac{d[P_1]}{dt} = k_i[I][M] - k_{p1}[P_1][M] + \sum_{n=1}^{x} k_{t,Mn}[P_n][M]$$

$$\frac{d[P_n]}{dt} = [M]\left\{ k_{p(n-1)}[P_{n-1}] - k_{pn}[P_n] - k_{t,Mn}[P_n] \right\}, \, n > 1$$

$$\frac{d[M_n]}{dt} = k_{t,Mn}[P_n][M], \, n \geq 1$$

If the initial conditions are assumed, then $t = 0$.

Thus the concentration terms will be

$$[M_n] = 0 \text{ and } [P_n] = 0, \, n > 1$$

$$[M] = [M_0]$$

$$[I] = [I_0]$$

Here $[M_0]$ and $[I_0]$ correspond to the initial concentration of monomer and initiator. The rate constants are dependent on the kinetic chain lengths of the polymerization reaction.

The aforementioned differential equations are evaluated and solved by Euler's method by using simulations. This was supported by finding the Poisson distribution's limiting case in a simulation. The simulation assumptions made were to take the termination constant as zero and the high magnitude of initiation rate constant. The degree of polymerization is taken as at least 5000 for this case. The number average and weight average molecular weights are dependent on various parameters like polymerization time and initiator to monomer ratio [32].

2.10 SUMMARY

PLA is a popular biodegradable polymer and welcomed in the domain of commodity plastics. The fabrication of this polymer promotes its end use and texture of applications. It is normally prepared in the bulk process of condensation polymerization systems. But condensation polymerization affects its quality and mechanical properties for the application. The monomer of PLA is lactic acid which is found as D, L or racemic mixtures. The internal physical properties and morphologies depend on their monomer choice. For racemic monomer feeds, the materials are tending to form amorphous domains. This polymer is also good for fiber formation and that's why melt spinning has been perfumed for this polymer also. Nowadays, high-molecular weight PLA is prepared by the coordination insertion method. This method requires catalysts that are especially metal-organic catalysts systems. The most common catalyst is tin and zinc-based catalysts systems. This chapter also discussed the effect of single catalysts and dual catalysts systems. The catalyst systems largely affect the molecular weight and tailor-made properties of the polymers. But still, there are several limitations to discuss the exact polymerization mechanisms where ring opening is involved. This is also a significant topic of research during these days.

REFERENCES

1. Hottle, T.A., M.M. Bilec, and A.E. Landis, *Sustainability assessments of bio-based polymers. Polymer Degradation and Stability*, 2013. **98**(9): pp. 1898–1907.
2. Das, P. et al., *Immobilization of heteroatom-doped carbon dots onto nonpolar plastics for antifogging, antioxidant, and food monitoring applications. Langmuir*, 2021. **37**(11): pp. 3508–3520.
3. Ganguly, S. and S. Margel, *Remotely controlled magneto-regulation of therapeutics from magnetoelastic gel matrices. Biotechnology Advances*, 2020. **44**: p. 107611.
4. Chen, G., T. Ushida, and T. Tateishi, *Scaffold design for tissue engineering. Macromolecular Bioscience*, 2002. **2**(2): pp. 67–77.
5. Peter, S. et al., *Polymer concepts in tissue engineering. Journal of Biomedical Materials Research*, 1998. **43**(4): pp. 422–427.
6. Ganguly, S. and N.C. Das, *Rheological properties of polymer–carbon composites, in carbon-containing polymer composites*, 2019. Springer. pp. 271–294.
7. Leicester, H.M. and H.S. Klickstein, *Tenney Lombard Davis and the history of chemistry. Chymia*, 1950. **3**: pp. 1–16.
8. Lunt, J., *Large-scale production, properties and commercial applications of polylactic acid polymers. Polymer Degradation and Stability*, 1998. **59**(1–3): pp. 145–152.
9. Garlotta, D., *A literature review of poly(lactic acid). Journal of Polymers and the Environment*, 2001. **9**(2): pp. 63–84.
10. Reshmy, R., et al., *Advanced biomaterials for sustainable applications in the food industry: Updates and challenges. Environmental Pollution*, 2021. **283**: p. 117071.
11. Ganguly, S., et al., *Microwave-synthesized polysaccharide-derived carbon dots as therapeutic cargoes and toughening agents for elastomeric gels. ACS Applied Materials & Interfaces*, 2020. **12**(46): pp. 51940–51951.
12. Santoro, M., et al., *Poly(lactic acid) nanofibrous scaffolds for tissue engineering. Advanced Drug Delivery Reviews*, 2016. **107**: pp. 206–212.
13. Bennett, R.G., *Selection of wound closure materials. Journal of the American Academy of Dermatology*, 1988. **18**(4): pp. 619–637.
14. Cai, C.-X., et al., *Stereoselective ring-opening polymerization of racemic lactide using alkoxy-amino-bis (phenolate) group 3 metal complexes. Chemical Communications*, 2004. (3): pp. 330–331.
15. Doherty, S., et al., *Dimeric aluminum chloride complexes of N-alkoxyalkyl-β-ketoimines: Activation with propylene oxide to form efficient lactide polymerization catalysts. Organometallics*, 2004. **23**(10): pp. 2382–2388.
16. McGuinness, D.S., et al., *Anionic iron (II) alkoxides as initiators for the controlled ring-opening polymerization of lactide. Journal of Polymer Science Part A: Polymer Chemistry*, 2003. **41**(23): pp. 3798–3803.

17. Hormnirun, P., et al., *Remarkable stereocontrol in the polymerization of racemic lactide using aluminum initiators supported by tetradentate aminophenoxide ligands. Journal of the American Chemical Society*, 2004. **126**(9): pp. 2688–2689.

18. Williams, C.K., et al., *A highly active zinc catalyst for the controlled polymerization of lactide. Journal of the American Chemical Society*, 2003. **125**(37): pp. 11350–11359.

19. Mehta, R., et al., *Synthesis of poly(lactic acid): a review. Journal of Macromolecular Science, Part C: Polymer Reviews*, 2005. **45**(4): pp. 325–349.

20. Chisholm, M.H. and K. Phomphrai, *Conformational effects in β-diiminate ligated magnesium and zinc amides: Solution dynamics and lactide polymerization. Inorganica Chimica Acta*, 2003. **350**: pp. 121–125.

21. Gibson, V.C., et al., *A well-defined iron (II) alkoxide initiator for the controlled polymerisation of lactide. Journal of the Chemical Society, Dalton Transactions*, 2002. **23**: pp. 4321–4322.

22. Kim, Y., G.K. Jnaneshwara, and J.G. Verkade, *Titanium alkoxides as initiators for the controlled polymerization of lactide. Inorganic Chemistry*, 2003. **42**(5): pp. 1437–1447.

23. Zhang, L., et al., *Ring-opening polymerization of D, L-lactide by rare earth 2, 6-dimethylaryloxide. Polymer International*, 2004. **53**(8): pp. 1013–1016.

24. Chisholm, M.H., J. Gallucci, and K. Phomphrai, *Lactide polymerization by well-defined calcium coordination complexes: Comparisons with related magnesium and zinc chemistry. Chemical Communications*, 2003. (**1**): pp. 48–49.

25. Sun, J., et al., *The ring-opening polymerization of D, L-lactide catalyzed by new complexes of Cu, Zn, Co, and Ni Schiff base derived from salicylidene and L-aspartic acid. Journal of Applied Polymer Science*, 2002. **86**(13): pp. 3312–3315.

26. Storey, R.F., et al., *Soluble tin (II) macroinitiator adducts for the controlled ring-opening polymerization of lactones and cyclic carbonates. Journal of Polymer Science Part A: Polymer Chemistry*, 2002. **40**(20): pp. 3434–3442.

27. Williams, C.K., et al., *Metalloenzyme inspired dizinc catalyst for the polymerization of lactide. Chemical Communications*, 2002. (18): pp. 2132–2133.

28. Myers, M., et al., *Phosphines: Nucleophilic organic catalysts for the controlled ring-opening polymerization of lactides. Journal of Polymer Science Part A: Polymer Chemistry*, 2002. **40**(7): pp. 844–851.

29. Chang, W.C. and W.-h. Sun, *Method of polymerization of lactide and polylactide homopolymer thereof*, 2002. Google Patents.

30. Nomura, N., et al., *Stereoselective ring-opening polymerization of racemic lactide using aluminum-achiral ligand complexes: Exploration of a chain-end control mechanism. Journal of the American Chemical Society*, 2002. **124**(21): pp. 5938–5939.

31. Baker, G.L. and M.R. Smith, III, *Process for the preparation of polymers of dimeric cyclic esters*, 2002, Google Patents.

32. Stolt, M. and A. Södergård, *Use of monocarboxylic iron derivatives in the ring-opening polymerization of L-lactide. Macromolecules*, 1999. **32**(20): pp. 6412–6417.

33. Kricheldorf, H.R. and S.-R. Lee, *Polylactones: 32. High-molecular-weight polylactides by ring-opening polymerization with dibutylmagnesium or butylmagnesium chloride. Polymer*, 1995. **36**(15): pp. 2995–3003.

34. Schwach, G., et al., *Zn lactate as initiator of DL-lactide ring opening polymerization and comparison with Sn octoate. Polymer Bulletin*, 1996. **37**(6): pp. 771–776.

35. Schwach, G., et al., *Stannous octoate-versus zinc-initiated polymerization of racemic lactide. Polymer Bulletin*, 1994. **32**(5): pp. 617–623.

36. Kasperczyk, J., *HETCOR NMR study of poly(rac-lactide) and poly(meso-lactide). Polymer*, 1999. **40**(19): pp. 5455–5458.

37. Kister, G., G. Cassanas, and M. Vert, *Effects of morphology, conformation and configuration on the IR and Raman spectra of various poly(lactic acid)s. Polymer*, 1998. **39**(2): pp. 267–273.

38. Schindler, A. and D. Harper, *Polylactide. II: Viscosity–molecular weight relationships and unperturbed chain dimensions. Journal of Polymer Science: Polymer Chemistry Edition*, 1979. **17**(8): pp. 2593–2599.

39. Van Dijk, J., et al., *Characterization of poly(d, l-lactic acid) by gel permeation chromatography. Journal of Polymer Science: Polymer Chemistry Edition*, 1983. **21**(1): pp. 197–208.

40. Dubois, P., et al., *Macromolecular engineering of polylactones and polylactides. 4: Mechanism and kinetics of lactide homopolymerization by aluminum isopropoxide. Macromolecules*, 1991. **24**(9): pp. 2266–2270.

41. Han, D.K. and J.A. Hubbell, *Lactide-based poly (ethylene glycol) polymer networks for scaffolds in tissue engineering. Macromolecules*, 1996. **29**(15): pp. 5233–5235.

42. Korhonen, H., A. Helminen, and J.V. Seppälä, *Synthesis of polylactides in the presence of co-initiators with different numbers of hydroxyl groups. Polymer*, 2001. **42**(18): pp. 7541–7549.

43. Zhang, X., et al., *Mechanism of lactide polymerization in the presence of stannous octoate: the effect of hydroxy and carboxylic acid substances. Journal of Polymer Science Part A: Polymer Chemistry*, 1994. **32**(15): pp. 2965–2970.

44. Hyon, S.-H., K. Jamshidi, and Y. Ikada, *Synthesis of polylactides with different molecular weights. Biomaterials*, 1997. **18**(22): pp. 1503–1508.

45. Jacobsen, S., et al., *New developments on the ring opening polymerisation of polylactide. Industrial Crops and Products*, 2000. **11**(2–3): pp. 265–275.

46. Rafier, G., et al., *Process for manufacturing homo-and copolyesters of Lactic acid*, 2003, Google Patents.

47. Möller, M., et al., *Stannous (II) trifluoromethane sulfonate: a versatile catalyst for the controlled ring-opening polymerization of lactides: Formation of stereoregular surfaces from polylactide 'brushes'. Journal of Polymer Science Part A: Polymer Chemistry*, 2001. **39**(20): pp. 3529–3538.

48. Abdel-Fattah, T.M. and T.J. Pinnavaia, *Tin-substituted mesoporous silica molecular sieve (Sn-HMS): Synthesis and properties as a heterogeneous catalyst for lactide ring-opening polymerization. Chemical Communications*, 1996. (5): pp. 665–666.

49. Kricheldorf, H.R., M. Berl, and N. Scharnagl, *Poly (lactones). 9: Polymerization mechanism of metal alkoxide initiated polymerizations of lactide and various lactones. Macromolecules*, 1988. **21**(2): pp. 286–293.

50. Stevels, W.M., et al., *Well defined block copolymers of ε-caprolactone and L-lactide using Y5 (μ-O)(OiPr) 13 as an initiator. Macromolecular Chemistry and Physics*, 1995. **196**(4): pp. 1153–1161.

51. Kasperczyk, J. and M. Bero, *Stereoselective polymerization of racemic DL-lactide in the presence of butyllithium and butylmagnesium. Structural investigations of the polymers. Polymer*, 2000. **41**(1): pp. 391–395.

52. Stere, C., et al., *Anionic and ionic coordinative polymerization of L-lactide. Polymers for Advanced Technologies*, 1998. **9**(6): pp. 322–325.

53. Köhn, R.D., et al., *Ring-opening polymerization of D, L-lactide with bis (trimethyl triazacyclohexane) praseodymium triflate. Catalysis Communications*, 2003. **4**(1): pp. 33–37.

54. Save, M. and A. Soum, *Controlled ring-opening polymerization of lactones and lactide initiated by lanthanum isopropoxide, 2. Mechanistic Studies: Macromolecular Chemistry and Physics*, 2002. **203**(18): pp. 2591–2603.

55. Barakat, I., et al., *Macromolecular engineering of polylactones and polylactides: X. Selective end-functionalization of poly(D, L)-lactide. Journal of Polymer Science Part A: Polymer Chemistry*, 1993. **31**(2): pp. 505–514.

56. Mukaiyama, T., J. Ichikawa, and M. Asami, *A facile synthesis of carboxylic esters and carboxamides by the use of 1, 1'-dimethylstannocene as a condensing reagent. Chemistry Letters*, 1983. **12**(5): pp. 683–686.

57. Schwach, G., et al., *More about the polymerization of lactides in the presence of stannous octoate. Journal of Polymer Science Part A: Polymer Chemistry*, 1997. **35**(16): pp. 3431–3440.
58. Schwach, G., et al., *Ring opening polymerization of D, L-lactide in the presence of zinc metal and zinc lactate. Polymer International*, 1998. **46**(3): pp. 177–182.
59. Kricheldorf, H.R. and C. Boettcher, *Polylactones, 24: Polymerizations of racemic and meso-d, l-lactide with Al-O initiators: Analyses of stereosequences. Die Makromolekulare Chemie: Macromolecular Chemistry and Physics*, 1993. **194**(6): pp. 1653–1664.
60. Kricheldorf, H.R. and R. Dunsing, *Polylactones, 8: Mechanism of the cationic polymerization of L, L-dilactide. Die Makromolekulare Chemie: Macromolecular Chemistry and Physics*, 1986. **187**(7): pp. 1611–1625.
61. Leenslag, J.W. and A.J. Pennings, *Synthesis of high-molecular-weight poly(L-lactide) initiated with tin 2-ethylhexanoate. Die Makromolekulare Chemie: Macromolecular Chemistry and Physics*, 1987. **188**(8): pp. 1809–1814.
62. Nijenhuis, A., D. Grijpma, and A. Pennings, *Lewis acid catalyzed polymerization of L-lactide. Kinetics and mechanism of the bulk polymerization. Macromolecules*, 1992. **25**(24): pp. 6419–6424.
63. Ganguly, S., P. Das, and N.C. Das, *Characterization tools and techniques of hydrogels*, in *Hydrogels based on natural polymers*, 2020. Elsevier. pp. 481–517.
64. Puaux, J.P., et al., *A study of L-lactide ring-opening polymerization kinetics*, in *Macromolecular symposia*, 2007. Wiley Online Library.
65. Odian, G., *Principles of polymerization*. 2004: John Wiley & Sons.
66. Higashimura, T., et al., *Rate constant of propagation in the cationic polymerization of styrene catalyzed by BF3 O (C2 H5) 2. Journal of Polymer Science Part B: Polymer Letters*, 1971. **9**(6): pp. 463–466.
67. Sawamoto, M. and T. Higashimura, *Stopped-flow study of the cationic polymerization of styrene derivatives. 1. Direct observation of the propagating species in the polymerization of p-methoxystyrene in 1, 2-dichloroethane. Macromolecules*, 1978. **11**(2): p. 328332.

3 Manufacturing Methods of PLA Composites

Guang-Way Bill Jang and Cheng-Han Hsieh
Industrial Technology Research Institute
Hsinchu, Taiwan

Allen Lai and Sagle Chan
Chiao Fu Enterprise
Taiwan

CONTENTS

3.1 INTRODUCTION

As polymer composites are primarily designed to achieve long-term durability, they are generally made from fossil-based synthetic polymers. However, greenhouse gas (GHG) emissions and microplastics are major concerns for plastic products. A potential alternative may be to utilize abundant, naturally occurring polymers produced by plants and animals. Included among them are cellulose, a polysaccharide comprising 40% of the organic matter on earth, and chitin, a plentiful polysaccharide found in the form of nanofibers in the exoskeleton of arthropods and various filamentous fungi. Biocomposites from renewable resources are receiving more attention than ever due to environmental, energy, and sustainability concerns. Cellulose-based materials, as renewable materials derived from abundant naturally occurring polymers, have been extensively investigated for polymer reinforcement. In addition to

DOI: 10.1201/9781003160458-3

sustainable feedstock supplies, recyclability and biodegradability are possible with suitable designs after employing biocomposite material.

The world's continued population growth has led to a spike in resource consumption, one that is environmentally unsustainable [1]. In the past few years, the use of natural reinforcement, such as that employing plant fibers, starch, lignin, and hemicellulose, has gained interest in order to produce entirely bio-based and biodegradable PLA composites for applications in food and cosmetics packaging, disposable items, toys, and other consumer products. Cellulose and chitin nanomaterials have been extensively investigated as additives in green nanocomposites because of their added benefits, including not only biodegradability, renewability, and biocompatibility, as well as the antibacterial properties of chitin and enhanced barrier properties and crystallization rate attributable to cellulose present in PLA matrices. These characteristics are particularly beneficial to packaging applications in extending food storage time and freshness. New product development needs to take into account sustainability to avoid traditional product production processes and applications with negative environmental impacts. Primary concerns are CO_2 emissions from production activities and material toxicity. A long trajectory approach based on a product life cycle assessment (LCA) perspective for product design and manufacturing is gradually being adopted by industries for sustainability concerns and to better understand the long-term impacts of products throughout their life cycles. Documentation of feedstocks, processing conditions, and byproduct-related information has to be thoroughly prepared for an extensive and representative assessment of product environmental impact. In addition to environmental considerations, social and economic factors are critical criteria for evaluating the sustainability of a product. Biocomposites are candidates for feasible alternatives due to their light weight, enhanced mechanical properties and recyclability, and/or biodegradability. Biocomposites have many industrial applications, including in the automotive and packaging industries. Key categories to consider in evaluating the carbon footprint of biocomposites include fossil depletion, climate change, agricultural land use, particulate matter formation, and human toxicity [1]. The lightweight nature of biocomposites provides improved energy efficiency (6.5%–16.4%) and lower greenhouse gasses emissions (16.0%–16.4%) [1, 2]. Economic considerations, such as mechanical performance, cost-effectiveness, raw material supply, and legislation, are crucial in automotive and consumer product applications. For certain applications, if the lifespan of the natural fiber (NF) biocomposites is limited, this may lead to safety or cost concerns, in which case alternative fibers or new designs may be required, which requires a compromise between environmental and social benefits. The balance between social, economic, and environmental criteria can be varied depending on a selected application.

For the fabrication of bionanocomposites and many other applications, NC can be isolated from plant fibers through several repeated processes to reach nanoscale dimensions. Cellulose's strong affinity for water renders its drying a challenge, a consideration that must be taken into account for aqueous suspensions in the manufacturing of composites. Success in developing nanocomposites with significant improvement in mechanical strength and low enforcement content relies on the uniform dispersion of cellulosic fiber in polymer matrices. This is achieved by solution casting of biodegradable polymers and nanocellulosic materials in aqueous media to obtain nanocomposites with improved mechanical strength [3–5]. Silylation, use of surfactants, and in situ polymerization approaches have been applied when using organic solvents for processing. Solution casting techniques are mostly carried out on a laboratory scale with limited industrial applications. Melt compounding for the processing of bionanocomposites poses a challenge due to the high temperature and viscosity of polymer melt relative to the solution process. Preparation of cellulose fiber-based nanocomposites requires innovative processing equipment set-up, optimization and controlling, and cellulose modification techniques in order to avoid degradation of cellulose whiskers and difficulty of separating fiber aggregates during the extrusion process.

There has been substantial growth in demand for plant-based plastics due to public awareness of the high-carbon footprint of petrochemical products and various environmental regulations. Manufacturers are constantly seeking sustainable feedstock alternatives to replace fossil resources. Poly(lactic acid)/polylactide (PLA) can be synthesized from renewable resources, and their inherent optical transparency and mechanical properties render biodegradable polyesters derived therefrom suitable for utilization in a broad range of applications, including biomedical, packaging, and textile fibers.

PLA is typically prepared by ring-opening polymerization (ROP) of lactide, usually referred to as polylactide, or prepared by polycondensation of lactic acid, also referred to as poly(lactic acid). The characteristics of the resulting polymers can vary widely based on the polymerization technique and the ratio of L- and D-form of monomers used. Owing to technological advances that have made cost reduction possible, at present lactic acid is synthesized by fermentation of sugar-containing material. The fermentation process produces specifically one major stereoisomer, L-isomer >99%, while chemically synthesized lactic acid consists of a racemic mixture consisting of 50% D and 50% L. Polymers with low D-isomer content are semicrystalline, whereas those with higher D-isomer content are required for amorphous polymers such as those used in heat-sealing applications. As crystallinity of PLA is essential in controlling its degradation rate, thermal resistance, optical, mechanical, and barrier properties, there are limitations on high-performance applications. The production of polymers requires considerable resources, including chemicals, water, and energy. Thus, it is important to assess the environmental impacts of a polymer matrix using emission factors evaluated in a cradle-to-gate life cycle. For environmental sustainability and ecological impact considerations, compostable or biodegradable matrixes are considered the most appropriate alternatives [1].

There are shortcomings of PLA in adjusting to a specific processing condition or end-use property requirements, such as low melting index/strength, crystallization rate, and impact strength. The tensile strength and elastic modulus of PLA are comparable to those of polyethylene terephthalate (PET). On the other hand, relative low ductility, impact resistance, heat distortion temperature (HDT), and resistance to hydrolysis limit applications of PLA polymers. Introduction of a selected additive, such as reinforcing fibers or nanofillers, into PLA matrix is considered a useful approach to significantly improve processability or other properties. Stability and dimensions of dispersed phases, their compatibility and interactions with PLA matrices, and manufacturing processes are critical factors in maximizing the performance of a resulting nanocomposite.

3.2 MODIFICATION OF CELLULOSIC FIBERS

The common challenge for using various types of nanomaterials (such as exfoliated clay, carbon nanotubes [CNTs], graphene, cellulose nanocrystals [CNCs], and nano-fibers) as reinforcements for the fabrication of nanocomposites is the poor dispersion of nanofillers in the polymer matrix. To enhance the compatibility between nanomaterials and hydrophobic polymers, different approaches, including silylation, esterification, grafting, and in situ polymerization, have been studied to improve dispersion. The surface properties of cellulose fibers (CFs) govern the fiber-fiber bonding within cellulose networks and the interfacial adhesion between fibers and polymers. Thus, these properties affect the performance of the resulting composite materials.

3.2.1 TYPES OF CELLULOSIC MATERIALS AND PRETREATMENT PROCESS

Due to rising resource sustainability and environmental impact concerns, rapid double-digit growth of market demand for NC material is expected to reach USD783 million by 2025, according to a MarketsAndMarkets study. Plant fibers mainly consist of cellulose, hemicellulose, lignin, structural water, as well as relatively small amounts of pectin, wax, or oil. The structural organization of a plant fiber, in general, consists of unidirectional cellulose microfibrils reinforcing elements in the matrix blend of hemicellulose and lignin. The mechanical performance of a plant fiber depends on its cellulose content, microfibrillar angle, fiber diameter, temperature, and water content, as well as the presence of defects. The properties of a given plant fiber may differ due to variations in chemical composition and structure (microfibrillar angle, crystallinity, and defects) due to environmental conditions during plant growth [6]. Commonly used NFs include flax, ramie, hemp, jute, sisal, coconut, cotton, nettle, kenaf, and bamboo. Agriculture residuals, such as rice straw, bagasse, and oil palm empty fruit bunches, are potential feedstocks for CFs. Cascade valorization approaches have been demonstrated by using residual fiber from grape pomace as filler for bioplastics after extraction of active ingredients [7, 8].

There is a broad range of physical and mechanical properties for various types of NFs derived from different sources [6]. These properties can vary based on the extraction method, harvesting season, and the parts of a plant from which the fibers are derived. Thus, it is critical to select a suitable type of fiber for a selected application. Plant fibers may be conditioned to multi-scale structures, including fabric yarn, fiber bundles, unit fibers, and even microfibrillated cellulose (MFC), nanocellulose fibers (NCF), microcrystalline cellulose (MCC) as well as cellulose nanocrystal (CNC), for the preparation of biocomposites. Crystalline cellulose has superior mechanical properties (Table 3.1) competitive with those of materials routinely used in engineering, such as steel, glass, and carbon fibers [9]. Typical dimensions of micro- and nano-cellulose materials are described in Table 3.2 [10].

The preparation of cellulose nanomaterials generally consists of purification steps using chemical treatments to remove the non-cellulosic components, followed by a mechanically induced destructuring process [11]. The size reduction process is often carried out with a diluted cellulosic fiber suspension using different shearing equipment (such as a homogenizer, a microfluidizer, or an ultra-fine friction grinder) to

TABLE 3.1

Mechanical and Physical Properties of NCC and Engineering Materials [9]

Material	Density (g/cm³)	Tensile Strength (MPa)	Tensile Modulus (GPa)	CTE (ppm/K)
Nanocrystalline cellulose	1.5	7500	145	3–22
Steel wire	7.8	4100	207	15
Kevlar	1.4	3800	130	−4
Graphite	2.2	21,000	410	2–6
Glass fiber	2.5	4800	86	13
CNT	1.0	11,000–73,000	270–970	–

release individually constitutive microfibrils. Pretreatments, mainly enzymatic, the introduction of charged groups through carboxymethylation, or TEMPO-mediated oxidation, were applied to facilitate the production route to improve the mechanical treatment efficiency. Cellulose nanomaterials can be divided into five categories based on their sources: cellulose nanofibers (CNF), CNCs, bacterial cellulose (BC), algal cellulose (AC), and tunicate cellulose [12]. Some specific properties of cellulosic materials, such as mechanical stiffness and specific surface and the available surface hydroxyl groups, are amplified upon downsizing. Nanocellulose (CNF & CNC) has superior intrinsic mechanical properties. Crystalline cellulose has tensile strength (7.5–7.7 GPa) greater than that of carbon fibers and an axial elastic modulus (110–220 GPa) comparable to that of Kevlar [13, 14]. Large-scale production of CNFs can be achieved by extensive mechanical fibrillation of wood pulp with or without acids, enzymes, or TEMPO oxidation treatment. CNFs possess a high aspect ratio, with the diameter ranging from 2 to 20 nm and lengths of up to several micrometers.

Conventional mechanical treatments of CFs include grinding, smashing, ball milling, beating, etc. CFs are subjected to treatments of various types, such as fiction, collision, impact, and shear, to reduce their crystallinity and interfacial bonding and to separate non-cellulosic materials. Mechanical treatment of CF results in enhancement of surface area and thus facilitates chemical modification of the surface. Ultrasonic treatment is an alternative mechanical treatment process, one in which high-vibration power-induced hydrodynamic force is used to separate CFs.

TABLE 3.2

Typical Dimensions of Micro- and Nano-Cellulose Materials [10]

Nanocellulose Type	Diameter (d, nm)	Length (L, nm)	Aspect Ratio (L/d)
Microfibril (MF)	2–10	>10,000	>1000
Microfibrillated cellulose (MFC)	10–40	>1000	100–150
Cellulose nanocrystal (CNC)	2–20	100–600	10–100
Microcrystalline cellulose (MCC)	>1000	>1000	<1

High-pressure homogenization is often applied for the mechanical processing of CFs by using a pressurized facility to deliver a cellulose suspension into a homogenizer. Cavitation, shear force, and the impact of fluid as well as the reactor internal wall causes delamination and size reduction of CFs.

High-pressure homogenization can be applied to produce cellulose nanomaterials. Physical treatment of wood cellulose pulps provided a smooth gel containing 2% MFC in water [15]. MFC gel has been demonstrated to have wide utility in the preparation of foods, such as helping ground meat retain moisture during cooking, cake frostings, and salad dressings. Extra-fine grinding method is based on the shear force generated by a disc to undermine cell wall/cellulose interaction and to obtain cellulose materials from wood pulp. Hardwood (birch and aspen) and softwood (pine) bleached sulfate pulps were ground in a ball mill followed by thermocatalytic treatment in the presence of hydrochloric acid to obtain MCC [16]. Investigation of the effect induced by a dry ball milling process on cellulose structure, morphology, and properties revealed a reduction in crystallinity from 0.53 to 0.15 and crystallite domain sizes from 4 nm down to 3.4 nm after 60-minute ball milling. The original fibrous structure vanished and yielded to a quasi-circular shape, and water absorbability increased from 7.3 wt.% to 11.6 wt.%.

Refining procedures in the pulp and paper manufacturing process help to swell and to delaminate CFs. The surface area of CFs increased after the refining process and thus facilitated the potential for surface modification by biological or chemical methods. NC fibers were prepared from a 0.5 wt.% bleached cotton fiber suspension by 30 passes through a disc refiner. The average diameter of the cotton fiber reduced from ~25 μm to 242 nm and the degree of polymerization (DP) decreased from 2720 to 740.

Acid hydrolysis of cellulose microfibrils results in transverse cleavage of the amorphous regions and releasing crystalline cellulose. Crystalline arrangement of the whiskers provides the resulting crystalline cellulose with high modulus and thus potentially acts as efficient reinforcement. Interest in nanostructured cellulose, NC, is growing due to its superior mechanical properties, high aspect ratio, biocompatibility, and renewability. Characteristics of an NC material, such as morphology and dimension, as well as physical and chemical properties, depend on the origin of the cellulose used, the isolation and processing conditions, and pre- and post-treatments used [17]. The hydrophilic behavior of NFs poses a challenge for composite materials intended for applications subject to variable climatic conditions. Absorption of moisture in wet environments results in structural modification of the reinforcing fiber, thus impacting the performance of composites. This hydrophilic behavior also causes poor adhesion between plant fibers and matrix polymers.

Typical treatment of MCC for separation of into whiskers for the preparation of composite is to swell by chemical and ultra-sonification methods. Oksman et al. [18] used DMAc with 0.5 wt.% LiCl solution for the treatment of 10 wt.% MCC particles by mechanical agitation at 70°C using a mechanical stirrer followed by sonification process. MCC was further swelled and separated into nanocellulose whiskers (NCWs) with dimensions of less than 10 nm in width and between 200 and 400 nm in length for the fabrication of bionanocomposite.

Biocomposites can be made from diverse types of NFs. Cellulose in several forms, including macro, micro, and nano, has been applied extensively as reinforcement for the preparation of PLA composites. It is a challenge to uniformly disperse cellulose materials in hydrophobic PLA matrices due to their fluffiness and hydrophilic characteristics, which lead to agglomeration and poor mechanical performance of the prepared composites. Approaches to improve interfacial bonding include physical treatments (plasma, corona, etc.) and chemical modification (maleic anhydride, organosilanes, isocyanates, sodium hydroxide, permanganate, peroxide, etc.) of the fiber surface.

3.2.2 Methods for the Modification of Cellulosic Materials

Substitution of hydroxyl groups and grafting techniques are often applied for tuning the surface polarity of CF. TEMPO oxidation pretreatment introduces carboxylic groups at CFs and facilitates their extraction in one step [12]. Methods for the modification of CFs can be divided into physical, chemical, and biological routes. Chemical approaches are most often used. Introducing side chains or components on the surface renders CFs with added functionality, thus widening the range of their potential applications. The aims of modification are providing CF with improved functionality, such as processability, adhesion, moisture resistance/absorbability, chemical resistance, and ion-exchange capacity [19].

3.2.2.1 Physical Modification

Polyelectrolytes bond with cellulose materials through ionic interaction. Thus, ionic density and distribution, as well as the presence of salt, will influence the efficiency of polyelectrolyte modification of cellulose. Electrostatic assembly of two negatively charged nanomaterials, nanosilver, and NC fibers, was mediated by subsequent adsorption of a positively charged polyelectrolyte, such as polydiallyldimethylammonium chloride, polyallylamine hydrochloride, and polyethyleneimine, followed by a negatively charged polyelectrolyte onto the surface of NCF. The adsorbed polyelectrolytes served as a linkage between NCF and Ag nanoparticles for the preparation of antibacterial paper [20].

Alternatively, modification of a cellulose surface can be achieved by using nonionic polyelectrolytes, including linear polymers, graft polymers, hyperbranched polymers, linear-dendritic hybrids, and dendritic polymers, through van der Waals or hydrogen bonding interaction.

3.2.2.2 Chemical Modification

Each anhydroglucose unit (AGU) of cellulose consists of three alcoholic hydroxyl groups, which are accessible for chemical modification reactions. Potential modification reactions for the primary alcohol on C-6 and the secondary alcohol on C-2 as well as C-3 are esterification, etherification as well as oxidation to be carried out in homogeneous or heterogeneous medium. However, due to high crystallinity, there are limited solvents for cellulose materials. Usually, chemical modification of cellulose materials is performed under heterogeneous conditions, and thus the cellulose structure can be maintained to a certain degree. Typical approaches for the chemical modification of cellulose are summarized below.

3.2.2.2.1 Carboxylation

A two-stage oxidation process with periodate followed by sodium chlorite treatments resulted in the transformation of secondary alcohol of cellulose to carboxyl group [21].

The pretreatment process turned the hardwood cellulose pulp into nanofibrillate oxidized cellulose gel with 0.38–1.75 mmol/g carboxyl contents. The resulting NCF had widths of 25±6 nm and approximately 40% crystallinity.

3.2.2.2.2 Sulfonation

Sulfonation introduces charge to a cellulose structure and thus enhances its dispersibility in a selected solvent. Many techniques have been developed for sulfonation of cellulose since 1800 and use the resulting cellulose materials as anticoagulants and flocculants [22, 23]. The consecutive pretreatment of cellulose with periodate and bisulfite was performed to facilitate nanofibrillation of hardwood pulp and to obtain sulfonated NCF using a homogenizer [24]. The periodate used can be efficiently recycled for sustainable consideration to avoid the production of halogenated wastes. The resulting sulfonated NCF had typical widths of 10–60 nm, crystallinity of 40%, and anionic charge densities of 0.18–0.51 mmol/g. A transparent film with a tensile strength of 164 MPa and Young's modulus of 13.5 MPa was obtained by solution casting.

3.2.2.2.3 Carboxymethylation

The esterification process for the production of carboxymethylcellulose typically involves swollen CFs in concentrated NaOH solution followed by reacting the hydroxyl groups of cellulose with the monochloroacetic acid. This approach, coupled with a microwave-assisted technique, has been successfully applied to modify cellulose extracted from Brewer's spent grain (BSG) with an average degree of substitution of 1.46 at 70°C for 7.5 min [25].

3.2.2.2.4 Quaternary Ammonium Salt Modification

Introducing quaternary amino groups onto the surface of cellulose provides a positive charge to the resulting material, thus promoting the separation of CF from its matrix. This can be achieved by the grafting of 2-(dimethylamino)ethyl methacrylate onto the surface of CFs by reversible addition-fragmentation chain transfer (RAFT) polymerization and subsequently quaternized with alky bromides of different chain lengths [26]. Antibacterial activity is added value for quaternary ammonium salt functionalized CFs. It was found that antibacterial activity was inversely correlated to chain length. Antibacterial hydrogel was prepared by functionalization of cellulose with glyoxypropyl trimethyl ammonium chloride followed by cross-linking reaction [27].

3.2.2.2.5 Etherification

Etherification of cellulose renders the material with improved water solubility. Cellulose was modified by the grafting of glycidyl methacrylate (GMA) using ceric ammonium nitrate as initiator, followed by reacting with polyethyleneimine for heavy metal remediation [28]. Layer-by-layer assembly of cellulose ether, such as hydroxyethyl cellulose, hydroxypropyl cellulose (HPC), and methylcellulose, with poly(acrylic acid) (PAA) followed by thermal treatment resulted in films with superior water adsorption ability [29]. Cellulose materials with improved water solubility, such as HPC, have great potential for application in biomedical, food, textile, and environmental fields [29–33].

3.2.2.2.6 Esterification

Cellulose acetate (CA) was synthesized in 1865 by Schuetzenberger and has wide applications in coating, consumer products, medicine, textile, adhesive, and films. The cellulose esters can be classified into inorganic, such as cellulose nitrate, sulfate, phosphate, and xanthate, as well as organic, such as CA, succinates, and many more. Reviews on chemistry and applications of cellulose esterification can be found elsewhere [34, 35]. Cellulose esters with a long side chain demonstrated superior moisture barriers (water vapor transmission rate [WVTR]) while maintaining oxygen permeability and thus are useful in applications for food packaging [36].

3.2.2.2.7 Silylation

Silane modification of cellulose can be carried out by a simple adsorption-curing process [37]. A quinolinium silane salt, (3-trimethoxysilylpropyl) quinolinium iodide (TMSQI), was synthesized by using (3-chloropropyl)trimethoxysilane (CPTMS) and quinolone in the presence of potassium iodide. Before it was subjected to a grafting process, TMSQI was pre-hydrolyzed in a mixture of 2-butanol/water (8:2) at room temperature to generate reactive silanol groups. Aqueous suspension of NCF was solvent exchanged from water to 2-butanol by several successive centrifugation and redispersion processes. Grafting reaction was conducted by stirring the mixture of pre-hydrolyzed TMSQI and NFC suspension at room temperature for 24 hours. Elementary analysis of the covalently grafted quinolinium groups of TMSQI on the surface of NFC revealed the degree of substitution (DS) was estimated to be 0.51. The grafted NFC demonstrated efficient antibacterial activity against both Gram-positive and Gram-negative bacteria.

3.2.2.2.8 Halogenation

Example of halogenation of cellulose is illustrated in Figure 3.1.

Chlorination of MCC was carried out by reacting with N-chlorosuccinimide-triphenylphosphine under homogeneous conditions in LiCl-N, dimethylacetamide. The maximum degree of substitution achieved is 1.86 [38]. On the other hand, the homogeneous bromination of cellulose with N-bromosuccinimide-triphenylphosphine in LiBr-DMA resulted in obtaining the degree of substitution of 0.9.

3.2.2.2.9 Pilot Production of Modified Cellulose

Pilot production of hydrophobic modification of MCC for the production of PET and PLA nanocomposites was carried out in our laboratories with a 70 L reactor. The 70 L reactor and the resulting modified MCC products are shown in Figure 3.2.

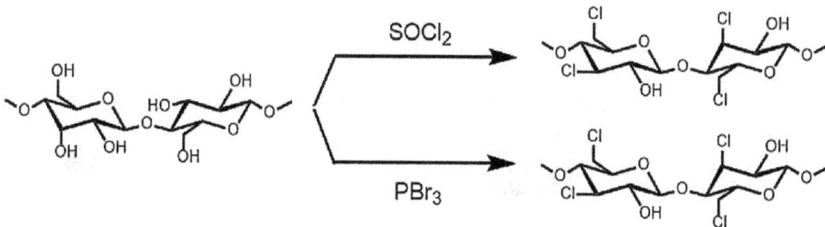

FIGURE 3.1 Halogenation of cellulose by $SOCl_2$ or PBr_3.

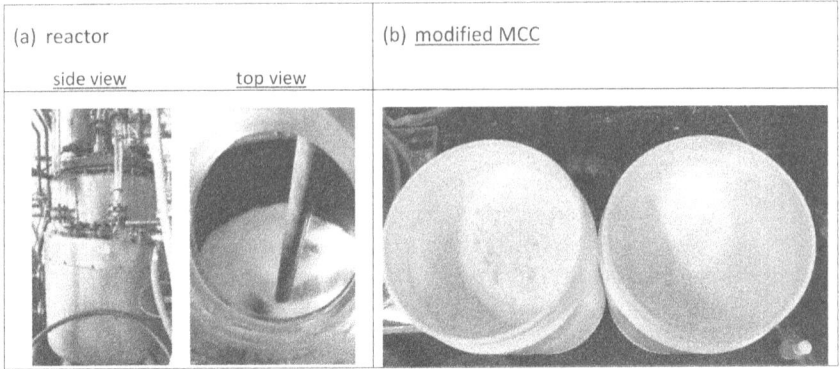

FIGURE 3.2 Hydrophobic modification of MCC in a (a) 70 L reactor and (b) the resulting modified MCC.

The compatibility of the modified MCC with PET matrix was evaluated by compounding 10 wt.% of the filler with the polymer at 245°C–260°C using a DSM microcompounder (Xplore™). Subsequent injection molding process resulted in "dog bone" specimens with 3.2 mm thickness for mechanical performance studies. Visual investigation of the specimens conducted with a backlight under the nanocomposites, as shown in Figure 3.3, indicates modified MCC is well dispersed in the PET matrix, while aggregates were present in the sample containing unmodified MCC. Mechanical properties of nanocomposites are summarized in Table 3.3. Nanocomposite-containing modified MCC demonstrated improved tensile strength and tensile modulus with a small decline in elongation at break value compared to those of PET blanks, but nonetheless superior to those of nanocomposites containing unmodified MCC.

3.2.2.3 Grafting

Conventional chemical and radiation modification technologies have the advantage of introducing a high degree of substitution on cellulose chains but have limited control of side chain growth. In recent years, living free radical polymerization approaches have been applied to the grafting of selected side chains to cellulose materials [39]. CFs were first functionalized with 2-bromoisobutyryl bromide followed by grafting of methyl acrylate (MA) with a targeted DP using the atom-transfer radical polymerization (ATRP) technique. This process rendered CFs with enhanced hydrophobicity. Subsequent grafting of 2-hydroxyethylmethacrylate (HEMA) on the modified CFs resulted in an improvement in hydrophilicity [40].

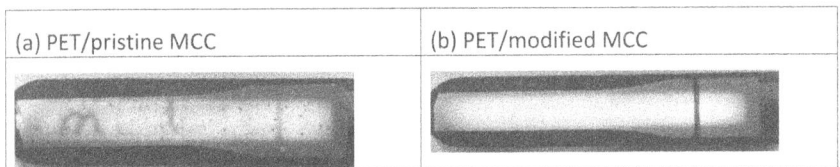

FIGURE 3.3 PET/MCC nanocomposites contain 10 wt.% (a) MCC without modification and (b) modified MCC.

TABLE 3.3

Mechanical Properties of PET and PET/MCC Nanocomposites

Nanocomposites	Tensile Strength (MPa)	Tensile Modulus (MPa)	Elongation at Break (%)
PET	55	1461	5.2
PET/pristine MCC	36	1573	2.6
PET/modified MCC	57	1604	4.7

Grafting of 2-(dimethylamino)ethyl methacrylate (DMAEMA) onto CFs modified with methoxycarbonylphenylmethyl dithiobenzoate was accomplished with the RAFT method [41].

Carboxymethyl cellulose (CMC) or HPC was modified with N-hydroxypyridine-2-thione ester (Barton reagent) for grafting of styrene to the surface using nitroxide mediated polymerization (NMP) approach in the presence of tetramethylpiperidine oxide (TEMPO) [42, 43].

Single-electron transfer living radical polymerization (SET-LRP) approach was applied to graft cellulose esters, cellulose diacetate (CDA), and cellulose acetate butyrate (CAB). The cellulose materials were functionalized by acylation of the backbones with 2-bromoisobutyryl and dichloroacetyl group before carrying out the grafting process [44, 45].

CNFs prepared by TEMPO-mediated oxidation of softwood CFs were grafted with L-lactide in DMSO using the ROP method for the preparation of PLA bionanocomposites [46].

3.2.2.4 Cross-linking

Cellulose materials can be cross-linked with other cellulose polymer chains or polymers, such as polyvinyl alcohol, chitosan, hyaluronic acid, using multi-functional cross-linking agents to improve mechanical properties or other desired application performance. Typical cross-linking agents include epichlorohydrin dicarboxylic acid, diisocyanate diacrylate, and many others.

Citric acid (CA) served as a cross-linking agent for sodium carboxymethylcellulose (CMCNa) and hydroxyethylcellulose (HEC) for the preparation of superabsorbents [47]. The presence of HEC is necessary to promote intermolecular cross-linking to overcome the electrostatic repulsion between CMCNa chains. Hydrogel with a swelling ratio (SR) of 900 and superior physical properties was obtained by cross-linking of the two cellulose materials.

Cross-linking treatment of cotton fabrics with four different carboxylic acids, namely 1,2,3,4-butane tetracarboxylic acid (BTCA), maleic acid (MA), succinic acid (SA), and CA, were studied in the presence of nanometer titanium dioxide (TiO_2) as a catalyst under UV irradiation. Improved dry and wet crease recovery as well as softness values were achieved for cross-linked cotton fabrics [48].

Cellulose films with improved mechanical properties were prepared by dissolving MCC in aqueous $ZnCl_2$ solution followed by introducing $CaCl_2$ into the Zn-cellulose solution [49]. This study concluded that Zn^{2+} weakened the hydrogen bonds between cellulose chains, while Ca^{2+} ions promoted interaction among the Zn-cellulose chains. Selected chemical modification methods are summarized in Table 3.4.

TABLE 3.4
Summary of Chemical Modification of Cellulose

Cellulose	Source	Modifier	Method	Results	References
NFC	Softwood pulp	Nano-Ag	Electrostatic	Antibacterial	[20]
Pulp	Hardwood cellulose	$NaIO_4 + NaClO_2$	Carboxylation	Carboxyl contents: 0.38–1.75 mmol/g; widths: 25 ± 6 nm; crystallinity index: 40%	[21]
Chemical wood pulp	*Betula verrucosa* and *B. pendula*	$NaIO_4 + NaHSO_3$	Sulfonation	Sulfonated groups content: 0.18–0.51 mmol/g; widths: 10–60 nm; crystallinity index: 40%	[24]
Bleached cellulose	Brewer's spent grain	$NaClO_2$ + monochloroacetic acid	Carboxymethylation	DS: 0.58–1.46; crystallinity index: 46.7%–56.5%	[25]
MCC	Aladdin-Reagent Inc.	3-chloro-2-hydroxypropyltrimethyl-ammonium chloride	Quaternary ammonium salt modification	Antibacterial cellulose hydrogels	[27]
MCC	Sigma-Aldrich	4-nitrobenzyl chloride + NH_4Cl	Etherification	Application as flame-resistant materials w/organophosphorus compounds	[28]
α-Cellulose	Sigma-Aldrich	C8-C18 fatty acid chlorides	Esterification	Low WVTR property	[36]
NFC	Masuko Sangyo	silane + quinoline	Silylation	Efficient antibacterial activity against both Gram-positive and Gram-negative bacteria	[37]

(Continued)

TABLE 3.4 (Continued)
Summary of Chemical Modification of Cellulose

Cellulose	Source	Modifier	Method	Results	References
MCC	Merck	Imidazole, etc.	Halogenation	DS values up to 1.6	[38]
Cellulose fiber		HMEA	ATRP	The hydrophilic/hydrophobic behavior of a cellulose surface can be controlled	[40]
Cellulose-CTA	Sigma-Aldrich	DMAEMA + C8-C16 alkyl bromide	RAFT	The PDMAEMA-grafted cellulose fiber with the highest degree of quaternization	[41]
CMC, hydroxypropyl cellulose (HPC)	Sigma-Aldrich	TEMPO + Barton reagent + styrene	NMP	To promote the controlled radical graft polymerization of styrene	[43]
CNC	Softwood, bagasse bleached pulp	L-lactide	ROP	The crystallinity of CNC-g-PLA: 59%–66%	[46]
CMCNa, hydroxyethyl cellulose (HEC)	Eigenmann e Veronelli S.p.A	Citric acid (CA)	Cross-linking	An optimal degree of swelling (SR = 900) was achieved	[47]
MCC	FMC Corporation	CaCl$_2$ + ZnCl$_2$	Cross-linking	The presence of Ca^{2+} ions cross-link the Zn-cellulose chains toward nanofibrils formation and yields transparent films	[49]
BC	Glucose+ microorganism	NaF	Hydrogen bonding	BC of different microstructures and morphologies and mechanical properties could be obtained	[50]

3.3 COMPOSITES PREPARATION

Surface modification enables the production of polymer/cellulose composites using conventional thermo-processing techniques. Industrial manufacturing and processing of PLA composites can be carried out in a similar fashion for convention plastics, such as melt compounding, injection molding, compression molding, extrusion, and spinning. As a polyester-based matrix, PLA is very sensitive to moisture; thus it is critical to dry the resin to the water content of less than 50~250 ppm prior to melt processing. Caution is recommended when using NFs as reinforcement, since the filler may contain crystallized water. Processing conditions, shear, temperatures, and mixing time need to be carefully controlled according to the type of equipment, nature of dispersed phases, composition, and end-use product properties [51].

Alternatively, solvent-casting methods may be used, especially for end-use products involved in the incorporation of heat-sensitive material, often for medical applications. Solvent-based processes require the use of organic solvents, which are not eco-friendly, and thus are limited in scale. These methods are capable of handling a small quantity of samples and quite suitable for early-stage laboratory-scale evaluation of composites. However, the properties of composites prepared by solvent-casting can be very different from those prepared by melt processing [51]. Study results showed thermo-compressed films have higher tensile strength and better thermal stability than solvent-casting films.

Various conventional processing techniques, including melt extrusion, melt spinning, and compression molding, have been applied to produce PLA/cellulose composites for eco-friendly consideration. The hydrophilic nature and thermal stability of cellulose nanomaterials are the key challenges in promoting their dispersion in various polymeric materials with melt-processing techniques. Although melt compounding is an established industrial process, improvements in fiber/polymer composites' mechanical properties are still limited due to difficulties in nanofiber dispersion, especially at high loadings, and to the insufficient formation of a stiff percolated nanofiber network [52]. It is difficult to feed NCs into extruders and achieve uniform dispersion in a polymer matrix and to achieve a percolation phenomenon like in ordinary paper with CFs mutually adhered by hydrogen bonds. Approaches to overcome these challenges include mixing NCs in a suitable medium in a liquid form before feeding into an extruder or drying followed by extrusion. Silane modification can preserve the integrity of CNCs and prevent thermal degradation of the nanomaterial during the extrusion processing step at 165°C [53]. NCs have a tendency to aggregate when dried because of their high surface and hydrophilic characteristics. Cellulose nanomaterials can be extracted from different sources or generated by bacteria and thus have various sizes, morphologies, and surface chemistries. In addition to pretreatment and modification of cellulose nanomaterials, the selection of a proper type of fiber material for a specific target application is critical to achieving optimal performance of the resulting composites.

3.3.1 SOLUTION PROCESSING

Cellulose nanomaterials are obtained in water suspension due to the hydrophilic nature thereof. The viscosity of the cellulose dispersion increases sharply as the concentration increases. Gelation occurs at only a few per cent solid content and the nanoparticles aggregate through hydrogen bonding upon drying. Solution casting has been a preferable approach to avoid irreversible agglomeration during drying. This approach is achieved by mixing the cellulose nanomaterial dispersion with a polymer/oligomer/monomer dispersed in the same or miscible liquid medium. The resulting mixture can be cast and freeze-dried to avoid the aggregation of NC. It is possible to achieve a percolation effect of fibers in composites by film casting of cellulose whiskers and latex resin. Wet processing methods generally lead to the highest mechanical properties, as the NC is well dispersed. Nanocomposites made from solution-casting methods demonstrated shear modulus at a rubbery state staying constant up to the temperature of cellulose decomposition [52]. This method is often applied to prepare masterbatches for melt processing of polymer cellulose composites [54–56].

Properties of the biocomposites can be increased by nanofibers alignment and improved dispersion. MFC and PLA were premixed in organic solvent followed by kneading after removal of the solvent to obtain a uniform dispersion of MCF in the polymer matrix [55]. Young's modulus and tensile strength of PLA increased without compromise yield strain with a fiber content of 10 wt.%. Alternatively, solvent evaporation technique commonly used for drug microencapsulation was utilized to deposit CNFs onto the surface of suspended PLA in water as microparticles [57]. Membrane filtration produced CNF sheets filled with PLA particles. Compression molding of the stacked sheets resulted in a biocomposite with improved modulus and strength up to 58% and 210%, respectively. The preparation methods and the resulting performance of a selected number of PLA/cellulose biocomposites are summarized in Table 3.5.

Animal protein, casein, was used as a compatibilizer and eco-friendly dispersant for the preparation of biocomposites comprising CNF and PLA to improve interfacial adhesion. [60] Proteins are amphoteric polymers and can display affinity with a wide range of hydrophilic and hydrophobic surfaces. The interactions between the hydroxyl groups of cellulose and the peptide bonds of protein promote the affiliation of the two materials. In fact, casein was used as an early wood adhesive. Bionanocomposite films were prepared by pressurized filtration of the mixture of CNF suspension and PLA latex in pH 8 buffer solution with or without the presence of casein followed by hot-pressing at 50°C. The tensile strength and Young's modulus of PLA/CNF (1/1) bionanocomposite films were reduced compared to those of the CNF film, while the presence of PLA led to a small increase of 0.5% in ductility. Incorporation of casein (1 wt.% to CNT) resulted in a significant increase of the film ductility by a factor of 2.3, and an enhancement of the tensile strength by 14% and Young's modulus by 12%. There is a 60% improvement in the PLA/casein/CNF film toughness. The distribution of CNFs in the polymer matrix is improved in the presence of casein. This also contributed to the improvement of the mechanical performance of the bionanocomposite. Protein-mediated adhesion between PLA and CNF reduced the moisture sensitivity of the resulting bionanocomposite films and improved the material stability for thermoforming applications.

Solution processing of a mixture of bacterial cell (BC) and PLA in 1,4-dioxane solution was applied to prepare porous composite microspheres as a masterbatch for the production of green PLA nanocomposites and self-reinforced PLA/NC

TABLE 3.5

Preparation Methods and Properties of PLA/Cellulose Biocomposites

PLA	Cellulose	Preparation	Performance	References
	CNF l-(-)-lactide	Solution casting	The crystallinity of TOCN-g-PLA products was 59–66% greater than the crystallinity of neat PLA	[46]
Landy PL	MFC Chitin nanofiber	Aqueous suspension mixture/hot press	PLA reinforced with cellulose and chitin nanofibers had higher tensile properties than its counterparts reinforced with one type of nanofiber alone	[52]
ESUN™	CNF (kenaf)	Masterbatch/solvent mixture + extrusion	Tensile modulus (2.9 GPa to 3.6 GPa) and tensile strength (58 MPa to 71 MPa) with 5 wt.% CNF	[54]
	CNC	Solution casting	The presence of 1% CNC in the PLA film resulted in a low WVTR	[58]
Ingeo 4032D	CNW	Melt extrusion	The mechanical properties of composites improved E-modulus about 90% compared to PLA, 800% enhancement of elongation to break	[59]
Ingeo 3052D Landy PL-3000 (PLA latex)	CNFs	CNF suspension + PLA latex/hot press	~50% increase in interfacial adhesion; an enhanced extensibility by 130% and tensile toughness by 60%	[60]
PLLA/Biomer	BC	mBC + PLA in dioxane, ppt in liq N2/extrusion	Tensile modulus of all the nanocomposites increased	[61]
Shanghai Yisheng	CNF	Solution casting	~60% enhancement of tensile strength and elongation at break	[62]
Ingeo 4032D	CNC	In situ PVAc/CNC and solvent casting/hot press	Increases in elastic modulus, yield strength, elongation to break and Tg	[63]
	MFC	Thin sheet of MCF and PLA fibers/compression molding	The modulus, strength, and strain at fracture increased linearly with the MFC content	[64]

(Continued)

TABLE 3.5 (Continued)
Preparation Methods and Properties of PLA/Cellulose Biocomposites

PLA	Cellulose	Preparation	Performance	References
PLLA/Biomer PDLLA/4060 D	BC	mBC + PLA in dioxane, ppt in liq N2/extrusion	PLLA/BC: tensile modulus (6 GPa) and strength (127 MPa) PLLA/PDLLA/CNC: 175% enhancement of viscoelastic	[57]
	CNC	Aqueous PVAc-GMA + melt extrusion	Epoxy groups of GMA accelerated biodegradation; effective dispersion of CNC	[65]
	CNC-g-PLLA	Extrusion	CNC as a nucleating agent; increase in Young's modulus	[66]
	CNWs	10 wt.% masterbatches/melt spinning	Fiber diameters 90–95 nm, drawing ratio 2; increased thermal stability, creep resistance; w/o affect stiffness and strength, significantly lower strain	[67]
Ingeo 2003D	CF	Solution casting/extrusion and injection molding	Improved elongation at break: PLA/PEO-15/CF-10 15.8% and PLA/PEO 450% compared to PLA 6.5%	[68]
Ingeo 2002D	CNC, MCF MMT	Compounded with a Brabender	All the composite materials exhibited higher storage modulus (E′) in the glassy region; PLA/CNC had improved elongation at break	[69]
Ingeo 2003D	Chitin nanofibers	Solution casting/extrusion	Chitin nanofibers could not reinforce the PLA matrix	[70]
Ingeo 4043D	CNFs	Compounded with a Brabender/extrusion	ESO can improve the ductility of the nanocomposites 5- to 10-fold (@ 10 wt.% CNF)	[71]
FUTERRO	CNC Chitin nanocrystals	Extrusion using liquid feeding	Chitin nanocomposites with slightly higher mechanical properties and higher transparency compare to CNC	[72]
Ingeo 4060D	MCF	Melt mixing/supercritical CO_2	The mechanical performance of bio-based foam initially increased with MCF and then decreased	[73]

composites with improved properties [61]. There is interest in BC for nanocomposites production due to its highly crystalline (~90%) structure and because it is free from wax, lignin, hemicellulose, and pectin (which are present in plant cellulose). Unlike plant fibers that have to be microfibrillated to produce NC, BC exists naturally as a nanosized material with diameters of 24–86 nm and lengths of several micrometers. Surface functionalization of BC using various organic acids (acetic, hexanoic, and dodecanoic acid) was evaluated to improve interfacial adhesion of BC and matrix PLA. Since the morphology of modified nanofibrils remained the same as that of neat BC, this indicates that the functionalization occurred essentially only on the surface and retained the highly crystalline bulk structure. The wettability between PLLA and the nanofiber increased with the chain length of the organic acid used to functionalize BC increased. To prepare a composite masterbatch for the extrusion process, BC was first dispersed in 1,4-dioxane with a homogenizer at 20,000 rpm, followed by dissolving PLA pellets at 60°C overnight. The resulting solution was introduced dropwise into liquid nitrogen for precipitation of composited micro-spheres. Enhancement of tensile modulus was observed for all the nanocomposites containing 2 wt.% and 5 wt.% of neat and modified BC. However, pounced negative effects on tensile strength and elongation at break were identified for nanocompos-ites containing neat and acetic acid-modified BC due to degradation of the polymer. It was suggested that acetic acid-modified BC is relatively hydrophilic in nature. The presence of water might cause the hydrolysis of the ester bonds in modified BC and degradation of PLLA, which was confirmed by GPC studies. Modification of BC with higher chain length, C6 and C12, led to improvement in both tensile modulus and tensile strength of nanocomposites. Enhancement of storage moduli of nanocomposites is evidence of improved interfacial adhesion between PLLA and the functionalized BC.

A chemo-mechanical approach was applied to prepare cellulose nanofibrils suspension in DMAc by treatment of pulpboard CF in a diluted sulfuric acid and subsequent dispersion of the pre-treated cellulose in DMAc with a homogenizer. Bionanocomposites were prepared by the casting of the solution containing PLA and PEO dissolved in DMAc with NCF suspension. In terms of mechanical properties, including tensile strength and elongation at break, an improvement of about 60% was achieved for PLA/PEO/NCF, as compared to those of bionanocomposites without PEO [62]. Polyethylene-oxide-assisted interfacial interaction was demonstrated here.

Dispersion of hydrophilic cellulose nanomaterials in a hydrophobic matrix was achieved by in situ polymerization of vinyl acetate in the presence of CNCs and subsequent mixing of the resulting in situ PVAc/CNC nanocomposite and PLA in DMF [63]. An aqueous suspension of CNCs was introduced to the vinyl acetate reaction mixture at the PVAc to CNC ratio of 80:20 for in situ emulsion polymeriza-tion to obtain in situ PVAc/CNC latex with 15 wt.% solid content. The high solid content of hybrid latex suspension was intended for potentially scaling up produc-tion. Equivalent suspension by mechanical mixing of CNCs suspension and PVAc latex was also prepared in comparative studies. Although both in situ and mechani-cal mixed suspensions were shown to be electrostatically stable with zeta potential measurements, the morphology of the two specimens is significantly different after drying. The unconsolidated in situ PVAc/CNC sample exhibits well dispersed PVAc particles of ~70.5 nm and CNCs of ~5 nm in diameter while the one prepared by a mixing method had large aggregates for both particles. Cross-linking of PVAc with sodium tetraborate decahydrate (borax) restricted the PVAc chains' mobility and strengthened interaction between the PVAc and CNCs. Nanocomposite films were prepared by casting in situ PVAc/CNC latex and PLA/DMF solution mixture with the addition of glyceryl triacetate (GTA) as a plasticizer in a Teflon Petri dish followed by hot pressing. A synergistic effect was demonstrated by significant increases in elastic modulus, yield strength, elongation to break, and glass transition temperature of PLA/in situ PVAc/CNC nanocomposites containing 0.1 wt.% CNC compared to the PLA/PVAc-only material.

3.3.2 Melt Processing

Solution-processing strategies face the limitation of scale for high-volume production, namely those concerning solvent handling and production efficiency. Extensive summaries of different nanocomposite systems prepared from cellulose nanomaterials by melt compounding can be found in the published literature [11]. The main governing factors influencing the performance of biocomposites prepared by melt processing are (i) irreversible aggregation, (ii) non-uniform dispersion, (iii) thermal stability, (iv) structural integrity, and (v) orientation of the NC fillers. Thus, melt-processing methods seem to be more challenging to apply to CNFs than to CNC due to the possibility of entanglement of CNFs.

3.3.2.1 Compression Molding

Compression molding is often applied to incorporate nanomaterials with relatively high cellulose content, up to more than 70%, into thermoplastic polymer matrices. In many cases, cellulose nanomaterials or their composites were first cast as films, spun into fibers, or woven into fabrics, following which PLA was introduced before compression molding to yield composite materials. For example, cellulose nanomaterial dispersions can be mixed with PLA in a selected solvent, followed by removal of the solvent and then compression molded into a composite sheet. The modulus, strength, and strain at fracture increased linearly with the MFC content for composites obtained by compression molding of stacked sheets made of uniformly dispersed MFC with PLA fibers [64, 74]. The manufacturing process for the preparation of MFC/PLA sheet is similar to paper-making and thus suitable for adoption at an industrial scale production. Instead of PLA fibers, a colloidal suspension of PLA can be mixed with a cellulose suspension in an aqueous medium and subsequently dried to form a paper-like sheet. A hot press process resulted in the melting of PLA to form a nanocomposite sheet.

Cellulose and chitin nanofibers were used as complementary reinforced fillers for PLA. Suspension mixtures of PLA, cellulose, and chitin nanofibers were cast into sheets and followed by compression molding to produce nanocomposites [52]. Reinforced fibers tend to concentrate on the bottom upon drying and lead to a gradient distribution along the thickness of the specimens. Chitin nanofibers served as a compatibilizer between hydrophobic PLA matrix and hydrophilic rigid CNFs. To a certain degree, this approach mimics the interaction of hemicellulose with cellulose and lignin. Polylactic acid reinforced with both nanofibers had higher tensile properties than its counterparts reinforced with cellulose or chitin nanofibers alone. Since chitin improved bonding between the hydrophobic PLA and hydrophilic MFC, HDT of nanocomposites, containing chitin to MFC nanofibers in the ratio of 1:4, 1:1, and 4:1 at 50% reinforcing phase loading, increased with an increase in chitin nanofibers content. Alternatively, nanocomposites can be prepared by surface modification of CNFs with lactic acid in an aqueous medium with the aid of sonication and subsequently polymerized by compression molding [75]. The modified nanopapers showed enhanced mechanical properties with an increase in elastic modulus and in yield strength of more than 30%. The PLA/NC composites also demonstrated improved thermal and moisture resistance.

Self-reinforced PLA/NC composites were fabricated by melt-spinning techniques and layered-filament winding followed by compression molding in an effort to improve the mechanical performance of PLA [57]. Self-reinforced thermoplastic polymer composites have the advantages of inherent compatibility between matrix and reinforced phases as well as the potential of recycling at the end of product life. Renewable reinforcement fillers are neat and modified BC in 2 wt.%. Strain-induced chain orientation led to improvement of tensile modulus (6 GPa) and strength (127 MPa) of poly(L-lactide) (PLLA). These improvements were accompanied by a reduced draw ratio and increased fiber diameters owing to an increase of the polymer melt viscosity in the presence of BC. Draw ratio can be improved, on the other hand, by surface modification of BC with hexanoic acid. BC also served as a nucleating agent to facilitate crystals growth in the polymer matrix.

Alternatively, PLLA fibers can be reinforced with a matrix of amorphous PDLLA containing 7 wt.% CNC with 175% enhancement of viscoelastic properties and 17% increase in strains to failure. Composite spheres were prepared as masterbatches using thermally induced phase separation (TIPS) by precipitation of 1,4-dioxane solution of BC and PLA in liquid nitrogen [61]. Amorphous PDLLA and semicrystalline PLLA (with or without BC/CNCs) were melt-spun on top of one another, followed by compression molding to produce aligned PLLA fiber-reinforced PDLLA composites. Consolidation temperatures need to be controlled to allow PDLLA/CNC matrices to melt and wetting the PLLA/CNC composite fibers is required to achieve desired performance. All fiber-reinforced composites demonstrated enhanced storage modulus. The composite of PLLA/CNC fiber-reinforced PDLLA/CNC achieved the highest storage moduli at 20°C of 11.6 GPa (bending) and 7.0 GPa (tension) among the samples evaluated, as compared to the values of 4.2 GPa and 2.8 GPa for neat PLA, respectively. Furthermore, significant improvements in both modulus (123%) and strength (35%) were observed due to the presence of PLLA/CNC fiber and CNC in the PDLLA matrix.

3.3.2.2 Extrusion

A typical extruder consists of three processing zones: (i) the feeding zone, (ii) the transition (kneading, compression) zone, and (iii) the metering zone. Variables to control for achieving desired final product performance include screw speed, screw configuration, screw length-to-diameter ratio (L/D), barrel temperature, feed rates, die shape/size, etc. All these variables need to be taken into account when selecting a set of PLA and CF. Polymers and fillers are conveyed to the center of the extruder barrel from the hopper at the feeding zone. Once the mixture enters the transition zone, the heat generated by the viscous shearing of polymer pellets and by the external heaters causes the melting of the polymers. Polymers melt gradually with a long kneading zone, which is suitable for compounding a thermally sensitive plastic. On the other hand, for a crystalline polymer that melts sharply, a short kneading is effective. For high filler content composite preparation, a high L/D ratio is required to ensure proper mixing of polymers and fillers in this section. Shear force screw arrangement designs range from high, medium, and low, which are based on the angle between two side-by-side adjacent screw threads, may be chosen based on the properties of the polymer and filler content. A combination of medium- and high-shear-force screw arrangements provided the best results in our laboratories when taking into account production efficiency and properties of the PLA/MCC masterbatch containing 50 wt.% filler. A metering zone with high shear temperatures with high pressure building up at the die was achieved. Compression molding is often applied to prepare masterbatches for extrusion process applications to produce bionanocomposites.

A masterbatch was prepared by mixing an acetone suspension of CNF extracted from kenaf pulp and PLA solution in a solvent mixture of acetone and chloroform in a 9:1 ratio. After evaporation of the solvent, the masterbatch was blended with PLA for the preparation of pellets using a twin-screw extruder [54]. The tensile strength and modulus were enhanced with increased CNF content from 1%, 3%, and 5%. Improvement in properties is less pronounced than that theoretically predicted.

There have been considerable efforts to overcome cellulose–cellulose interactions and to improve interfacial compatibility between cellulose and PLA by surface functionalization of cellulose materials or by using compatibilizers to afford the use of melt processing route to prepare biocomposites. Maleic anhydride-grafted PLA was used as a compatibilizer to enhance PLA/CNWs interfacial adhesion [76]. However, processing conditions, such as temperature and mixing time, need to be optimized to avoid PLA and CNW degradation as well as incomplete dispersion. Cellulose nanowhiskers (CNWs) were extracted from MCC and used as filler for improved mechanical properties of the resulting nanocomposites. In the presence of maleic anhydride-grafted PLA as a compatibilizer and nanoclay, the tensile strength doubled and the glass transition temperature was 23°C higher for the resulting nanocomposites compared to that of PLA/MCC composite.

CNCs were dispersed in aqueous poly(vinyl acetate) (PVAc) followed by melt extrusion with PLA [65]. Grafting of epoxy groups onto PVAc was performed by copolymerization of vinyl acetate and GMA by using ammonium cerium(IV) nitrate as initiator. It was observed that the epoxy groups accelerated the biodegradation rate of PLA/PVAc-GMA/CNC(3 wt.%) through the formation of radicals.

CNCs, obtained from acid hydrolysis of MCC, were grafted with L-lactide by ROP initiated from the hydroxyl groups available at the surface of the nanomaterials. Incorporation of the resulting CNC-g-PLLA into PLA matrix was carried out by extrusion process to obtain biocomposites containing 1 wt.% and 3 wt.% CNC [66]. A higher graft ratio improves the dispersion of CNC in PLA matrices and thus facilitates the polymer's crystallization rate. The best efficiency was revealed by the increase in Young's modulus of a biocomposite containing 3 wt.% of CNC-g-PLLA.

Melt compounded masterbatches of PLA with 10 wt.% CNWs were prepared for melt spinning of PLA/CNW nanocomposite fibers [67]. The presence of CNWs restricted the drawing ratio of the fibers to a factor of two without affecting fiber stiffness or strength but resulted in a significantly lower strain. A restriction of PLA chain mobility by CNWs filler resulted in improvement of thermal stability, creep resistance, and reduction in thermal shrinkage of PLA fiber. Other observations include increased surface roughness, aggregations in the fibers containing CNWs, and slightly increased crystallinity.

Several studies have indicated that cellulose materials can be well dispersed in poly(ethylene oxide) (PEO), an amphiphilic polymer, due to the formation of strong hydrogen bonds. The use of PEO as a compatibilizer for the fabrication of PLA/CF composites could represent an environmentally friendly alternative due to its non-toxicity, water solubility, biodegradability, and biocompatibility. Treatment of CF in an aqueous solution of PEO improved the interfacial interaction among the components, reduced the fluffiness of CF, and allowed a continuous as well as uniform feeding in the extruder [68]. The synergistic effect of PEO and CF led to the enhancement of the physical-chemical properties of PLA biocomposites. The masterbatch of PEO/CF was prepared by solution casting followed by hot pressing at 90°C. Hot-pressed PEO/CF blends of different compositions, CF and PEO, were cut into small pieces and fed to a co-rotating twin-screw extruder equipped with six temperature zones with a screw speed of 250 rpm and a temperature profile ranging from 175°C to 190°C. The PLA/PEO/CF biocomposites consisted of 10, 15, and 30 wt.% of CF while maintaining a PEO content at 15%. Injection molding of the dried extruded pellets was carried out at 175°C with an injection pressure of 60 bar to obtain testing specimens. Rheological investigation of the apparent shear viscosity and shear stress characteristics indicated that all biocomposites demonstrated the typical shear-thinning behavior. To ensure the integrity of the extruded composite pellet and injected product, a high shear viscosity at a low shear rate is critical during extrusion, whereas a low shear viscosity at a high sheet rate is positive for the injection molding process. A higher shear viscosity of PLA/PEO/CF compared to PEO plasticized PLA is an indication of a good reinforcing effect of CF. A stronger interfacial adhesion of the cellulosic fibers at a content of 10 wt.% was speculated to result in a higher shear viscosity than that of neat PLA for PLA/PEO/CF-10 biocomposite at a fixed PEO content of 15 wt.%. The storage modulus of PLA/PEO and PLA/PEO/CF was significantly higher than that of neat PLA at temperatures above 80°C. The value increased with increasing CF content in the biocomposites. Since the storage modulus in rubbery state depends on the aspect ratio of the filler, crystallinity, and interaction between the phases, the result indicates PEO modified CF was well dispersed in the PLA matrix. The biocomposite, PLA/PEO/CF, had comparable tensile strength and modulus, while the elongation at break improved by 73%–143% compared to neat PLA. The tensile modulus and tensile strength of the biocomposites were also remarkably improved by 52%–85% and 46%–54%, respectively, when compared to those of PLA/PEO.

Polylactide reinforced with organo-modified mineral nanoparticles and CFs were prepared for competitive studies [69]. It was observed that the filler intensified PLA cold crystallization due to a nucleating effect without affecting the glass transition temperature of the polymer. It has been widely reported that exfoliated and oriented organo-modified montmorillonite (o-MMT) have significantly improved barrier properties with respect to neat PLA. The nanocomposite specimens were prepared by melt mixing in an internal mixer Brabender Plasti-Corder at 180°C at the constant rotation speed of 50 rpm for 10–15 minutes. To obtain three-component hybrid composites containing microcellulose fibers, PLA was first melt mixed with nanofillers in the mixer for 5 minutes, followed by the addition of 15 wt.% MCF for an additional 5-minute mixing at 170°C. The three-component system was prepared for studies of the effect of MCF on the properties of PLA nanocomposites with different types of nanofillers. Thermomechanical properties studies revealed that all the composite materials exhibited higher storage modulus (E′) in the glassy region in the following order: PLA/MMT/MCF > PLA/NCC/MCF > PLA/MMT > PLA/MCF > PLA/NCC. During uniaxial drawing, PLA/MMT exhibited properties comparable to those of neat PLA in terms of yield stress and stress at break values but had drastically diminished elongation at break percentage. On the other hand, both yield stress and stress at break decreased for the PLA/CNC nanocomposite while elongation at break improved. However, the tensile behavior of the hybrid composites was negatively influenced by the presence of MCF due to poor interfacial bonding between the fiber and PLA matrix.

A pre-composite of poly(ethylene glycol) (PEG) and chitin nanocrystals were prepared by water evaporation before combining with PLA in an extruder to obtain transparent nanocomposites [70]. Transparent specimens and cryofractured surfaces without aggregates compared to 10-micrometer aggregates observed in the PLA/chitin film prepared without pre-composites made clear the role of PEG as a dispersing agent. The SEM micrographs studies of cryofractured surfaces revealed that the dimensions of the PEG domains decreased as the content of chitin nanofibers increased. This suggested that chitin nanofibers acted as interfacial stabilizers and are present in both the plasticized PLA matrix as well as in the PEG domains. The Young's modulus decreased when plasticizer was introduced to the PLA matrix and it decreased further by the addition of chitin nanofibers to the plasticized PLA. The stress-at-break and the stress-at-yield measurements resulted in a similar trend. Thus, chitin nanofibers did not show a clear reinforcing action in the presence of 10 wt.% plasticizer. The stress at yield of nanocomposites containing PEG 8000 was higher than those containing PEG 400. This is due to the system with a lower molecular plasticizer allowing for easier sliding of macromolecules in correspondence with the beginning of the yield. Studies of the influence of chitin nanofiber content on the mechanical properties of plasticized PLA at 10 wt.% PEG 400 indicated the composite at 5 wt.% of the filler was most rigid, and the elastic modulus slightly decreased initially by adding 2 wt.% chitin nanofibers and remain about the same up to 12 wt.%. The stress at yield and the Young's modulus decreased as a function of PEG content. The linear reduction of the Young's modulus as a function of PEG content allowed us to estimate the optimal value of 3.45 GPa for the composite consisting of PLA and 2 wt.% chitin nanofibers at the intercept. This value is still lower than the value determined for pure PLA (3.5 GPa), and thus, Coltelli et al. concluded that chitin nanofibers could not reinforce the PLA matrix. For the earlier studies with improved Young's modulus by Chitin nanofiber, it can be attributed to the low affinity of the dispersing agent, triethyl citrate (TEC), and the PLA matrix. However, the presence of 2 wt.% chitin nanofibers did not affect the 180% increase in the elongation at break of PLA plasticized with 10 wt.% PEG.

A tertiary system consisting of PLA, CNF, and epoxidized soybean oil (ESO) was developed to maintain tensile strength and to increase the ductility of the biocomposite [71]. The mixture of PLA, CNF, and ESO was melt blended with a Brabender at 160°C and 60 rpm rotor speed. It was followed by extrusion and compress molding at 220°C for sample preparation. For a two-component system, the tensile strength and modulus were improved in the PLA-CNF biocomposites with increasing CNF content, while the values decreased in the PLA-ESO cases with the increase of ESO content. By combining CNF and ESO to the PLA matrix, the biocomposite retained comparable tensile strength and 20% increased modulus values to that of PLA while the ductility and toughness reached at least three times of neat PLA. According to the study results, it was suggested that 10 wt.% CNF was responsible for maintaining the tensile strength and modulus and ESO contributed to the improvement of ductility of the biocomposite. Further increase of CNF to 20 and 30 wt.% did not lead to improvement in strength, modulus, or strain in the presence of ESO (5 wt.%) compared to the case of the PLA-CNF binary system. Superior properties at high volume fractions of CNFs were attributed to the formation of strong hydrogen bonds among NCs, which results in geometrical and mechanical percolation. Force transfer occurring between CNF and PLA matrix became the dominant factor at 20 wt.% CNF content. The percolation threshold range depended on the dispersion of the filler, aspect ratio, and filler/matrix interaction. In a binary system, at low filler content, both CNF (10 wt.%) and ESO (2 and 5 wt.%) enhanced the PLA crystallization rate. At higher content, the fillers may serve as impurities that interfere with nucleation or crystal growth. Irregular degrees of crystallization were observed for the tertiary system due to hydrogen bonding interaction between CNFs and ESO. Thus, it is critical to optimally control the amount of CNF and ESO to obtain a biocomposite with desired mechanical performance.

Liquid feeding with suitable processing aid is an alternative route to disperse cellulose nanomaterials in polymer matrices without drying before compounding. The NC suspension was fed into polymer melt directly during extrusion to avoid aggregation. A venting system is thus critical for the liquid feeding extrusion process [18]. A co-rotating twin-screw extruder coupled with a gravimetric feeder, a peristaltic pump, and steam removal accessory was applied for the preparation of PLA/CNCs nanocomposites using liquid-feeding technique [72]. The extrusion operation conditions for the preparation of PLA nanocomposites were a screw speed of 300 rpm and a temperature profile of 170°C at the feeding zone to 200°C at the die. Suspensions for liquid feeding consisted of CNC or chitin nanomaterials dispersed in a 1:1 ratio of water and ethanol solution with the addition of TEC as a plasticizer. According to AFM measurement, chitin nanocrystal had a high aspect ratio and almost doubled maximum diameter and length, as compared to those of CNC. Chitin nanocrystal also had higher crystallinity (91% vs 80%) and thermal stability (onset at 280°C vs 219°C). Plasticized PLA nanocomposite films were prepared by extrusion and followed by compression molding at two different cooling rates. Fast cooling rates produced ductile transparent films with an elongation to break of ~300%, while slower cooling rates resulted in nanocomposite films with haziness and crazing characteristics. The yield strength and the Young's modulus of the fast-cooled plasticized PLA film were significantly increased by 316% (from 3.7 to 15.4 MPa) and by 267% (from 0.3 to 1.1 GPa), respectively, with the addition of 1 wt.% CNC. In the presence of 1 wt.% chitin nanocrystal, the yield strength improved by 478% (from 3.7 to 21.4 MPa) and the Young's modulus improved by 300% (from 0.3 to 1.2 GPa). In the case of the slow cooling process, the ultimate strength of nanocomposites containing CNC or chitin nanocrystal increased by 26% (from 15.8 to 19.9 MPa) and 28% (from 15.8% to 20.3 MPa), respectively. The Young's modulus increased by 50% (from 0.6 to 0.9 GPa) for the plasticized PLA nanocomposite containing 1 wt.% CNC and by 67% (from 0.6 to 1.0 GPa) for the sample with the addition of chitin nanocrystal. However, the yield strength of the slow-cooled specimens was slightly increased by 9% (from 19.6 to 21.4 MPa) in the presence of CNC and was not significantly increased with the addition of chitin nanocrystal. Since the chitin nanocomposites showed a tendency to be more amorphous, PLA crystallization did not impact the mechanical properties of the chitin nanocomposites due to the internal migration of the plasticizer from the crystalline to the amorphous region.

MFCs were introduced into the PLA matrix to improve mechanical properties and regulate cell size as well as structure for the fabrication of bio-resourced foams for building insulation application [73]. The glass transition temperature of PLA decreases in the presence of MCF and thus leads to a decrease in cell ball thickness. Fiber orientation, concentration, and dispersion affect the voids, porosity, and pore structure of a fiber/PLA composite foam and thus have a great influence on the insulation properties. Studies found that foams prepared by the supercritical CO_2 method using acetylated cellulose had a higher bulk density with improved interfacial properties and tensile strength up to 9 wt.% loading while foams made with unmodified CFs had smaller average cell sizes and exhibited higher flexibility and toughness.

A masterbatch of PET nanocomposite containing 30 wt.% modified MCC was prepared by using an extruder at temperatures ranging between 235°C and 245°C at a production rate of 10 Kg/h (Figure 3.4). Barrier properties of PET/MCC nanocomposites containing 10 wt.% and 30 wt.% MCC with and without modification were evaluated. Nanocomposite sheets with thicknesses of about 500 micrometers were obtained by using a T-die extruder. A nanocomposite sheet containing 30 wt.% MCC is shown in Figure 3.5. The resulting oxygen barrier properties, (oxygen transmission rate [OTR]), of the tested specimens are summarized in Table 3.6. Thorough dispersion of modified MCC in the polymer matrix was clearly demonstrated by a one-order magnitude reduction of OTR value, as compared to that of nanocomposites prepared with MCC without modification.

A much higher MCC loading to PLA was achieved than that to PET in our laboratories for the production of masterbatches. The operating temperature range for the production of PLA/modified MCC masterbatches is between 155°C and 170°C. The resulting masterbatch containing 50 wt.% MCC has a melting index of 10 at 200°C under 2.16 kg loading and specific gravity of 1.25. The resulting masterbatch revealed onset degradation at about 300°C based on a thermalgravimetric analysis (TGA) thermogram study, as shown in Figure 3.6. A compostable PLA/cellulose nanocomposite with improved barrier properties will have both environmental and social benefits. Further optimization of fabrication techniques including cellulose modification and processing conditions are required to achieve this goal.

FIGURE 3.4 Pilot production of PET nanocomposite masterbatch containing 30 wt.% modified MCC.

FIGURE 3.5 PET/30 wt.% modified MCC sheet.

TABLE 3.6
OTR of PET/MCC Nanocomposites

	PET/Unmodified MCC		PET/Modified MCC	
MCC (wt.%)	30	10	30	10
OTR (c.c/m² day atm)	16.27	20.1	1.09	7.1

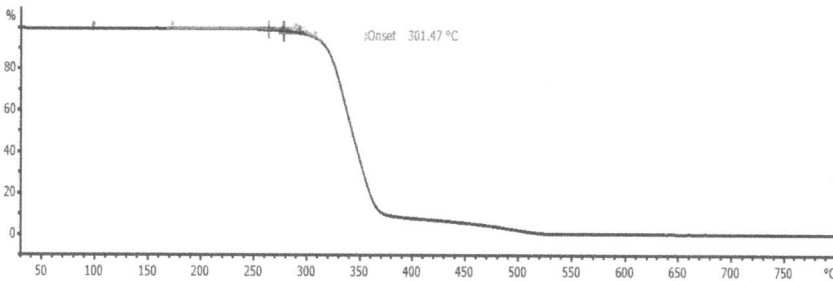

FIGURE 3.6   TGA thermogram of PLA/MCC (50 wt.%) masterbatch.

BSG accounts for about 85% of all waste generated during beer production. Proper management of the food processing industrial residue provides economic and environmental benefits. Brewery wastes are often used as animal feed additives and food ingredients in the food industry [77]. In our laboratories, BSG was dried and ground to less than 15–30 micrometers, with finer particles preferred for thinner blown films, using a mill for the production of PLA/cellulose biocomposites. The resulting ground BSG was then blended with polybutylene adipate terephthalate (PBAT) and PLA at a temperature range of between 150°C and 180°C in a twin-screw extruder. Typical BSG loading was about 5–10 wt.% for the production mulch films (Figure 3.7). For selected applications, such as strawberry and watermelon

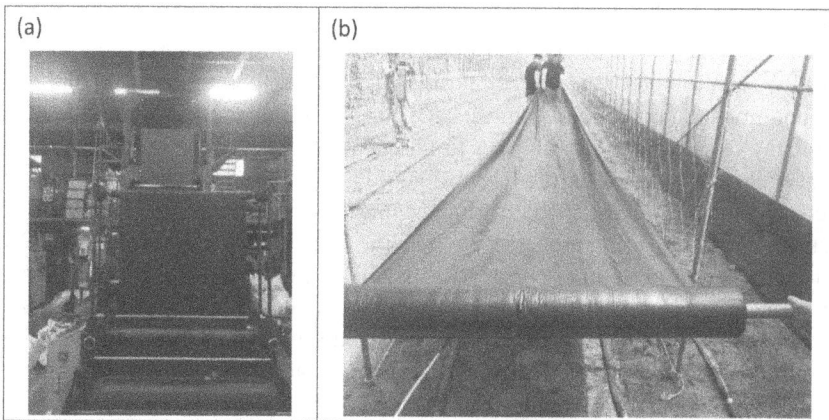

FIGURE 3.7   PLA/BSG biocomposite mulch film (a) blown film, (b) at plantation.

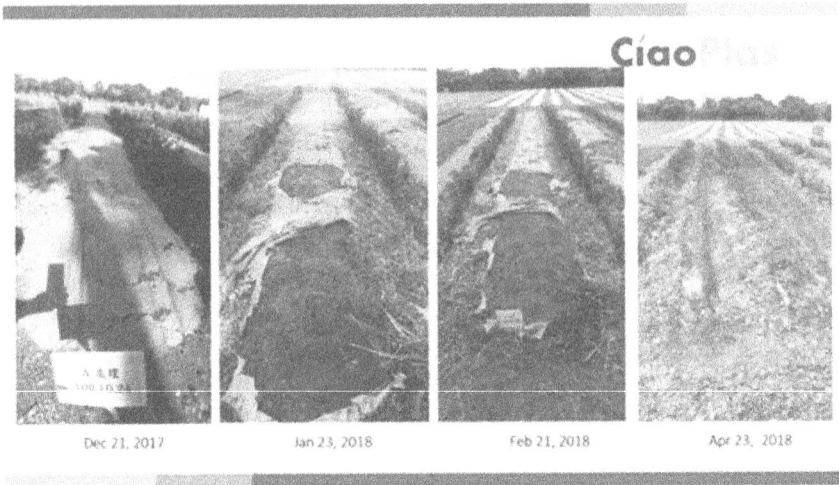

Dec 21, 2017   Jan 23, 2018   Feb 21, 2018   Apr 23, 2018

FIGURE 3.8 Degradation of biocomposite mulch film in the farm.

cultivation, a thermally stable bacterial component was incorporated into the biocomposites to control the degradation rate. The resulting film disintegrated in a few months in an exterior field, as shown in Figure 3.8.

3.4 SUMMARY AND SUGGESTIONS

Polylactic acid/CF composites are potentially compostable and recyclable. There has been intensive investigation into the preparation technologies and properties of PLA cellulose composites in recent years. Current efforts are focused on overcoming the interfacial adhesion between hydrophobic PLA and hydrophilic cellulose materials. Approaches include chemical and physical modification of cellulose materials and choice of suitable and optimal composite preparation conditions to both preserve the integrity of the structure of cellulose materials and avoid degradation of the polymer. Improved performance in various properties, including barrier, mechanical, and thermal properties, was achieved for biocomposites obtained by reinforcing PLA with a broad range of cellulosic materials, such as wood fibers and flour, bamboo fibers, microcellulose fibers, MCC fibers, NC fibers, nanocrystalline CFs, and many others. However, in order to make the best use of the numerous renewable cellulosic resources for the fabrication of high-performance biocomposites and to contribute to achieving global carbon neutrality, it is still crucial to continuously perfecting manufacturing processes for both the preparation of fiber-reinforced PLA and modification of cellulose materials. There is growing interest in plant-based composite materials that are both environmentally friendly and high-performance. For this emerging market, there is an urgent need for clear labeling and end-of-use management schemes for biodegradable and compostable products. The strategic use of biocomposites will have both social and environmental benefits. The involvement of all stakeholders is necessary to achieve a closed-loop sustainable cycle for renewable materials.

REFERENCES

1. Correa, J.P., Montalvo-Navarrete, J.M., Hidalgo-Salazar, M.A., "Carbon footprint considerations for biocomposite materials for sustainable products: a review", *J. Clean. Prod.* 2019, 208, 785–794.
2. Boland, C.S., Kleine, R., Keoleian, G.A., Lee, E.C., Kim, H.C., Wallington, T.J., "Life cycle impacts of natural fiber composites for automotive applications: effects of renewable energy content and lightweighting", *J. Ind. Ecol.* 2016, 20(1), 179–189.
3. Mathew A.P., Dufresne A., "Morphological investigation of nanocomposites from sorbitol plasticized starch and tunicin whiskers", *Biomacromolecules* 2002, 3(3), 609–617.
4. Morin A., Dufresne A., "Nanocomposites of chitin whiskers from Riftia tubes and poly(caprolactone)", *Macromolecules* 2002, 35(6), 2190–2199.
5. Helbert W., Cavaille C.Y., Dufresne A., "Thermoplastic nanocomposites filled with wheat straw cellulose whiskers. Part I: processing and mechanical behaviour", *Polym. Comp.* 1996, 17(4), 604–611.
6. Célino, A., Fréour, S., Jacquemin, F. Casari P., "The hygroscopic behavior of plant fibers", *Front. Chem.* January 24, 2014,1, 43.
7. Monari, S., Ferri, M., Vannini, M., Sisti, L., Marchese, P., Ehrnell, M., Xanthakis, E., Celli, A., Tassoni, A., "Cascade strategies for the full valorisation of Garganega white grape pomace towards bioactive extracts and bio-based materials", *PLoS One.* September 18, 2020, 15(9), Article no. e0239629.
8. Gontard, N., Sonesson, U., Birkved, M., Majone, M., Bolzonella, D., Celli, A., Angellier-Coussy, H., Jang, G.W., Verniquet, A., Broeze, J., Schaer, B., Batista, A.P., Sebok, A. "A research challenge vision regarding management of agricultural waste in a circular bio-based economy", *Crit. Rev. Environ. Sci. Technol.* 2018, 48(6), 614–654.
9. Kim, J.H., Shim, B.S., Kim, H.S., Lee, Y.J., Min, S.K., "Review of nanocellulose for sustainable future materials", *Int. J. Precis. Eng. Manuf-Green-Technol.* 2015, 2, 197–213.
10. Paunonen, S., "Cellulose derivatives, review", *BioResources* 2013, 8, 3098–3121.
11. Dufresne, A., "Cellulose nanomaterials as green nanoreinforcements for polymer nanocomposites", *Phil. Trans. R. Soc. A.* 2018, 376, 20170040.
12. Mokhena, T.C., Sefadi, J.S., Sadiku, E.R., John, M.J., Mochane, M.J. Mtibe, A., "Thermoplastic processing of PLA/cellulose nanomaterials composites", *Polymers* 2018, 10, 1363.
13. Moon, R.J., Martini, A., Nairn, J., Simonsen, J., Youngblood, J., "Cellulose nanomaterials review: structure, properties and nanocomposites", *Chem. Soc. Rev.* 2011, 40(7), 3941–3994.
14. Habibi, Y., Lucia, L.A., Rojas, O.J., "Cellulose nanocrystals: chemistry, self-assembly, and applications", *Chem. Rev.* 2010, 110(6), 3479–3500.
15. Turbak, A.F., Snyder, F.W., Sandberg, K.R., "Microfibrillated cellulose, a new cellulose product: properties, uses, and commercial potential", *J. Polym. Sci. Polym. Symp.* 1983, 37, 815–827.
16. Laka, M., Chernyavskaya, S., "Obtaining of microcrystalline cellulose from softwood and hardwood pulp", *BioResources.* 2007, 2, 583–589.
17. Trache, D., Tarchoun, A.F., Derradji, M., Hamidon, T.S., Masruchin, N., Brosse, N. Hussin, M.H., "Nanocellulose: from fundamentals to advanced applications", *Front. Chem.* 2020, 8, 392.
18. Oksman, K., Mathew, A.P., Bondeson, D., Kvien, I., "Manufacturing process of cellulose whiskers/polylactic acid nanocomposites", *Comp. Sci. Technol.* 2006, 66, 2776–2784.
19. Pang, L., Gao, Z., Feng, H., Wang, S., Wang, Q., "Cellulose based materials for controlled release formulations of agrochemicals: a review of modifications and applications", *J. Control. Release* 2019, 316, 105–115.

20. Martins, N.C.T., Freire, C.S.R., Pinto, R.J.B., Fernandes, S.C.M., Neto, C.P., Silvestre, A.J.D., Causio, J., Baldi, G., Sadocco, P., Trindade, T., "Electrostatic assembly of Ag nanoparticles onto nanofibrillated cellulose for antibacterial paper products", *Cellulose* 2012, 19, 1425–1436.

21. Limatainen, H., Visanko, M., Sirvi, J.A., Hormi, O.E.O., Niinimaki, J., "Enhancement of the nanofibrillation of wood cellulose through sequential periodate-chlorite oxidation", *Biomacromolecules* 2012, 13, 1592–1597.

22. Nagasawa, K., Tohira, Y., Inoue, Y., Tanoura, N., "Reaction between carbohydrates and sulfuric acid: Part I: depolymerization and sulfation of polysaccharides by sulfuric acid", *Carbohydr. Res.* 1971, 18, 95–102.

23. Luo, J., Semenikhin, N., Chang, H., Moon, R.J., Kumar, S., "Post-sulfonation of cellulose nanofibrils with a one-step reaction to improve dispersibility", *Carbohydr. Polym.* 2018, 181, 247–255.

24. Liimatainen, H., Visanko, M., Sirviö, J., Hormi, O., Niinimäki J., "Sulfonated cellulose nanofibrils obtained from wood pulp through regioselective oxidative bisulfite pretreatment", *Cellulose* 2013, 20, 741–749.

25. Martins, D., Bukzem, A., Ascheri, D., Signini, R., Aquino, G.L.B., "Microwave-assisted carboxymethylation of cellulose extracted from brewer's spent grain", *Carbohydr. Polym.* 2015, 131, 125–133.

26. Nagalakshmaiah, M., Kissi, N.E., Dufresne, A., "Ionic compatibilization of cellulose nanocrystals with quaternary ammonium salt and their melt extrusion with polypropylene", *ACS Appl. Mater. Interfaces* 2016, 8, 8755–8764.

27. Yang, S., Fu, S., Li, X., Zhou, Y., Zhan, H., "Preparation of salt-sensitive and antibacterial hydrogel based on quaternized cellulose", *Bioresources* 2010, 5, 1114–1125.

28. Chang, S., Condon, B., Edwards, J.V., "Preparation and characterization of aminobenzyl cellulose by two step synthesis from native cellulose", *Fibers Polym.* 2010, 11, 1101–1105.

29. Zhang, X., Lin, F., Yuan, Q, Zhu, L., Wang, C., Yang, S., "Hydrogen-bonded thin films of cellulose ethers and poly (acrylic acid)", *Carbohydr. Polym.* 2019, 215, 58–62.

30. Nau, M., Trosien, S., Seelinger, D., Boehm, A. K., Biesalski, M., "Spatially resolved crosslinking of hydroxypropyl cellulose esters for the generation of functional surface-attached organogels", *Front. Chem.* 2019, 7, 367.

31. Godinho, M.H., Gray, D.G., Pieranski, P., "Revisiting (hydroxypropyl) cellulose (HPC)/ water liquid crystalline system", *Liq. Cryst.* 2017, 44, 2108–2120.

32. Weissenborn, E., Braunschweigb, B. "Hydroxypropyl cellulose as a green polymer for thermo-responsive aqueous foams", *Soft Matter.* 2019, 15, 2876–2883.

33. Wallmeier, M., Hauptmann, M., Majschak, J.P., "New methods for quality analysis of deep-drawn packaging components from paperboard", *J. Packag. Technol. Sci.* 2015, 28, 91–100.

34. Fox, S.C., Li, B., Xu, D., Edgar, K.J., "Regioselective esterification and etherification of cellulose: a review", *Biomacromolecules* 2011, 12, 1956–1972.

35. Wang, Y., Wang, X., Xie, Y., Zhang, K., "Functional nanomaterials through esterification of cellulose: a review of chemistry and application", *Cellulose* 2018, 25, 3703–3731.

36. Bras, J., Garcia, C.V., Borredon, M.E., "Oxygen and water vapor permeability of fully substituted long chain cellulose esters (LCCE)", *Cellulose* 2007, 14, 367–374.

37. Hassanpour, A., Asghari, S., Lakouraj, M.M. "Synthesis, characterization and antibacterial evaluation of nanofibrillated cellulose grafted by a novel quinolinium silane salt", *RSC Adv.* 2017, 7, 23907–23916.

38. Furuhata, K.I., Aoki, N., Suzuki, S., Sakamoto, M., Saegusa, Y., Nakamura, S., "Bromination of cellulose with tribromoimidazole, triphenylphosphine and imidazole under homogeneous conditions in LiBr-dimethylacetamide", *Carbohydr. Polym.* 1995, 26, 25–29.

39. Roy, D., Semsarilar, M., Guthrie, J.T., Perrier, S., "Cellulose modification by polymer grafting: a review", *Chem. Soc. Rev.* 2009, 7, 2046–2064.
40. Carlmark, A., Malmstrom, E.E., "ATRP grafting from cellulose fibers to create block-copolymer grafts", *Biomacromolecules* 2003, 4, 1740–1745.
41. Roy, D., Knapp, J.S., Guthrie, J.T., Perrier, S., "Antibacterial cellulose fiber via RAFT surface graft polymerization", *Biomacromolecules* 2008, 9, 91–99.
42. Lizundia, E., Meaurio, E., Vilas, J.L., *Multifunctional Polymeric Nanocomposites Based on Cellulosic Reinforcements* 1st Edition, Elsevier, 2016.
43. Daly, W.H., Evenson, T.S., Iacono, S.T., "Recent developments in cellulose grafting chemistry utilizing Barton ester intermediates and nitroxide mediation", *Macromol. Symp.* 2001, 174, 155–164.
44. Rosen, B.M., Jiang, X., Wilson, C.J., Nguyen, N.H., Monteiro, M.J., Percec, V., "The disproportionation of Cu(I)X mediated by ligand and solvent into Cu(0) and Cu(II)X2 and its implications for SET-LRP", *J. Polym. Sci. A Polym. Chem.* 2009, 47, 5606–5628.
45. Vlček, P., Raus, V., Janata, M., Kříž, J., Sikora, A., "Controlled grafting of cellulose esters using SET-LRP process", *J. Polym. Sci. A Polym. Chem.* 2011, 49, 164–173.
46. Chuensangjun, C., Kitaoka, K., Chisti, Y., Sirisansaneeyakul, S., "Surface-modified cellulose nanofibers-graft-poly (lactic acid) s made by ring-opening polymerization of L-Lactide", *J. Polym. Environ.* 2019, 27, 847–861.
47. Demitri, C., Del Sole, R., Scalera, F., Sannino, A., Vasapollo, G., Maffezzoli, A., Ambrosio, L., Nicolais, L., "Novel superabsorbent cellulose-based hydrogels cross-linked with citric acid", *J. Appl. Polym. Sci.* 2008, 110, 2453–2460.
48. Wang, C.C., Chen, C.C., "Physical properties of crosslinked cellulose catalyzed with nano titanium dioxide", *J. Appl. Polym. Sci.* 2005, 97, 2450–2456.
49. Xu, Q., Chen, C., Rosswurm, K., Yao, T., Janaswamy, S., "A facile route to prepare cellulose-based films", *Carbohydr. Polym.* 2016, 149, 274–281.
50. Sun, B., Zhang, L., Wei, F., AL-Ammari, A., Xu, X., Li, W., Chen, C., Lin, J., Zhang, H., Sun, D., "In situ structural modification of bacterial cellulose by sodium fluoride", *Carbohydr. Polym.* 2020, 231, 115765.
51. Murariu, M., Dubois, P., "PLA composites: From production to properties", *Adv. Drug Deliv. Rev.* 2016, 107, 17–46.
52. Nakagaito, A., Kanzawa, S., Takagi, H., "Polylactic acid reinforced with mixed cellulose and chitin nanofibers: effect of mixture ratio on the mechanical properties of composites", *J. Compos. Sci.* 2018, 2, 36.
53. Raquez, J.M., Murena, Y., Goffin, A.L., Habibi, Y., Ruelle, B., DeBuyl, F., Dubois, P., "Surface-modification of cellulose nanowhiskers and their use as nanoreinforcers into poly-lactide: a sustainably-integrated approach", *Compos. Sci. Technol.* 2012, 72, 544–549.
54. Jonoobi, M., Harun, J., Mathew, A.P., Oksman, K., "Mechanical properties of cellulose nanofiber (CNF) reinforced polylactic acid (PLA) prepared by twin screw extrusion", *Compos. Sci. Technol.* 2010, 70, 1742–1747.
55. Iwatake, A., Nogi, M., Yano, H., "Cellulose nanofiber-reinforced polylactic acid", *Compos. Sci. Technol.* 2008, 68, 2103–2106.
56. Wang, T., Drzal, L.T., "Cellulose-nanofiber-reinforced poly(lactic acid) composites prepared by a water-based approach", *ACS Appl. Mater. Interfaces* 2012, 4(10), 5079–5085.
57. Blaker, J.J., Lee, K.-Y., Walters, M., Drouet, M., Bismarck, A., "Aligned unidirectional PLA/bacterial cellulose nanocomposite fibre reinforced PDLLA composites", *React. Funct. Polym.* 2014, 85, 185–192.
58. Hubbe, M.A., Ferrer, A., Tyagi, P., Yin, Y., Salas, C., Pal, L., Rojas, O.J., "Nanocellulose in thin films, coatings, and plies for packaging applications: a review", *Ind. Crop. Prod.* 2017, 95, 574–582.
59. Kalia, S., Dufresne, A., Cherian, B.M., Kaith, B.S., Avérous, L., Njuguna, J., Nassiopoulos, E., "Cellulose-based bio- and nanocomposites: a review", *Int. J. Polym. Sci.* 2011, 1–35.

60. Khakalo, A., Filpponen, I., Rojas, O.J., "Protein-mediated interfacial adhesion in composites of cellulose nanofibrils and polylactide: enhanced toughness towards material development", *Compos. Sci. Technol.* 2018, 160, 145–151.

61. Lee, K.Y., Blaker, J.J., Bismarck, A., "Surface functionalization of bacterial cellulose as the route to produce green polylactide nanocomposites with improved properties", *Compos. Sci. Technol.* 2009, 69, 2724–2733.

62. Qu, P., Gao, Y., Wu, G.F., Zhang, L.P., "Nanocomposites of Poly(lactic acid) reinforced with cellulose nanofibrils", *BioResources* 2010, 5(3), 1811–1823.

63. Geng, S., Wei, J., Aitomäki, Y., Noël M., Oksman, K., "Well-dispersed cellulose nanocrystals in hydrophobic polymers by in situ polymerization for synthesizing highly reinforced bionanocomposites", *Nanoscale* 2018, 10, 11797–11807.

64. Nakagaito, A.N. Fujimura, A. Sakai, T.; Hama, Y. Yano, H., "Production of microfibrillated cellulose (MFC)-reinforced polylactic acid (PLA) nanocomposites from sheets obtained by a papermaking-like process", *Compos. Sci. Technol.* 2009, 69, 1293–1297.

65. Haque, M.M.-U., Puglia, D., Fortunati, E., Pracella, M., "Effect of reactive functionalization on properties and degradability of poly(lactic acid)/poly(vinyl acetate) nanocomposites with cellulose nanocrystals", *React. Funct. Polym.* 2017, 110, 1–9.

66. Lizundia, E., Fortunati, E., Dominici, F., Vilas, J.L., León, L.M., Armentano, I., Torre, L., Kenny, J.M., "PLLA-grafted cellulose nanocrystals: role of the CNC content and grafting on the PLA bionanocomposite film properties", *Carbohydr. Polym.* 2016, 142, 105–113.

67. John, M.J., Anandjiwala, R., Oksman, K., Mathew, A.P., "Melt-spun polylactic acid fibers: effect of cellulose nanowhiskers on processing and properties", *J. Appl. Polym. Sci.* 2013, 127, 274–281.

68. Singh, A., Genovese, M.E., Mancini, G., Marini, L., Athanassiou, A. "Green processing route for polylactic acid: cellulose fiber biocomposites", *ACS Sustain. Chem. Eng.* 2020, 8, 4128–413.

69. Piekarska, K., Sowinski, P., Piorkowska, E., Haque, Md.M.-Ul., Pracella, M., "Structure and properties of hybrid PLA nanocomposites with inorganic nanofillers and cellulose fibers", *Compos. A Appl. Sci. Manuf.* 2016, 82, 34–41.

70. Coltelli, M.-B., Cinelli, P., Gigante, V., Aliotta, L., Morganti, P., Panariello, L., Lazzeri, A., "Chitin nanofibrils in poly(lactic acid) (PLA) nanocomposites: dispersion and thermo-mechanical properties", *Int. J. Mol. Sci.* 2019, 20, 504.

71. Meng, X., Bocharova, V., Tekinalp, H., Cheng, S., Kisliuk, A., Sokolov, A.P., Kunc, V., Peterd, W.H., and Ozcan, S., "Toughening of nanocelluose/PLA composites via bioepoxy interaction: mechanistic study", *Mater. Des.* 2018, 139, 188–197.

72. Herrera, N., Salaberria, A.M., Mathew, A.P., and Oksman, K., "Plasticized polylactic acid nanocomposite films with cellulose and chitin nanocrystals prepared using extrusion and compression molding with two cooling rates: effects on mechanical, thermal and optical properties", *Compos. A Appl. Sci. Manuf.* 2016, 83, 89–97.

73. Oluwabunmi, K., D'Souza, N., Zhao, W., Choi, T.Y. Theyson, T., "Compostable, fully biobased foams using PLA and micro cellulose for zero energy buildings", *Sci. Rep.* 2020, 10, 17771.

74. Robles, E., Kánnár, A., Labidi, J., Csóka, L., "Assessment of physical properties of self-bonded composites made of cellulose nanofibrils and poly (lactic acid) microfibrils", *Cellulose* 2018, 25, 3393–3405.

75. Sethi, J., Farooq, M., Sain, S., Sain, M., Sirviö, J.A., Illikainen, M., Oksman, K., "Water resistant nanopapers prepared by lactic acid modified cellulose nanofibers", *Cellulose* 2018, 25, 259–268.

76. Hong, J., Kim, D.S., "Preparation and physical properties of polylactide/cellulose nanowhisker/nanoclay composites", *Polym. Compos.* 2013, 34, 293–298.

77. Rachwa, K., Wasko, A., Gustaw, K., Polak-Berecka, M., "Utilization of brewery wastes in food industry", *Peer J.* 2020, 8, e9427. http://doi.org/10.7717/peerj.9427

4 Chemical Modification of Cellulose and Its Reinforcement Effects on the Properties of PLA-Based Composites

Jesús Rubén Rodríguez-Núñez
Universidad de Guanajuato
Celaya, Mexico

Heidy Burrola-Núñez
Universidad Estatal de Sonora
Hermosillo, Mexico

Luis Ángel Val-Félix and Tomás Jesús Madera-Santana
Centro de Investigación en Alimentación y Desarrollo
Hermosillo, Mexico

CONTENTS

4.1 INTRODUCTION

Our society has taken an enormous consciousness about the resilience of plastics in the environment, concerns about the toxic emissions during incineration of these, and the pollution produced in lagoons, rivers, and seas produced by macro and microplastics. The abundance of plastics in the environment (land, sea, and air), a lack of landfill space, and the dangers of entrapment or ingestion of these materials to human health, as well as animals, birds, and fish, have prompted efforts to identify more environmentally acceptable alternative materials.

The capability of natural fibers (NFs) to be incorporated into plastic materials for common and engineering applications is vast. According to their origin, natural fibers are classified as plant, animal, and mineral. Cellulose, hemicellulose, lignin, pectins, and waxes are components of natural plant fibers (Bandhu et al., 2010). Stems (soft sclerenchyma) fibers, leaf (hard fibers), fruits, seeds, wood, cereal straw, and grass fibers are examples of natural plant fibers (John and Thomas, 2008; Huber et al., 2011). NFs are considered composite materials, and they are made up of cellulose fibrils that are embedded into a matrix of lignin. Along the length of the fiber, the cellulose fibrils are aligned, which imparts its mechanical characteristics (tensile and flexural strength, rigidity, etc.). The reinforced efficiency that these fibers impart to the composite is due to the structure of cellulose. It consists of small semicrystalline crystals and microfibrils that are distributed randomly into the matrix (Jedvert and Heinze, 2017). Cellulose is a linear-homopolysaccharide that consists of D-anhydro-glucopyranose ($C_6H_{11}O_5$) repeated units, joined by 1,4-β-D-glycosidic links in positions C1 and C4 (Eyley and Thielemans, 2014), as is observed in Figure 4.1. The three hydroxyl groups in the monomeric repeating unit can produce hydrogen bonds, which are responsible for the crystalline packaging and the physical properties of cellulose. This biopolymer is resistant to oxidizing agents and too strong alkali, although it can be hydrolyzed by acids to produce water-soluble sugars (John and Thomas, 2008).

NFs are now arising as feasible alternatives to substitute traditional fibers (glass, carbon, etc.); these are used in composite materials for a variety of applications in automobile parts, building constructions, and rigid packaging materials, either alone or in combination. The main advantages of NFs over synthetic or man-made fibers, such as glass, are their low cost, low density, competitive specific mechanical properties; carbon dioxide sequestration; sustainability; recyclability; and biodegradability (John and Thomas, 2008).

On the other hand, biocomposites comprise reinforced NFs and a polymeric matrix. The natural fiber is biodegradable, but common thermoplastics (polyethylene or polypropylene) and thermosets (epoxy or unsaturated polyester resins) are not biodegradable. These biocomposites are named "partially biodegradable" (Mohanty et al., 2005). However, if the matrix is a biopolymer (starch, chitosan, agar, etc.) or

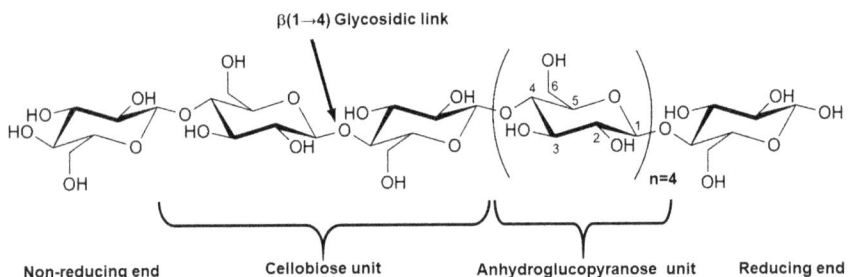

FIGURE 4.1 Molecular structure of the cellulose biopolymer.

a bioplastic (polylactic acid, polycaprolactone, etc.), the biocomposite is classified as "completely biodegradable". In this sense, the mixture of biofibers and bioplastics has the ability to operate either alone or in conjunction with petroleum-based polymers, and they provide an alternate option to create environmentally acceptable materials in the 21st century. Polylactic acid (PLA) is a bioplastic made from renewable resources. The commercial accessibility of PLA has motivated the development of biocomposites contained as a reinforcement NFs. These biocomposites are completely biodegradable and have shown a beneficial ecological, economic, and technological balance. The monomeric unit of PLA is 2-hydroxy-propanoic acid (lactic acid), where the asymmetric carbon atom has two optical configurations (D and L). There are two routes to synthesize PLA, one is via polycondensation polymerization of D- or L-lactic acid, and the other is through ring-opening polymerization (ROP) of the lactide molecule. The last process is used to produce high-molecular-weight PLA. PLA synthesized by polycondensation polymerization has low molecular weight and this bioplastic does not have suitable mechanical properties for several applications (Södergård and Stolt, 2010). The physical properties of PLA are related to the number of chiral repeating units and stereoregularity of the polymer. Varying the crystallinity, molecular structure and weight, and additives (coupling agents, plasticizers, processing aids, etc.), the PLA can range from soft and elastic bioplastic to a hard, high strength material. Among the thermal properties, PLA has a glass transition temperature (Tg) between 50 and 72°C, a melting temperature (Tm) 170–190°C, and crystallinity around 35% (Södergård and Stolt, 2002; Tsusj, 2002). Other physicochemical properties of the PLA are that it is soluble in chlorinated and fluorinated organic solvents, and acetone, pyridine, tetrahydrofuran xylene, ethyl acetate, ethyl lactate, and others. The barrier properties of PLA show that this bioplastic has a medium permeability level to oxygen and water.

This chapter is an overview of methods used to perform the chemical modification of cellulose and its use as a reinforcing agent in PLA-based biocomposites. Other aspects that will be covered in this chapter are the sources of NFs for cellulose production, the procedures for cellulose production, and the classification of cellulose fiber modification methods. It also covers an overview of PLA-based biocomposites containing modified cellulose fibers, their trends and future perspectives in special applications.

4.2   SOURCES FOR CELLULOSE PRODUCTION

The NFs from the vegetable origin are structured by macrofibrils, a single NF has millions of macrofibrils. Each macrofibril is composed of microfibrils, where cellulose, hemicellulose, lignin, pectins, waxes, and extractives are among the primary components (Hao et al., 2018). Hemicellulose, which is made up of carbohydrates and has branched chains, is the second most prevalent substance in the NFs after cellulose. Lignin is an amorphous complex structure, high molecular weight, and less polar than cellulose. It contains aromatic polymers of phenyl propane units. The function of lignin in plant fibers is an adhesive agent in the cellulose-hemicellulose matrix (Komuraiah, et al., 2014). The percentage of cellulose depends on the source, as shown in Table 4.1. It is decisive when choosing a source to produce cellulose.

TABLE 4.1

The Chemical Composition of Representative Natural Fibers

Source	Cellulose (wt.%)	Hemicellulose (wt.%)	Lignin (wt.%)	Waxes (wt.%)
Cotton	85–90	5.7	-	0.6
Pineapple	81	-	12.7	-
Ramie	68.6–76.2	13–16	0.6–0.7	0.3
Isora	74	-	23	1.09
Kenaf	72	20.3	9	-
Flax	71	18.6–20.6	2.2	1.5
Jute	61–71	14–20	12–13	0.5
Hemp	68	15	10	0.8
Sisal	65	12	9.9	2
Banana	63–64	19	5	-
Henequen	60	28	8	0.5
Abaca	56–63	20–25	7–9	3

Sources: Kumar et al. (2011) and Geremew et al. (2021).

The production of cellulose can be performed by chemical, biological, and physical treatments. The selection of treatment will depend on the source fiber, the desired final shape (microfiber, nanofiber, nanocrystal, and others), and its application. The main processes implicated in the production of cellulose and examples are shown in Figure 4.2.

The production of cellulose involves, in most cases, immersing the NF in a chemical solvent in a process called "lyocell". However, many of these solvents are highly toxic to the environment, such as sodium hydroxide and sodium sulfide. Among these, N-methylmorpholine N-oxide (NMMO), an oxidizing agent capable of producing large amounts of cellulose at low cost, was first used in 1994 (Borbély 2008). Another alternative that has been recently investigated is the use of ionic

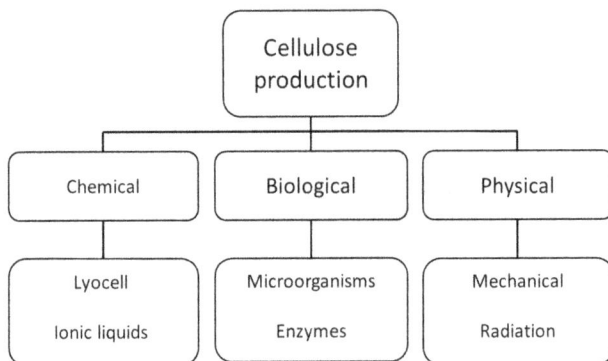

FIGURE 4.2   Main processes to produce cellulose fibers.

liquid solvents. These are defined as salts or electrolytes whose melting point is less than 100°C and can be cations or anions. Ionic liquids have the advantage of being chemically and thermally more stable than the aforementioned techniques. Besides, the process protects the environment since it is carried out at a low vapor pressure. At a low temperature, the solvents can be reused (>99.5%), and they are soluble in water and other organic compounds (Hermanutz et al. 2019). Regarding the desired final shape, cellulose fibers are in the micro-scale, although cellulose nanofibers attract the attention for advanced materials applications. However, it is important to point out that the shape and properties of the micro and nanofibers are dependent on the origin of the cellulose (crystallinity, purity, and degree of polymerization) and the treatment applied to the fibers (surface treatment and chemical modification) (Shanghaleh et al. 2018). He et al. (2018) found that these are easier methods to obtain cellulose nanofibrils from soft pine woods. Non-bleached cellulose nanofibers with low lignin and high hemicellulose are more viscous and retain more water than those that were bleached. Melikoglu et al. (2019) reported the use of alkaline treatment to extract cellulose nanofibers from apple pomace. The optimal variables were a 10.23% sodium hydroxide solution, 69.82°C for 161.54 min. Subsequently, acid hydrolysis and ultrasonication treatment were applied to produce cellulose nanocrystals.

Biological treatments often use cellulase to hydrolyze the cellulose chains and allow the formation of nanofibers. Cellulases are enzymes that break down cellulose's β-1,4-glycosidic linkages. Cellulases are a multi-component enzyme system that usually consists of three enzymes that work together to hydrolyze cellulose: endoglucanase (EC 3.2.1.4), cellobiohydrolase (EC 3.2.1.91), and cellobiase (-glucosidase, EC 3.2.1.21). The first two enzymes operate directly on cellulose, converting it to cellobiose and glucose. Soluble cellodextrins and amorphous cellulose are degraded by endoglucanases and cellobiohydrolases. Cellobiohydrolases, on the other hand, are best degrading crystalline cellulose (Howard et al., 2003; Wanmolee et al. 2016). However, enzymatic treatment is expensive and requires very specific conditions and stability (Durham and Sastry, 2020). Banvillet et al. (2021) reported the use of enzymatic treatment in combination with alkaline treatment to obtain cellulose nanofibers from eucalyptus. However, they found that the mechanical properties were affected when combining the treatments compared to the nanofibers that only received alkaline treatment.

Another important product is bacterial cellulose (BC) which can be produced by certain species of bacteria, such as *Aerobacter, Acromobacter, Komagataeibacter, Gluconacetobacter, Rhizobium, Salmonella, Escherichia, Vallonia, Saprolegnia,* and others as part of their mechanisms of adaptation to the environment (Chen et al. 2010). BC is used in the food industry, but its applications have been expanding to the pharmaceutical, biotechnology, biomedical, cosmetic, papermaking, and electronics industries. Compared to plant-based cellulose, BC has a unique nanostructure, high degree of polymerization, crystallinity, purity, and biocompatibility. BC has high water uptake capacity, aspect ratio, Young's modulus, total surface area, among other properties (Wang et al., 2019).

To obtain bacterial cellulose, the common methods are static, agitated-shaking, and bioreactor cultures. The method used to produce the BC depends on the final application. Moreover, it is important to consider the bacterial stain, nutrition medium, oxygen content, pH, among other variables during the production. A commercial growth medium is Hestrin–Schramm, although its cost is an inconvenience to produce cellulosic fibers on a large scale. Recently, researchers have been focused on studying the feasibility of agro-industrial waste as a growing medium to reduce costs; for instance, suitable nutrients can be found locally through citrus fruit peel waste at low costs. Guzel and Akpinar (2019) found that lemon and orange peels allow bacterial cellulose to be obtained from *Komagataeibacter hansenii* GA2016, with thermal properties and moisture absorption superior to the obtained from the commercial growth medium.

Several technologies have been used to improve the applications of BC, among them the production of BC nanofibers with chemical treatments (bleaching, alkaline, acid hydrolysis, insolation, and drying. Some authors have varied the solvent concentration and temperature of cellulose extraction and nanofiber formation (Agwuncha et al. 2020). A current trend is the acquisition of cellulose from the recycled fabrics made with NF. Studies show that cotton fabric waste, with or without dye, can be hydrolyzed in sulfuric acid to reduce the molecular weight of the NF and can regenerate it with the wet spinning technique. The fibers obtained have mechanical and thermal properties comparable to commercial rayon fabric (Liu et al. 2019).

4.3　CLASSIFICATION OF MODIFICATION PROCEDURES OF CELLULOSE FIBERS

NFs are frequently subjected to physicochemical methods or treatments to extract cellulose, exposing functional groups, improving their adherence with polymeric matrices, and thereby improving the properties of the biocomposites formed. These methods can be chemical, physical, and biological. The choice of each will depend on the applications, properties, and conditions of the cellulose that is expected to be obtained and used (Durham and Sastry 2020; Tibolla et al. 2020).

NFs (jute, henequen, hemp, sisal, etc.) are very polar because of the presence of the hydroxyl groups (Madera-Santana et al. 2013). However, pectin and waxes coat these fibers, preventing the hydroxyl groups from interacting with polar matrices and generating mechanical interlocking adhesion with them (Mwaikambo and Ansell 2002). Cellulosic fibers are treated by physical or chemical treatments that modify the surface and structure to produce reactive hydroxyl (-OH) groups and rough surfaces for mechanical interlocking or adhesion with polymeric matrices (Bisanda and Ansell 1991; Mwaikambo and Ansell 1999), illustrated in Figure 4.3. Physical treatments lead to proton abstraction and create unstable radicals, which are converted into chemical groups. Ultraviolet (UV) and γ-rays produce reactive free macroradicals in the fibers causing their deterioration and changes in some properties (Zaman et al. 2010; Mahzan et al. 2017). Etching, chemical implantation, crystallization, and polymerization, are all caused by electrical discharge techniques (corona discharge, cold plasma, etc.). As a result, the mechanical bonding of the filler-polymer matrix is enhanced on the surface characteristics of cellulose fibers (Roy et al. 2009; Shi et al. 2013; Cichosz et al. 2020).

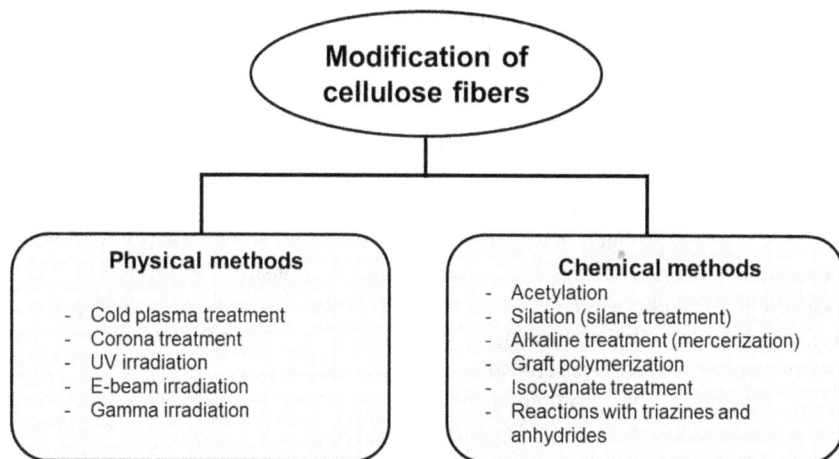

FIGURE 4.3   Modification methods of cellulose fibers.

The common chemical methods to modify the cellulose surface reported in the literature are acetylation, silylation (e.g. formation of trimethylsilyl derivatives), alkaline treatment (mercerization), and treatments with isocyanates, anhydrides, organosilanes, and triazines; as well as graft polymerization using hydrophobic monomers (Frone et al. 2011; Khalil et al. 2012; Le Duigou et al. 2012; Larsson et al. 2012; Cichosz et al. 2020).

The cellulose has hydroxyl groups that are the reactive groups for a variety of chemical modifications. These modifications are conducted by nucleophilic reactions by hydroxyl groups of the anhydro-glucopyranose units. Another alternative method of modifying the cellulose is by oxidation, which introduces carbonyl and carboxyl groups to the biopolymer (Heinze et al. 2012). The introduction of functional groups imparts new physicochemical properties to the cellulose without altering desirable natural properties. The foremost common derivatization reactions of cellulose are esterification and etherification. The cellulose esterification reactions may give rise to important cellulose esters, such as cellulose acetate (CA), CA propionate, CA butyrate, etc. Another reaction to modify the structure of the cellulose chemically, is etherification, which gives rise to important commercial cellulose ether products such as methylcellulose (MC), carboxymethyl cellulose (CMC), hydroxyalkyl cellulose (HAC) (Roy et al. 2009). The cellulose graft copolymerization is an alternative way to modify cellulose, incorporating it into its main chain (backbone) branches of long sequences of a monomer. Several authors have used this process to improve the properties (adhesion) of cellulose in polymeric composites (Hebeish and Guthrie 1981; Kang et al. 2013).

4.3.1 SURFACE TREATMENTS FOR CELLULOSE FIBERS

Surface treatment through non-ionizing (UV) and ionizing (gamma – γ) radiation has been used to generate physicochemical changes in natural fibers, leading to structural changes, thus improving some properties. On the one hand, UV radiation produces reactive free radicals in the fibers causing their deterioration, since the chemical compounds of the fiber are sensitive to UV rays (cellulose, hemicellulose, and lignin) (Mahzan et al. 2017). Lignin is the main component that absorbs UV radiation because its repeating phenylpropane unit is rich in aromatic rings. It also contains phenolic units, ketones, and other chromophores (Sadeghifar et al. 2017). UV radiation causes photodegradation of polysaccharides, generating scission of gly-cosidic chains and formation of the carbonyl group (Sionkowska et al. 2006). On the other hand, γ-rays produce macrocellulosic radicals and supply energy to cellulose by Compton scattering. The radicals produced are responsible for altering the fibers' physical-chemical characteristics (Zaman et al. 2010). The main effects of γ-rays on cellulosic fibers are chain scission, which reduces the molecular weight, and chain crosslinking, which produces cross-linked structures (Belgacem and Gandini, 2005). UV and γ-radiation can promote chemical reactions in the molecular structures of NFs, mainly modifying their hydrophilic character and their mechanical properties.

UV radiation generates changes in the surface polarity of NFs, improving adhesion with polymeric matrices (Sionkowska et al. 2006; Asha et al. 2017), so mechanical properties are improved. As shown in Table 4.2, upon NFs, like coir and banana, UV radiation at low doses increases mechanical properties (tensile strength, elongation at break, and Young's modulus or elastic modulus); due to crosslinking between neighboring cellulose molecules. In composites, with different fibers like hemp, wood, or ramie in a PLA matrix, UV radiation generates a decrease in tensile strength, elastic modulus, flexural strength, and modulus, among others. Continuous chain scissions in the polymer and the fiber decreased mechanical properties of com-posite materials and it is a dose-dependent effect (Islam et al. 2010). The interaction between the fiber and the matrix is weak; moreover, swelling and shrinkage of the fiber produce cracks, which also generates a reduction of stress transfer between the fiber and the polymeric matrix (Yatigala et al. 2018).

In NFs like cotton, flax, and silk, γ-radiation does not affect tensile strength at a dose of 15 kGy, while it is reduced at higher doses (100 kGy). However, in com-posites, the mechanical properties generally improve with exposure to γ-radiation between doses of 2.5 and 100 kGy. Tensile (tensile strength, elongation at break, elastic modulus), impact, and flexural parameters increase in some composites with exposure to γ-radiation (Table 4.2). When subjected to γ-radiation, photodegrada-tion and photocrosslinking can occur simultaneously (Rahman et al. 2019), with the latter being predominant. The development of reticulated structures is linked to an increase in mechanical properties, it is due to the formation of C–C bonds, and improved adhesion fiber-matrix (Rahman et al. 2019; EL-Zayat et al. 2019; Lenfeld et al. 2020; Xia et al. 2020). It has been reported that when exposed to γ-radiation, cotton fibers' crosslinks depend on water content (Borsa et al. 2003). In addition to crosslinking effects, increased crystallinity is related to improved mechanical prop-erties. A gamma radiation dosage of 25 kGy has been reported to increase crystal-linity to 71.5% from 64% in hemp fibers (Olaru et al. 2016).

Comparing both types of radiation, γ-radiation has been reported to cause less significant degradation in fibers compared to UV radiation. It has been reported that UV radiation (500 h) affects intra- and inter-molecular H-bond networks, while exposure to γ-radiation of up to 10 kGy has a less clear effect in hemp fibers (Olaru

TABLE 4.2

Mechanical Properties of Natural Fibers and Composites Subjected to Radiation Surface Treatment

Radiation Surface Treatment	Natural Fiber or Composites	Influence on Mechanical Properties	References
UV	Coir	An increase in tensile strength, elongation at break, and elastic modulus	Rahman and Khan (2007)
	Banana	An increase in tensile strength and elastic modulus at 168 h exposure to UV	Benedetto et al. (2015)
	Hemp/PLA	After exposure to UV (250–1000 h), tensile strength, Young's modulus, flexural strength, and flexural modulus decrease, while impact strength increase	Islam et al. (2010)
	Wood/PLA	2000 h of exposure to UV generated a decrease in flexural strength and modulus	Yatigala et al. (2018)
	Ramie/PLA	672 h of exposure to UV generated a decline in shear, tensile and flexural strength	Chen et al. (2011)
γ-radiation	Cotton, flax, and silk	Tensile strength was stable at a dose of 15 kGy, while at 100 kGy it was reduced to 26–33%	Machnowski et al. (2013)
	Flax/PLA	A dose of 20 kGy increased impact and tensile strength, while elongation-at-break and Young's modulus remained unchanged	Xia et al. (2020)
	Pineapple leaf fiber/LDPE	Tensile strength, flexural strength, and impact strength were optimum at a dose of 7.5 kGy	Rahman et al. (2019)
	Sugarcane bagasse/HDPE	A dose of 100 kGy provided an increase in tensile strength, elastic modulus, and elongation-at-break	EL-Zayat et al. (2019)
	Okra or Jute/PP	A dose between 2.5 and 10 kGy resulted on an improvement in tensile strength, tensile modulus, and elongation-at-break	Rahman et al. (2018)
UV/γ-radiation	Jute	The combined effect gave an improvement on tensile strength of 108%	Khan et al. (2006)

et al. 2016). UV radiation affects mainly lignocellulosic fibers (Sadeghifar et al. 2017), resulting in the scission of the macromolecular chains. Aromatic polymers like lignin have been shown to have higher radiation resistance than aliphatic chains due to the resonance of phenyl rings providing a protective at intra- and intermolecular levels (Pentimalli et al. 2000). It has been found that the main functional groups in the molecular structure of alkaline lignin are not destroyed after gamma irradiation (Yuting and Xiansu 2008). Likewise, the combined effect of UV and gamma radiation generates a confident effect on the mechanical properties of the fibers. Jute fibers subjected to UV and γ-radiation can show an improvement upon the mechanical properties was observed, since cross-linked structures are predominant (Khan et al. 2006).

Other physical treatments that modify the surfaces of cellulosic fibers include the use of plasma, electron beams, and corona treatment, among others. The synthesis of superhydrophobic materials obtained with cellulose that has been subjected to plasma treatment. Yao et al. (2020) stated that the application of plasma treatment on cellulose promotes compatibility with hydrophobic compounds such as polymers, since they modify their surface and improve adhesion with the matrix. This is because the plasma treatment ionizes the functional groups on the cellulose surface, producing the creation of Van der Waals-type interactions, and condensation reactions with the matrix. Çaglar et al. (2016) reported the use of plasma treatment on cellulose microfibers to improve their adherence to PLA in a multi-layer biocomposite. The reported mechanical and barrier properties make it a sustainable food packaging material. The biocomposite obtained can maintain an oxygen barrier even in environments where the relative humidity is 60%. The electron beam effect, in cellulosic pulp, generates an increase in oxidized groups in the crystalline areas, and weakens the network of hydrogen bonds, and also increases the formation of carbonyl groups (Shin and Sung, 2008; Henniges et al. 2013). Meanwhile, the corona effect on *Aloe vera* fibers generates a decrease in tensile strength and elastic modulus. This decrease is caused by the modification in the surface and composition of the fibers (Hassani et al. 2020).

4.3.2 CHEMICAL TREATMENTS FOR CELLULOSE FIBERS

ROP, free radical copolymerization, atom-transfer radical polymerization, reversible addition-fragmentation chain-transfer polymerization, radiation process (high-energy radiation and photoirradiation), and enzymatic grafting are some of the techniques used to graft different monomers on the cellulose backbone (Roy et al. 2009; Wojnarovits et al. 2010; Tosh and Routray 2014). Traditionally, free-radical grafting polymerization (FRP) redox systems are widely used to modify cellulose fibers; however, the limitation of FRP is the broad molecular weight distribution of graft-polymer and the formation of homopolymer. In contrast, procedures such as ring-opening polymerization and living free-radical polymerization (LFRP) allow obtaining narrow molecular distributions.

The sort of initiator used has been found to have a significant impact on grafting efficiency (grafting percentage) of monomers grafted into the cellulose fibers, chemical initiators (redox type), and free radical's initiators are described in Table 4.3.

TABLE 4.3

Chemical and Free Radical Initiators Used in Grafting Cellulose Fibers

Chemical Initiators	References
Ceric ammonium nitrate (CAN) or ceric (IV) ion	Routray and Tosh (2012, 2013)
Ceric (IV) sulfate (CS)	Wohlhauser et al. (2018)
Ceric ammonium sulfate (CAS)	Ibrahim et al. (2002); Gupta and Khandekar (2006)
Cobalt (III) acetylacetonate complex salts	Gupta and Sahoo (2001a)
Cobalt (II) – potassium monopersulfate	Sahoo et al. (1986); Pinho and Soares (2018)
Sodium sulfite – ammonium persulfate	Yang et al. (2009); Khalilzadeh et al. (2020)
Iron (II) – hydrogen peroxide	Gupta and Sahoo (2001b); Jideonwo and Adimula (2006)
Free Radical Initiators	**Reference**
Potassium persulfate ($K_2S_2O_8$) or KPS	Ibrahim et al. (2008); Wohlhauser et al. (2018)
Ammonium persulfate (($NH_4)_2S_2O_8$) or APS	Sutirman et al. (2017)
Azobisisobutyronitrile ($C_8H_{12}N_4$) or AIBN	Ouajai et al. (2004); Gürdağ and Sarmad (2013)
Benzoyl peroxide ($C_{14}H_{10}O_4$) or BPO	Tosh and Routray (2014); Anah et al. (2015)
Sodium bisulfite ($NaHSO_3$) or SBS	Wohlhauser et al. (2018)

Source: Tosh and Routray (2014).

4.4 MODIFIED CELLULOSE FIBERS IN PLA-BASED BIOCOMPOSITES

In recent years, the development of biocomposites has increased the preferred use of biopolymers rather than petroleum-derived polymers, due to this biodegradability, renewability, and low manufacturing cost. PLA is one of the most promising bioplastics to substitute petroleum-derived polymers, which is obtained from lactic fermentation and is an ecofriendly plastic showing suitable mechanical properties and easy processability (Kyutoku et al. 2019). However, PLA has lower strength and thermal stability, requiring the use of natural fibers (cellulose, hemicellulose, lignin, pectin) for its reinforcement.

The industrial use of NFs has serious disadvantages, for example, low thermal and dimensional stability, as well as hydrophilic properties, and poor interfacial adhesion with several polymer matrices. For this reason, several physical and chemical modifications of natural fibers have been described in the literature (Doineau et al. 2021). Zhang et al. (2020) described that plasma treatments, UV radiation, alkalis, permanganate, acetylation, or coupling agent represent the main methods to recover the interfacial properties of the natural fibers/polymer matrix. It has also been reported that using nanotechnology (nanoparticles), the interfacial properties between the natural fibers and polymer matrix can be improved significantly.

In this sense, Bendorou et al. (2021) reinforced recycled PLA with cellulosic microfibers (75 µm–7 mm) from hemp fibers and paper solid wastes. The biocomposite obtained with PLA and hem fibers showed the best values for elastic modulus (324.53 ± 3.10 MPa), impact strength (27.61 ± 2.94 kJ/m^2), and biodegradation rate (1.97%). Kyutoku et al. (2019) improved the durability of PLA by adding cellulose fiber (CF) treated with epoxides as surface modifiers, showing that above-accelerated degradations conditions (60°C and 70% RH) the thermal and mechanical properties of PLA/CF composites did not present degradation. However, the melting point and the storage modulus showed a reduction. Also, Kowalczyk et al. (2011) reported that the reinforcement of PLA with nanofibers and standard fibers of cellulose from spruces, showed a decrease of the elongation at break at 25°C using nano or standard cellulose fibers; yet, the storage modulus increased using nanofibers (14–25%) compared with the neat PLA. Furthermore, the glass transition and crystallization of the PLA matrix were not modified by the cellulose fibers. Table 4.4 shows more recent studies of biocomposites of PLA with modified cellulose fibers.

Sudamrao and Patel (2020) stated out that biocomposites with renewables fibers are equivalent to composites of synthetic fibers, and the use of these technologies will begin to grow at the global level in areas such as automotive, packaging, medical, and textiles (e.g. the Mercedes-Benz E-class use natural fibers). Besides, the use of NFs to enhance the mechanical and thermal characteristics of PLA, is an environmentally friendly way. The biocomposites of PLA reinforced with modified cellulose fibers represent an interesting option to develop new biodegradable materials, however, it is important to use suitable methodologies by industrial and biocomposites fields for the performance of engineering products (Kian et al. 2019).

Whiskers of microcrystalline cellulose (MCC) has been used to reinforce the PLA matrix (Mathew et al. 2005). The particle size of MCC around 10–15 µm was obtained from crystalline cellulose derived from high-quality wood pulp and disintegrates into cellulose whiskers by acid hydrolysis (Bandhu et al. 2010). However, the MMC showed aggregation of crystalline cellulose entities within the PLA matrix, and therefore, the biocomposite showed poor mechanical properties (tensile stress and elongation at break); despite the composites having slightly greater stiffness than pure PLA. The MMC particles showed no adhesion with the PLA matrix. The alternative is the modification of MCC by grafting or sulfonation (Li et al. 2016; Zhai et al. 2021). As was previously mentioned, grafting polymerization is a free radical polymerization process, where a modification of cellulose's surface properties is performed. Zhu et al. (2020) reported the grafting polymerization method of methacrylic acid (MA) on MCC and the preparation of MA-MCC/PLA composites. The authors reported that the modification of MCC improves the dispersion and compatibility of MCC in the PLA matrix; tensile and impact strength showed values of 52.6 MPa and 8.16 kJ/m^2, respectively; which were much higher than the PLA matrix. The incorporation of MA in the MCC improves the fluidity of the PLA chains and facilitates its crystallization, and produces a lower cold crystallization temperature. The modification of MCC by the sulfonation process uses sulfuric acid, and can be improved by ultrasound aid. Zhao et al. (2020) reported the modification of MCC by sulfonation and ultrasound and prepared PLA composites by casting. The FTIR analysis showed many hydroxyl groups and the tensile strength reached values of 505 MPa in the modified MCC/PLA composites. These composites showed an increase in decomposition temperature from 260 to 320°C.

TABLE 4.4

Modification of Cellulose Fibers to Improve the Physicochemical Properties of PLA Biocomposites

Biocomposite/Cellulose Treatment	Improved Property	Reference
Cellulose fiber from sorghum bagasse and PLA. The cellulose was treated with KOH (5%), then was bleached with sodium chlorite solution in acetate buffer (2.3%) and finally, was treated with laccase enzyme from Trametes versicolor to degrade lignin and xylanase.	– The nanocellulose fibers improved the maximum strain (6.4%). – The biodegradability was improved with 15% of nanocellulose fibers.	Yulianto et al. (2020)
Wood fibers from spruce and PLA. The alkenyl succinic anhydride treated calcium carbonate was used to surface modification of fibers.	– Decreased the melt viscosity of the polymer. – Improved the processability because of decreased flowability. – Improved the fiber adhesion into PLA matrix.	Ozyhar et al. (2020)
Bamboo cellulose nanowhiskers and PLA. The bamboo cellulose was treated with NaOH and bleached with NaClO$_2$, finally was hydrolyzed with H$_2$SO$_4$ (65%).	– The addition of bamboo cellulose nanowhiskers modified the homocrystallites of PLA and the crystallite size was increased. – The addition of bamboo cellulose nanowhiskers increased the Tg, Tc, and Tm, producing a development in its thermal properties. – With the increase of bamboo cellulose nanowhiskers content, the tensile strength of the biocomposites decreased.	Qian et al. (2018)
Cellulose nanofibrils from hardwood Kraft pulp and PLA. The cellulose pulp was treated using FiberCare R type enzyme, an enzymatic treatment, which is an endoglucanase able to hydrolyze the 1,4-β-D-glycosidic bonds of cellulose.	– The cellulose nanofibrils increased the brittleness of the materials and decreased the tensile strength. – Tensile tests increased Young's modulus (37%). – The thermal stability did not show significant differences; however, higher concentrations of cellulose nanofibrils (10%) had an unfavorable effect on crystallization behavior of PLA.	Gazzotti et al. (2019)
Hydrophobic-modified nanocellulose fiber and PLA. The nanocellulose fiber was modified via grafting with butyl acrylate and ammonium ceric (IV) nitrate (CAN) as initiator.	– The water vapor transmission rate (WVTR) decreased with the addition of hydrophobic-modified nanocellulose fiber into PLA. – The lowest WVTR value was 34 g/m^2/d, it was obtained by adding 1% of modified nanocellulose fiber to PLA.	Song et al. (2014)
Micro and nanofibers from eucalyptus using ball milling, followed by surface treatment (2.5% NaOH and detergents [5%]). These fibers were incorporated in PLA matrix.	– Decreased the hydrophobicity of the film surface. – Increased the moisture absorption (%). – Concomitant decrease of the crystallinity index of the films. – Improve the mechanical properties (Young modulus).	Silva et al. (2019)

4.5 SUMMARY AND FUTURE TRENDS

Cellulose is the most abundant biopolymer on our planet, and it is one of the most available renewable alternatives for reducing and replacing the enormous amount of petroleum-derived plastic materials. Cellulose is polar and not inherently compatible with hydrophobic polymeric matrices (polyolefins or PLA). Consequently, its surface needs to be modified by various coupling agents to improve matrix-polymer adhesion, particularly at the interface. Currently, different treatment routes have been reported, which are grouped into physical and chemical treatments. Protons and unstable radicals are produced as a result of physical treatments, which are converted into chemical groups. Physical treatments, such as electrical discharge (cold plasma, corona discharge), can lead to chemical grafting, polymerization, crystallization, or etching. With these, the surface properties of the cellulose fibers are modified and the mechanical interaction between the matrix and the cellulose fiber is enhanced. Surface treatment through non-ionizing (UV) and ionizing γ-radiation has shown that γ-rays cause less significant degradation on cellulose fibers in comparison to UV radiation.

Coupling agents promote optimal stress transfer at the matrix-filler interface via cellulose surface hydrophobization (fiber-fiber interaction decrease), or by establishing a chemical bond between the matrix and the fiber. Polymer grafting is a technique that has permitted the insertion of well-defined polymer chains' highly regulated structure, is another noteworthy example of chemical modification. Polymer grafting improves the cellulose fiber properties such as elasticity, acoustic absorption, thermal resistance, resistance to abrasion and wear, and resistance to water absorption. In the last decade of the 20th century, car manufacturers in Germany are striving to make all components recyclable and biodegradable. For instance, plastics reinforced with flax fibers were used to make Mercedes door panels. Flax fibers in a polypropylene matrix are also being used by Canadian firms to produce the rear-shelf panel of the 2000 Chevrolet Impala.

Recently, the use of cellulose nanofibers, nanocrystals, and/or nanowhiskers in biodegradable and renewable materials used in PLA matrices have produced nanocomposites with improved properties, which represents a promising technology in the development of new biomaterials. The biodegradability and cytocompatibility of nanocellulose biocomposites have increased their use in the biomedical field.

All-cellulose composites (ACCs) have attracted a lot of research attention in recent years and are considered to be a class of biocomposites that have a matrix of dissolved and regenerated cellulose and undissolved cellulose fibers as reinforcement. ACCs are made of cellulose, which improves phase compatibility and the mechanical properties of the biocomposite. Research on different cellulose sources and their properties in composites is ongoing. The recent development of ACCs may open up new opportunities for industries to generate new applications for ACCs, as they are biodegradable biocomposites (in soil, compost, etc.); however, their degradation potential consequently limits its uses in some applications.

REFERENCES

Agwuncha, S. C., S. Owonubi, D. P. Fapojuwo, A. Abdulkarim, T. P. Okonkwo, and E. M. Makjatha. 2020. Evaluation of mercerization treatment conditions to extracted cellulose from shea nut shell using FTIR and thermogravimetric analysis. *Mater Today: Proc* 38(2):958–963.

Anah, L., N. Astrini, and A. Haryono. 2015. The effect of temperature on the grafting of acrylic acid onto carboxymethyl cellulose. *Macromol Symp* 353(1):178–184.

Asha, A. B., A. Sharif, and M. E. Hoque. 2017. Green biocomposites: design and applications. *Interface interaction of jute fiber reinforced PLA biocomposites for potential application.* In: Jawaid M., Salit M. S., Alothman O. Y. (eds). 285–307. Springer.

Bandhu G. S., S. Bandyopadhyay-Ghosh, and M. Sain. 2010. *Composites. Poly(lactic acid): Synthesis, Structures, Properties, Processing, and Applications.* Auras R., Lim L.T., Selke S E M., and Tsuji H. (eds). 293–310. John Wiley & Sons, Inc.

Banvillet, G., G. Depres, N. Belgacem, and J. Bras. 2021. Alkaline treatment combined with enzymatic hydrolysis for efficient cellulose nanofibrils production. *Carbohydr Polym* 255:117383.

Belgacem, M. N., and A. Gandini. 2005. The surface modification of cellulose fibers for use as reinforcing elements in composite materials. *Compos Interf* 12(1–2):41–75.

Bendourou, F. E., G. Suresh, M. A. Laadila, P. Kumar, T. Rouissi, G. S. Dhillon, K. Zied, S. K. Brar, and R. Galvez. 2021. Feasibility of the use of different types of enzymatically treated cellulosic fibres for polylactic acid (PLA) recycling. *Waste Manag* 121:237–247.

Benedetto, R. M. D., M.V. Gelfuso, and D. Thomazini. 2015. Influence of UV radiation on the physical-chemical and mechanical properties of banana fiber. *Mater Res* 18:265–272.

Bisanda, E. T. N. and M. P. Ansell. 1991. The effect of silane treatment on the mechanical and physical-properties of sisal-epoxy composites. *Compos Sci Technol* 41(2):165–178.

Borbély, E. 2008. Lyocell, the new generation of regenerated cellulose. *Acta Polytec Hun* 5(3):11–18.

Borsa, J., T. Toth, E. Takacs, and P. Hargittai. 2003. Radiation modification of swollen and chemically modified cellulose. *Radiat Phys Chem* 67(3–4):509–512.

Çaglar, M., M. Minelli, M. G. De Angelis, M. G. Baschetti, A. Stancampiano, R. Laurita, M. Gherardi, V. Colombo, J. Trifol, P. Szabo, and T. Lindstrom. 2016. Atmospheric plasma assisted PLA/microfibrillated cellulose (MFC) multilayer biocomposites for sustainable application. *Ind Crops Prod* 93:235–243.

Chen, D., J. Li, and J. Ren. 2011. Influence of fiber surface-treatment on interfacial property of poly (l-lactic acid)/ramie fabric biocomposites under UV-irradiation hydrothermal aging. *Mat Chem Phys* 126(3):524–531.

Chen, P., S. Y. Cho, and H. J. Jin. 2010. Modification and applications of bacterial celluloses in polymer science. *Macromol Res* 18(4):309–320.

Cichosz, S., A. Masek, and A. Rylski. 2020. Cellulose modification for improved compatibility with the polymer matrix: Mechanical characterization of the composite material. *Materials* 13(23), 5519.

Doineau, E., G. Coqueugniot, M. F. Pucci, A.-S. Caro, B. Cathala, J.-C Bénézet, J. Bras, and N. L. Moigne. 2021. Hierarchical thermoplastic biocomposites reinforced with flax fibres modified by xyloglucan and cellulose nanocrystals. *Carbohydr Polym* 254:117403.

Durham, E. K. and S. K. Sastry. 2020. Moderate electric field treatment enhances enzymatic hydrolysis of cellulose at below-optimal temperatures. *Enzyme Microb Technol* 142:109678.

EL-Zayat, M. M., A. Abdel-Hakim, and M. A. Mohamed. 2019. Effect of gamma radiation on the physico mechanical properties of recycled HDPE/modified sugarcane bagasse composite. *J Macromol Sci A* 56(2):127–135.

Eyley, S. and W. Thielemans. 2014. Surface modification of cellulose nanocrystals. *Nanoscale* 6:7764–7779.

Frone, A. N., S. Berlioz, J.-F. Chailan, D. M. Panaitescu, and D. Donescu. 2011. Cellulose fiber-reinforced polylactic acid. *Polym Compos* 32(6):976–985.

Gazzotti, S., R. Rampazzo, M. Hakkarainen, D. Bussini, M. A. Ortenzi, H. Farina, G. Lesma, and A. Silvani. 2019. Cellulose nanofibrils as reinforcing agents for PLA-based nanocomposites: an in situ approach. *Compos Sci Technol* 171:94–102.

Geremew, A., P. De Winne, T. Adugna, and H. De Backer. 2021. An overview of the characterization of natural cellulosic fibers. *Key Eng Mat* 881:107–116.

Gupta, K. and S. Sahoo. 2001a. Graft copolymerization of 4-vinylpyridine onto cellulose using Co (III) acetylacetonate complex in aqueous medium. *Cellulose* 8(3):233–242.

Gupta, K. and S. Sahoo. 2001b. Co(III) acetylacetonate-complex-initiated grafting of *N*-vinyl pyrrolidone on cellulose in aqueous media. *J Appl Polym Sci* 81(9):2286–2296.

Gupta, K.C. and K. Khandekar. 2006. Graft copolymerization of acrylamide onto cellulose in presence of comonomer using ceric ammonium nitrate as initiator. *J Appl Polym Sci* 101(4):2546–2558.

Gürdağ, G. and S. Sarmad. 2013. Cellulose graft copolymers: synthesis, properties, and applications. *Polysaccharide Based Graft Copolymers*. In: Kalia S., Sabaa M. (eds). 15–57. Springer.

Guzel, M. and O. Akpinar. 2019. Production and characterization of bacterial cellulose from citrus peels. *Waste Biomass Valor* 10(8):2165–2175.

Hao, L. C., S. M. Supuan, M. R. Hassan, and R. M. Sheltami. 2018. Natural fiber reinforced vinyl polymer composites. *Natural Fibre Reinforced Vinyl Ester and Vinyl Polymer Composites*. In: S. M. Sapuan, H. Ismail, and E. S. Zainudin (eds). 27–70. Woodhead Publishing Elsevier.

Hassani, F. O., N. Merbahi, A. Oushabi, M. H. Elfadili, A. Kammouni, and N. Oueldna. 2020. Effects of corona discharge treatment on surface and mechanical properties of Aloe Vera fibers. *Mater Today Proc* 24:46–51.

He, M., G. Yang, J. Chen, X. Ji, and Q. Wang. 2018. Production and characterization of cellulose nanofibrils from different chemical and mechanical pulps. *J Wood Chem Tech* 38(2):149–158.

Hebeish, A. and J. T. Guthrie. 1981. *The Chemistry and Technology of Cellulosic Copolymers*. Springer-Verlag, Berlin.

Heinze, T., A. Koschella, T. Liebert, V. Harabagiu, and S. Coseri. 2012. The European Polysaccharide Network of Excellence (EPNOE) – Research initiatives and results. *Cellulose: Chemistry of Cellulose Derivatization*. In: P. Navard (ed). 283–327. Springer-Verlag, Wein.

Henniges, U., M. Hasani, A. Potthast, G. Westman, and T. Rosenau. 2013. Electron beam irradiation of cellulosic materials – opportunities and limitations. *Materials* 6(5):1584–1598.

Hermanutz, F., M. P. Vocht, N. Panzier, and M. R. Buchmeiser. 2019. Processing of cellulose using ionic liquids. *Macromol Mater Eng* 304(2):1800450.

Howard, R.L., E. Abotsi, E. L. van Rensburg, S. Y. Howard. 2003. Lignocellulose biotechnology: issue of bioconversion and enzyme production. *Afr J Biotechnol* 2:602–619.

Huber, T., J. Müssig, O. Curnow, S. Pang, S. Bickerton, and M. P. Staiger. 2011. A critical review of all-cellulose composites. *J Mat Sci* 47(3):1171–1186.

Ibrahim, Md., E. M. Flefel, and W. K. El-Zawawy. 2002. Cellulose membranes grafted with vinyl monomers in homogeneous system. *J Appl Polym Sci* 84(14):2629–2638.

Ibrahim, Md., H. Mondal, Y. Uraki, M. Ubukata, and K. Itoyama. 2008. Graft polymerization of vinyl monomers onto cotton fibres pretreated with amines. *Cellulose* 15(4):581–592.

Islam, M. S., K. L. Pickering, and N. J. Foreman. 2010. Influence of accelerated ageing on the physico-mechanical properties of alkali-treated industrial hemp fibre reinforced poly(lactic acid) (PLA) composites. *Polym Degrad Stabil* 95(1):59–65.

Jedvert, K. and T. Heinze. 2017. Cellulose modification and shaping-a review. *J Polym Eng* 37(9):845–860.

Jideonwo, A. and H. A. Adimula. 2006. The graft copolymerisation of acrylamide onto cellulose using enhanced Fe^{2+}/H_2O_2 redox initiator system. *J Appl Sci Environ Mgt* 10(3):151–155.

John, M. J. and S. Thomas. 2008. Biofibres and biocomposites. *Carbohydr Polym* 71(3):343–364.

Kang, H., R. Liu, and Y. Huang. 2013. Cellulose derivatives and graft copolymers as blocks for functional materials. *Polym Int* 62(3):338–344.

Khalil, H. A., A. Bath, and A. Yusra. 2012. Green composites from sustainable cellulose nanofibrils: A review. *Carbohydr Polym* 87(2):963–979.

Khalilzadeh, M. A., S. Hosseini, A. S. Rad, and R. A. Venditti. 2020. Synthesis of grafted nanofibrillated cellulose-based hydrogel and study of its thermodynamic, kinetic, and electronic properties. *J Agric Food Chem* 68(32):8710–8719.

Khan, M.A., N. Haque, A. Al-Kafi, M. N. Alam, and M. Z. Abedin. 2006. Jute reinforced polymer composite by gamma radiation: effect of surface treatment with UV radiation. *Polym Plast Technol Eng* 45(5):607–613.

Kian L.K., N. Saba, M. Jawaid, and M.T.H. Sultan. 2019. A review on processing techniques of bast fibers nanocellulose and its polylactic acid (PLA) nanocomposites. *Int J Biol Macromol* 121:1314–1328.

Komuraiah, A., N. Shyam Kumar, and B. D. Prasad. 2014. Chemical composition of natural fibers and its influence on their mechanical properties. *Mech Compos Mater* 50(3):359–376.

Kowalczyk, M., E. Piorkowska, P. Kulpinski, and M. Pracella. 2011. Mechanical and thermal properties of PLA composites with cellulose nanofibers and standard size fibers. *Compos Part A Appl Sci Manuf* 42:1509–1514.

Kumar, R., S. Obrai, and A. Sharma. 2011. Chemical modifications of natural fiber for composite material. *Der Chemica Sinica* 2(4):219–228.

Kyutoku, H., N. Maeda, H. Sakamoto, H. Nishimura, and K. Yamada. 2019. Effect of surface treatment of cellulose fiber (CF) on durability of PLA/CF bio-composites. *Carbohydr Polym* 203:95–102.

Larsson, K., L. A. Berglund, M. Ankerfors, and T. Lindström. 2012. Polylactide latex/nanofibrillated cellulose bionanocomposites of high nanofibrillated cellulose content and nanopaper network structure prepared by a papermaking route. *J Appl Polym Sci* 125(3):2460–2466.

Le Duigou, A., A. Bourmaud, E. Balnois, P. Davies, and C. Baley. 2012. Improving the interfacial properties between flax fibres and PLLA by a water fibre treatment and drying cycle. *Ind Crops Prod* 39:31–39.

Lenfeld, P., P. Brdlík, M. Borůvka, L. Běhálek, and J. Habr. 2020. Effect of radiation crosslinking and surface modification of cellulose fibers on properties and characterization of biopolymer composites. *Polymers* 12(12):3006.

Li, H., Z. Cao, D. Wu, G. Tao, W. Zhong, H. Zhu, P. Qiu, and C. Liu. 2016. Crystallisation, mechanical properties and rheological behaviour of PLA composites reinforced by surface modified microcrystalline cellulose. *Plast Rubber Compos* 45(4):181–187.

Liu, W., S. Liu, T. Liu, J. Zhang, and H. Liu. 2019. Eco-friendly post-consumer cotton waste recycling for regenerated cellulose fibers. *Carbohydr Polym* 206:141–148.

Machnowski, W., B. Gutarowska, J. Perkowski, and H. Wrzosek. 2013. Effects of gamma radiation on the mechanical properties of and susceptibility to biodegradation of natural fibers. *Textile Res J* 83(1):44–55.

Madera-Santana, T. J., H. Soto Valdez, and M. O. W. Richardson. 2013. Influence of surface treatments on the physicochemical properties of short sisal fibers: Ethylene vinyl acetate composites. *Polym Eng Sci* 53(1):59–68.

Mahzan, S., M. Fitri, and M. Zaleha. 2017. UV radiation effect towards mechanical properties of natural fibre reinforced composite material: a review. *IOP Conf Ser: Mater Sci Eng* 165:012021.

Mathew, A. P., K. Oksman, and M. Sain. 2005. Mechanical properties of biodegradable composites from polylactic acid (PLA) and microcrystalline cellulose (MCC). *J Appl Polym Sci* 97(5):2014–2025.

Melikoglu, A. Y., S. E. Bilek, and S. Cesur. 2019. Optimum alkaline treatment parameters for the extraction of cellulose and production of cellulose nanocrystals from apple pomace. *Carbohydr Polym* 215:330–337.

Mohanty, A. K., M. Misra, L. T. Drzal, S. E. Selke, B. R. Harte, and G. Hinrichsen. 2005. Natural fibers, biopolymers, and biocomposites: an introduction. *Natural Fibers, Biopolymers, and Biocomposites.* In: Mohanty, A. K., M. Misra, and L. T. Drzal (eds). 1–35. CRC Press.

Mwaikambo, L. Y. and M. P. Ansell. 1999. The effect of chemical treatment on the properties of hemp, sisal, jute and kapok for composite reinforcement. *Die Angew Makromol Chem* 272(1):108–116.

Mwaikambo, L. Y. and M. P. Ansell. 2002. Chemical modification of hemp, sisal, jute, and kapok fibers by alkalization. *J Appl Polym Sci* 84(12):2222–2234.

Olaru, A., T. Măluțan, C. M. Ursescu, M. Geba, and L. Stratulat. 2016. Structural changes in hemp fibers following temperature, humidity and UV or gamma-ray radiation exposure. *Cell Chem Technol* 50(1):31–39.

Ouajai, S., A. Hodzic, and R. A. Shanks. 2004. Morphological and grafting modification of natural cellulose fibers. *J Appl Polym Sci* 94(6):2456–2465.

Ozyhar, T., F. Baradel, and J. Zoppe. 2020. Effect of functional mineral additive on processability and material properties of wood-fiber reinforced poly(lactic acid) (PLA) composites. *Compos Part A Appl Sci Manuf* 132:105827.

Pentimalli, M., D. Capitani, A. Ferrando, D. Ferri, P. Ragni, and A. L. Segre. 2000. Gamma irradiation of food packaging materials: an NMR study. *Polymer* 41(8):2871–2881.

Pinho, E. and G. Soares. 2018. Functionalization of cotton cellulose for improved wound healing. *J Mater Chem B* 6(13):1887–1898.

Qian, S., H. Zhang, W. Yao, and K. Sheng. 2018. Effects of bamboo cellulose nanowhisker content on the morphology, crystallization, mechanical, and thermal properties of PLA matrix biocomposites. *Compos B Eng* 133:203–209.

Rahman, A. M., S. Alimuzzaman, R. A. Khan, and J. Hossen. 2018. Evaluating the performance of gamma irradiated okra fiber reinforced polypropylene (PP) composites: comparative study with jute/PP. *Fash Text* 5(1):1–17.

Rahman, H., S. Alimuzzaman, M. A. Sayeed, and R. A. Khan. 2019. Effect of gamma radiation on mechanical properties of pineapple leaf fiber (PALF)-reinforced low-density polyethylene (LDPE) composites. *Int J Plastics Technol* 23(2):229–238.

Rahman, M. M. and M. A. Khan. 2007. Surface treatment of coir (*Cocos nucifera*) fibers and its influence on the fibers' physico-mechanical properties. *Compos Sci Technol* 67(11–12):2369–2376.

Routray, C. and B. Tosh. 2012. Controlled grafting of MMA onto cellulose and cellulose acetate. *Cellulose* 19(6):2115–2139.

Routray, C. and B. Tosh. 2013. Graft copolymerization of methyl methacrylate (MMA) onto cellulose acetate in homogeneous medium: effect of solvent, initiator and homopolymer inhibitor. *Cellulose Chem Technol* 47(3–4): 171–190.

Roy, D., M. Semsarilar, J. T. Guthrie, and S. Perrier. 2009. Cellulose modification by polymer grafting: a review. *Chem Soc Rev* 38(7):2046–2064.

Sadeghifar, H., R. Venditti, J. Jur, R. E. Gorga, and J. J. Pawlak. 2017. Cellulose-lignin biodegradable and flexible UV protection film. *ACS Sustain Chem Eng* 5(1):625–631.

Sahoo P. K., H. S. Samantaray, and R. K. Samal. 1986. Graft copolymerization with new class of acidic peroxo salts as initiators. I. Grafting of acrylamide onto cotton–cellulose using potassium monopersulfate, catalyzed by Co(II). *J Appl Polym Sci* 32(7):5693–5703.

Shanghaleh, H., X. Xu, and S. Wang. 2018. Current progress in production of biopolymeric materials based on cellulose, cellulose nanofibers and cellulose derivatives. *RSC Adv* 8(2):825–842.

Shi, J., L. Lu, W. Guo, Y. Sun, and Y. Cao. 2013. An environment-friendly thermal insulation material from cellulose and plasma modification. *J Appl Polym Sci* 130(5):3652–3658.

Shin, S. J. and Y. J. Sung. 2008. Improving enzymatic hydrolysis of industrial hemp (Cannabis sativa L.) by electron beam irradiation. *Radiat Phys Chem* 77(9):1034–1038.

Silva, C. G., A. L. P. Campini, D. B. Rocha, and D. S. Rosa. 2019. The influence of treated eucalyptus microfibers on the properties of PLA biocomposites. *Compos Sci Technol* 179:54–62.

Sionkowska, A., H. Kaczmarek, M. Wisniewski, J. Skopinska, S. Lazare, and V. Tokarev. 2006. The influence of UV irradiation on the surface of chitosan films. *Surf Sci* 600(18):3775–3779.

Södergård A. and M. Stolt. 2002. Properties of lactic acid based polymers and their correlation with composition. *Prog Polym Sci* 27(6):1123–1163.

Södergård A. and M. Stolt. 2010. Industrial production of high molecular weight poly(lactic acid). *Poly(lactic acid): Synthesis, Structures, Properties, Processing, and Applications.* In: R. Auras, L.-T. Lim, S. E. M. Selke, and H. Tsuji (eds). 293–310. John Wiley & Sons, Inc.

Song, Z., H. Xiao, and Y. Zhao. 2014. Hydrophobic-modified nano-cellulose fiber/PLA biodegradable composites for lowering water vapor transmission rate (WVTR) of paper. *Carbohydr Polym* 111:442–448.

Sudamrao A. G. and B. Patel. 2020. A review: bio-fiber's as reinforcement in composites of polylactic acid (PLA). *Mater Today* 26:2116–2122.

Sutirman, Z. A., M. M. Sanagi, A. A. Naim, K. J. A. Karim, and W. A. W. Ibrahim. 2017. Ammonium persulfate-initiated graft copolymerization of methacrylamide onto chitosan: synthesis, characterization and optimization. *Sains Malaysiana* 46(12):2433–2440.

Tibolla, H., A. Czaikoski, F. M. Pelissari, F. C. Menegalli, and R. L. Cunha. 2020. Starch-based nanocomposites with cellulose nanofibers obtained from chemical and mechanical treatments. *Int J Biol Macromol* 161:132–146.

Tosh, B. and C. R. Routray. 2014. Grafting of cellulose based materials: a review. *Chem Sci Rev Lett* 3(10):74–92.

Tsusj, H. 2002. Polylactide. *Biopolymers. Polyesters III. Applications and Commercial Products.* In: Y. Doi, A. Steinbuchel (eds). 129–177. Wiley-VCH Verlag GmbH.

Wang, J., J. Tavakoli, and Y. Tang. 2019. Bacterial cellulose production, properties and applications with different culture methods: a review. *Carbohyd Polym* 219: 63–76.

Wanmolee, W., W. Sornlake, N. Rattanaphan, S. Suwannarangsee, N. Laosiripojana, and V. Champreda. 2016. Biochemical characterization and synergism of cellulolytic enzyme system from *Chaetomium globosum* on rice straw saccharification. *BMC Biotechnol* 16(1):82.

Wohlhauser, S., G. Delepierre, M. Labet, G. Morandi, W. Thielemans, C. Weder, and J. O. Zoppe. 2018. Grafting polymers from cellulose nanocrystals: synthesis, properties, and applications. *Macromol* 51(16):6157–6189.

Wojnarovits, L., C. M. Foldvary, and E. Takacs. 2010. Radiation-induced grafting of cellulose for adsorption of hazardous water pollutants: a review. *Rad Phys Chem* 79(9):848–862.

Xia, X., X. Shi, W. Liu, S. He, C. Zhu, and H. Liu. 2020. Effects of gamma irradiation on properties of PLA/flax composites. *Iran Polym J* 29(7):581–590.

Yang, F., G. Li, Y. G. He, F. X. Ren, and J. X. Wang. 2009. Synthesis, characterization, and applied properties of carboxymethyl cellulose and polyacrylamide graft copolymer. *Carbohydr Polym* 78(1):95–99.

Yao, M. Z., Y. Liu, C. N. Qin, X. J. Meng, H. Zhao, S. F. Wang, and Z. Q. Huang. 2020. Facile fabrication of hydrophobic cellulose-based organic/inorganic nanomaterial modified with POSS by plasma treatment. *Carbohydr Polym* 253:117193.

Yatigala, N.S., D. S. Bajwa, and S. G. Bajwa. 2018. Compatibilization improves performance of biodegradable biopolymer composites without affecting UV weathering characteristics. *J Polym Environ* 26(11):4188–4200.

Yulianto, D. N. Putri, M. S. Perdani, R. Arbianti, L. Suryanegara, and H. Hermansyaha. 2020. Effect of cellulose fiber from sorghum bagasse on the mechanical properties and biodegradability of polylactic acid. *Energy Rep* 6:221–226.

Yuting, C. and C. Xiansu. 2008. Radiation effect of γ-rays on lignin. *J Radiat Res Radiat Proces* 26(5):275–279.

Zaman, H.U., M. A. Khan, R. A., Khan, M. Z. I. Mollah, S. Pervin, and M. D. Al-Mamun. 2010. A comparative study between gamma and UV radiation of jute fabrics/polypropylene composites: effect of starch. *J Reinf Plast Compos* 29(13):1930–1939.

Zhai, S., Q. Liu, Y. Zhao, H. Sun, B. Yang, and Y. Weng. 2021. A review: research progress in modification of poly(lactic acid) by lignin and cellulose. *Polymers* 13:776–790.

Zhang, Z., Y. Li, K. Fu, and Q. Li. 2020. Determination of interfacial properties of cellulose nanocrystal-modified sisal fibre in epoxy by cyclic single-fibre pull-out. *Compos Sci Technol* 193:108142.

Zhao, B., Y. Zhang, and H. Ren. 2020. Effects of microcrystalline cellulose surface modification on the mechanical and thermal properties of polylactic acid composite films. *Plast Rubber Compos* 49(10):450–455.

Zhu, T., J. Guo, B. Fei, Z. Y. Feng, and X. Y. Gu, 2020. Preparation of methacrylic acid modified microcrystalline cellulose and their applications in polylactic acid: Flame retardancy, mechanical properties, thermal stability and crystallization behavior. *Cellulose* 27, 2309–2323.

5 PLA/Cellulose Composites and Their Hybrid Composites

Martin A. Hubbe
North Carolina State University
Raleigh, NC

Warren J. Grigsby
Scion
Rotorua, New Zealand
Present employer: Henkel NZ, Ltd

CONTENTS

DOI: 10.1201/9781003160458-5

5.1 INTRODUCTION: DRIVING FORCES AND BACKGROUND

Poly(lactic acid) or "PLA" has emerged as the most widely produced and utilized bioplastic that can be considered as a substitute for petroleum-based plastics in a wide range of applications. There has been much attention to PLA and its composites in the scholarly literature. Because of that, the present chapter will focus on a series of pivotal questions that are likely to be important in determining the most successful future paths for the implementation of PLA, with an emphasis here on PLA's composite structures with cellulosic particles. Additionally, it makes sense to consider the inclusion of a second type of filler or reinforcement, along with the cellulosic particles, *i.e.* for the formation of hybrid composites.

5.1.1 DRIVING FORCES

The packaging of a food product is often the first thing seen by a potential customer. Corporations selling the products hope that the packaging sends an attractive message. Though the bright colors and printing messages on a package can attract customers, there is an increasing likelihood that discerning customers may be influenced by the eco-friendly nature of the packaging. Due to the high prominence of packaging in the retailing of food items at grocery stores, one can expect that renewable and biodegradable packaging will attract increasing attention from food distributors.

Increasing sophistication of consumers can be expected in the coming years, so it is necessary to consider some tough questions. For instance, corporations involved with packaging will want to know whether it is really worth their time and funding to develop a whole new or expanded reliance on a renewable-based supply chain, with all of the risks and effort associated with developing something new. Consumers may have heard some conflicting messaging about whether or not PLA-containing packaging really degrades if these materials become cast into the environment or placed in a landfill. If the idea is to promote a very high percentage of recovery and recycling of the packaging materials, then there will be concerns about the level of degradation of molecular mass and other polymer attributes with each successive generation of processing.

In the grand scheme, one can expect general agreement and support in future years for the concept that packaging materials ought to be largely prepared from plant-based materials, *i.e.* coming from materials that are photosynthetically renewable (Gandini and Lacerda 2015). Such materials can be justified in terms of the long-term sustainability of food-distribution practices. Given the high value of packaging in reducing the percentage of wasted food items due to spoilage, there does not seem to be much threat that food packaging will "disappear" in coming years. Rather, it can be expected to evolve as a result of a competition among a variety of factors, some of which are highlighted in the present review article. To lay groundwork for that discussion, the next section provides some general technical background pertaining to the use of PLA in food packaging and some other applications.

5.1.2 Technical Background

5.1.2.1 Fundamentals and Sources

Poly(lactic acid) (PLA) is a hydrophobic thermoplastic that can achieve favorable elastic modulus (about 3500 MPa) and a relatively high melting point (*e.g.* 165°C), depending on its purity and molecular mass (Farah *et al.* 2016). Although PLA is hydrophobic, it contains the polar functionality common to polyesters, and this confers upon it a relatively high cohesive energy density, giving a Hildebrand parameter near to 10 cal$^{0.5}$•cm$^{-1.5}$ (Siemann 1992). The cohesive forces can contribute to its barrier performance against various permeants (Lagarón *et al.* 2004), which makes it possible for PLA to be considered for such applications as packaging films and beverage containers. Among the various polymers that can be manufactured starting with plant materials, *i.e.* bioplastics, PLA is currently manufactured in the highest amount, with a reported annual production of over 600,000 tons (Jem and Tan 2020). PLA has become widely used especially in 3D printing.

In principle, it would be possible to produce PLA from petroleum-derived lactic acid; however, such products would tend to have inferior properties (Eiteman and Ramalingam 2015). By using biologically sourced material, it is practical to achieve a high degree of optical purity, given that only the L-isomer of PLA is present (Abdel-Rahman *et al.* 2013). The optically pure PLA generally has a higher melting point, a greater crystallinity after cooling of the melt, and a higher Young's modulus.

Presently PLA is primarily manufactured from the starch or sugar content of plants such as corn (Vaidya *et al.* 2005; Reddy *et al.* 2008), and this situation can raise concerns regarding competition for resources with food products. Many people think that industrial products ought to instead be manufactured starting with non-food plant materials, including the residues left over after harvesting and processing of food items (Doran-Peterson *et al.* 2008; Alves De Oliveira *et al.* 2020; Mazzoli 2020). By the use of such "second-generation" lactic acid sources to manufacture PLA and related polymers, there is less tendency for competition to drive up the price of food items or to increase their scarcity. In addition, the expected adverse environmental impacts are decreased relative to other sourcing of plastic materials (Adom and Dunn 2017).

5.1.2.2 PLA-Cellulose Composites

The technical properties of PLA films and structures can be enhanced by the formation of composites in which solid particles can serve as reinforcements or "fillers". In particular, there have been useful review articles focusing on the usage of cellulose-based reinforcing particles in the formulation of PLA-based composites (Saba *et al.* 2017; Hubbe *et al.* 2021). Thus, only a brief overview of the main issues is provided here, and the main attention is paid to critical questions that may be expected to influence the paths of future implementation in the field of food packaging.

An earlier review of literature established that by forming PLA-cellulose composites, the modulus of elasticity often can be increased by a factor as great as two compared to the PLA by itself (Hubbe and Grigsby 2020). However, there was a wide scatter of outcomes from the many studies considered in that review. One factor that consistently tended to increase a wide range of physical properties, including tensile and flexural moduli and breaking strength, was a treatment or additive enhancing the interfacial compatibility. For example, it is possible to render the cellulosic surfaces more hydrophobic by esterification with alkyl groups (Yin *et al.* 2020), where this treatment had a favorable effect on the properties of PLA composites. In the cited work, hydrophobization of the surfaces of nanocellulose particles was achieved by eco-friendly lipase-catalyzed esterification, using lauric acid. Although there has been much emphasis on the use of nanoparticles for the reinforcement of bioplastics, there does not appear to be a significant dependency of composite strength properties on the size of the reinforcing particles (Hubbe and Grigsby 2020).

It is well known that the incorporation of reinforcing particles into a plastic matrix tends to make the material more brittle. Thus, in order to tune the resulting properties, it is common to employ various plasticizing agents or other polymers blended with the PLA (Liu and Zhang 2011; Ferri *et al.* 2016; Koh *et al.* 2018). By such approaches, in combination with the reinforcing particles, formulators are able to tune the composite properties to achieve favorable combinations of toughness, modulus of elasticity, and other desired properties.

5.1.2.3   PLA-Cellulose Hybrid Composites

A hybrid composite can be defined as a material comprised of a continuous matrix that is filled with two or more different kinds of filler particles. Such formulations can be justified when each type of filler provides a desirable attribute to the final product (Hubbe 2017). For example, PLA composites have been prepared with a combination of talc and newsprint paper fibers (Huda *et al.* 2007). The composites were prepared by melt-compounding and molding. The talc addition contributed to greater flexural modulus, an effect that was attributed to the greater stiffness of the mineral. As a further example, plastic hybrid nanocomposites prepared with either nanofibrillated cellulose or cellulose nanocrystals, in addition to platy nanoclay particles, exhibited improved resistance to permeation of a film by both oxygen and water vapor (Trifol *et al.* 2016). In another approach to forming a hybrid composite, a hybrid filler is first prepared, in which two components are coupled together before their addition to a polymer matrix. Following such a procedure, a silica-lignin combination was used as a filler for PLA (Grzabka-Zasadzinska *et al.* 2016); and the resulting hybrid formulation provided better nucleation of the matrix polymer. The word "hybrid" can also be applied when blends of different matrix polymers are reinforced with one kind of filler. For instance, Zhang *et al.* (2010) filled a blend of PLA and polypropylene with bamboo fibers. The incorporation of the polypropylene, in combination with a coupling agent, improved the toughness of the material. Such approaches can be important because the neat PLA matrix was relatively brittle and prone to breakage.

5.1.2.4   Current and Potential Applications

The combination of melt-formability, relatively high modulus of elasticity, and hydrophobic character allow PLA, as well as its blends and composites, to be considered as replacements for some widely used petroleum-based plastics in many applications. Two main categories of products that are of interest for food packaging are beverage containers and films. Work by Auras *et al.* (2006) showed that the properties of PLA generally lie between those of oriented polystyrene (OPS) and polyethylene terephthalate (PET). In other words, the properties fall within the range of some petroleum-based plastics commonly used for the containment of liquids. The topic of bio-based barrier-type films, including PLA-cellulose composite films, has been widely studied and reviewed (Arrieta *et al.* 2017; Sun *et al.* 2018; Hubbe *et al.* 2021; Khosravi *et al.* 2020).

5.2 PIVOTAL QUESTIONS

In light of the driving forces (environmental issues, technical requirements) and technical background just covered, there are some serious concerns that are likely to determine the degree of uptake, success, and the extent of implementation of PLA and its cellulose composites in the coming years. Briefly stated, some questions addressed in the subsections that follow are (1) whether PLA-cellulose composites ought to be regarded as a boutique or a volume product category, (2) should PLA be excluded from litter-prone applications, (3) can PLA-cellulose composites be trusted in demanding environments, (4) why blends of PLA and a second polymer often yield synergistic effects, (5) when should relatively large cellulose particle be used in PLA-cellulose composites, (6), when is it best to use a second type of filler in addition to a cellulosic particle in PLA, (7) should pilot-scale processing be used when developing PLA composites, and, lastly, (8) where should researchers be placing an emphasis with respect to directed applications of PLA?

5.2.1 PLA-Cellulose Composites: Boutique or Volume?

Arguably, research on PLA and PLA-cellulose materials is an active area, in which the literature has steadily built up over the last decade. However, an important question remains: Does this research match the scale and breadth needed for industry and global market uptakes? In other words, as illustrated in Figure 5.1, the material might be considered either for boutique applications, such as figurines and logo displays, or volume applications, such as beverage containers and furniture. Globally, volumes of PLA are *ca.* 600,000 tonnes per annum, and although these are currently dwarfed by petrochemically-derived plastics, PLA growth rates and projected volumes over the next decades suggest that PLA alongside other emerging bioplastics will be a significant proportion of polymer production by 2040 (Stuchtey and Dillion 2021; MacArthur *et al.* 2021). Primarily this growth in PLA production will be in the regions of SE Asia, Europe, and the Americas.

Moreover, generally, this existing and new PLA production can be matched to the locations of cellulose materials and their availability. Cellulose is globally traded, and cellulose can be accessed in a wide range of forms and volumes over nano-, micro-, and macro-dimensions, with these materials available as native, modified, and regenerated celluloses. Based on existing manufacturing bases, it can be imagined that micro- and macro-scale cellulosic materials in traditional fiber and particle forms can be readily matched to PLA volumes in key manufacturing markets. This includes traditional markets of SE Asia and North America together with emerging capacity such as Brazil that has evolved over the last decades. However, while highly valued, the volumes of nanoscale celluloses presently produced can be considered as

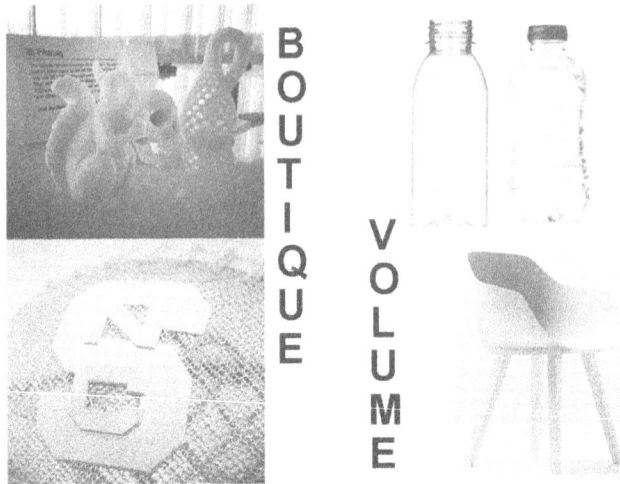

FIGURE 5.1   Examples of boutique (left) and volume (right) applications of bioplastics such as PLA.

niche materials, limited to 100s tonnes per annum and at least another decade away before manufacturing capacity grows to make these materials more available in any commercial volumes.

There are also commercial-scale equipment issues when considering potential production scales and practicalities of producing PLA-cellulose composites. With petrochemical-derived plastics, the inclusion of cellulose particles is preferably achieved via first forming the composite material into pre-formed pellets as a starting feedstock for further processing (Mertens *et al.* 2017). For example, celluloses in the powdered form, such as wood flour, have a history of being processed into plastics as construction materials and other applications (Bledzki *et al.* 1998). With PLA, it is envisaged that the initial processing routes to cellulose composites will be similar. Cellulose in the form of kraft pulp or wood fibers or as other native lignocellulosic fibers is available in suitable volumes, and processes for integrating these cellulose forms with PLA have been well reported (Chan *et al.* 2018). With similarities in composites processing between PLA and petrochemical plastics, it is reasonable to expect that cellulose materials in these or larger particle size ranges will not be constrained in uptake at large scales, as materials handling and processing equipment is already widely available. PLA substitution of existing plastics and their manufacturing practice is also expected to realize this large-scale uptake.

Beyond their availability, the current forms of nanocelluloses may offer challenges to their integration with PLA and the scaling of these processes when compared with larger cellulose particles (Hubbe *et al.* 2008). Nanocellulosics are typically available as low-solids suspensions in wet form, which undergo agglomeration or congeal on drying (Calvino *et al.* 2020; Wang *et al.* 2019a). Practically, plastic composites are often marred by nanocellulosic particles clumping during processing with the plastic. This is also the case for PLA, and presently it can be considered detracting from the wider uptake and applications of PLA-nanocellulose composites. This situation also presents opportunities for research and development. Researchers have undertaken chemical and surface modifications to promote dispersions of nanocellulose in plastics (Rol *et al.* 2019). However, undertaking such modifications in a production environment has not been reported. Nonetheless, this provides opportunities to investigate routes alongside combining PLA with nanocellulosics in aqueous form prior to their dehydration and formation of the hydrophobic PLA composite. Furthermore, wet-processing with PLA emulsions is still in its infancy (Lee *et al.* 2016), but may offer new routes to producing PLA composites with nanocellulose. It is envisaged that such wet-processing may be done with nanocelluloses in which surface modifications are the first step prior to introducing PLA.

Overall, when one considers the global scale and opportunity for PLA-cellulose composites, it can be stated that "size does matter" alongside the ability to adapt and scale fundamental research activities. Each is equally relevant to whether PLA-cellulose composites remain boutique or become produced in volumes projected by the wider availability of PLA and associated polyhydroxyalkanoates. This is matched with suitable pathways that already exist for the large-scale production of PLA composites with cellulose forms with larger dimensions. In contrast, for nano-sized cellulosics, a new process or processing alternatives will need to be available. Otherwise, the integration of nanocelluloses into PLA will remain the limiting factor to scaling these materials.

5.2.2 Should PLA Be Excluded from Litter-Prone Applications?

The harmful effects and unsightly appearance of litter, as well as the adverse effects of plastic-based litter persisting in the environment, provide much of the motivation behind the push for bioplastics. The idea is that, in addition to the fact that the bioplastics can be produced from renewable resources, they also are expected to be biodegradable. Relative to polyethylene (PE), PLA can be judged to be highly biodegradable. PE has been estimated to require 1000 years to break down in the natural environment, whereas PLA has been estimated to require only several weeks at elevated temperatures similar to industrial composting conditions (Hakkarainen *et al.* 2000). As depicted in Figure 5.2, an ideal goal would be for any PLA discarded as litter to completely decompose in the environment soon after its undesired disposal. However, much longer times of degradation are required under typical conditions of soils or seawater (Wadsworth *et al.* 2013; Wan *et al.* 2019).

It has been proposed that PLA can be rendered substantially more biodegradable by preparing it as a composite with the inclusion of cellulosic reinforcing particles (Hubbe *et al.* 2021). The idea is that the water-loving cellulosic materials render the material susceptible to the actions of water. In principle, water that enters a PLA composite by way of cellulosic reinforcing particles will tend to swell those particles, and this can further open up the material to more water and induce further swelling. In addition, where the water is able to go, one might propose that enzymes can go as well. Because PLA is a polyester, it seems reasonable to expect that PLA

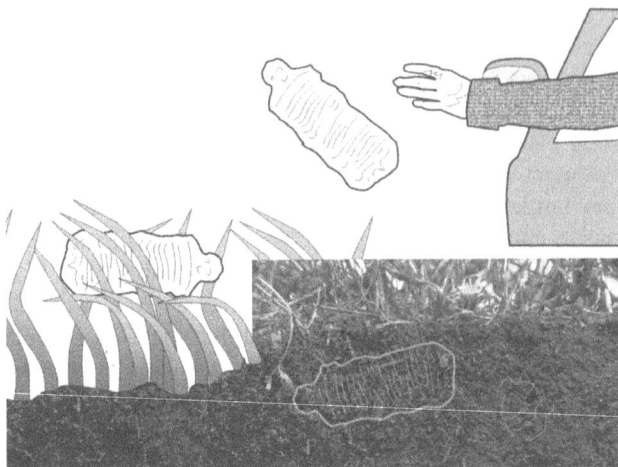

FIGURE 5.2   Due to the prevalence of littering, a suitably high rate of biodegradation of bioplastic food containers or packaging is needed under typical outdoor conditions.

will be susceptible to increased rates of hydrolysis in the presence of ester-cleaving enzymes such as lipases and proteinases (Lee *et al.* 2014). However, even when there is enhanced contact between the aqueous phase and the PLA, there is reason to question whether enzyme-catalyzed reactions will proceed fast enough to satisfy consumer expectations. The combination of a hydrophobic nature, a high density, and a high level of crystallinity conspire to render PLA resistant to enzymatic attack (Hubbe *et al.* 2021). In contrast, much faster enzymatic hydrolysis rates are observed for some other polyesters, *e.g.* poly(ethyleneterephthalate) (PET) (Austin *et al.* 2018) and triglyceride fats (Hermansyah *et al.* 2007).

Studies suggest that abiotic reactions play a rate-determining role in PLA degradation and that such reactions are highly temperature-dependent (Agarwal *et al.* 1998). Industrial-scale composting involves the heating up of the material to over 58°C (Gorrasi and Pantani 2013), and such heating is sufficient for the abiotic breakdown processes to proceed. Ester-targeting enzymes also have been shown to play a significant role in this process (Lee *et al.* 2014). However, various other studies have shown unsatisfactorily slow biodegradation of PLA in ground burial tests or in seawater testing, which generally involve temperatures well below 58°C (Wan *et al.* 2019). A recent review article considered various ways that PLA composites can be formulated to promote more rapid biodegradation, which includes strategies such as blending the PLA with other matrix polymers such as starch, using cellulosic reinforcing particles, and adding carboxylic acid-containing compounds (Hubbe *et al.* 2021). Moreover, high rates of biodegradation have been reported for various PLA blends or composites involving such materials as starch (Koh *et al.* 2018). In general, however, the degree of biodegradability of ordinary PLA and its composites does not appear to be high enough to merit its inclusion in certain items such as fast-food containers, which have a high likelihood of ending up as litter in the environment.

5.2.3 CAN PLA/CELLULOSE BE TRUSTED IN DEMANDING ENVIRONMENTS?

This chapter has so far focused on near-term consumer applications and opportunities for PLA-cellulose composites such as packaging. More broadly, olefin plastics find applications across a wide range of sectors as engineering plastics, with many examples in construction and automobiles. If engineering requirements can be met, then is it worth asking whether PLA-cellulose composites can be similarly adapted into applications such as a 3D-printed facade for a building (Köhler-Hammer *et al.* 2016). Moreover, it is often a consumer request for a bioplastic to be first used in a challenging environment, with the expectation this material can then biodegrade at the end of its useful life. Figure 5.3 illustrates some competing expectations that consumers may hold, sometimes simultaneously, with respect to such issues. This creates a further tension regarding the appropriateness of using PLA in exterior or demanding environments versus retaining desired end-of-life attributes.

As a starting point for this consideration, PLA has well-known deficiencies, including being prone to hydrolysis, brittle, and deforming under load or at a heat deflection temperature (Nakajima *et al.* 2017; Raquez *et al.* 2013). Positioning PLA-cellulose composites in any outdoor applications will need to overcome or minimize these deficiencies. PLA-wood composites have been evaluated in prototyping building and construction materials, including cladding (Dong *et al.* 2018; Dahy 2019). It can be expected that such composites are well placed to be utilized as extruded profiles in applications such as decking, like other WPCs. PLA has also been considered for automotive applications (Raquez *et al.* 2013). It would be expected that PLA-cellulose composites might be employed in automobiles, depending on whether PLA's susceptibility to brittleness can be overcome or via dampening of vibrations. While specific PLA deficiencies may be overcome through co-polymer addition, blending, or via cellulose reinforcement in engineering applications, the available research suggests that

FIGURE 5.3 It will be important to work on societal expectations with respect to (a) maximizing the service life of durable goods made from bioplastics and (b) the ability of those same items to rapidly biodegrade if inappropriately cast out or at the end of their service life.

any PLA use in demanding environments should be limited to non-structural, low impact uses, while also avoiding heat sources or high temperatures, which may be expected in direct sunlight.

Building on the above, if PLA-cellulose composites applications include their placement or utilization in exterior environments, then their exposure to weathering will prove challenging, as for any plastic (Köhler-Hammer *et al.* 2016). Additives and stabilizers are routinely added to plastics to protect and aid longevity in exterior applications. There have been decades of development of plastic additives for use in synthetic plastics such as PVC and PP, and the results of such efforts provide insights to adapting their suitability for PLA and cellulose composites. In the case of PLA, additives and stabilizers have been extensively investigated for enhancing processing thermal stability and UV protection (González-López *et al.* 2020), with this also including adaptation of a variety of bio-derived materials for these functions (Grigsby *et al.* 2013; Masek and Latos-Brozio 2018). Furthermore, it has been reported that the inclusion of nanocelluloses in modified forms can also provide protection roles for the PLA (Chang *et al.* 2013). Research is still emerging on whether native and or modified nanocelluloses provide greater efficacy. In contrast, while promoting PLA longevity on UV exposure, ultimately, it is the susceptibility to hydrolysis and biodegradability that will direct exterior longevity and product performance (Scaffaro *et al.* 2019). Overall, the current state of literature suggests there is potential for PLA-cellulose composites to be used in outdoor applications, but not all research and development is yet in place to be confident regarding exterior exposure and durability (Gardner *et al.* 2015) for wider commercial uptake of these PLA materials. As a final comment, consumer expectations regarding biodegradability at the end of life for composites in exterior applications will need to be tempered. At this stage, research has not been progressed to the point where it is possible to offer both PLA plastic and PLA-cellulose composite durability and also on-demand biodegradation.

The predominant use of PLA and PLA-cellulose composites will always be in consumer-facing applications such as packaging. Sustainability, branding, and consumer preference are at the forefront of this sector. However, in packaging applications, exposure to temperature extremes can be expected. Examples could include PLA use in packaging containers, films, and laminated forms stored at sub-ambient temperatures, or the filling of this PLA packaging with hot liquids. The mechanical properties of PLA will be variable over these processing and storage regimes, being dependent on temperature and moisture content, including the PLA Tg, melt, and heat deflection (Auras *et al.* 2004). Each of these PLA attributes is key to processing, use, and performance of PLA and PLA-cellulose composites as packaging materials and products. These attributes may also impact the suitability and use of these materials in temperature extremes, particularly in the frozen state or with hot water. PLA in film form has been evaluated in freeze-thaw cyclic testing, which can variously impact PLA hydrolysis (Lee and Rhim 2006). Research has tended to focus on containment and spoilage within PLA, with examples including plasticized PLA composites demonstrating reduced brittleness, retention of transparency, and mechanical properties (Râpă *et al.* 2016). With emerging research into the inclusion of modified celluloses in micro- and nanoforms, it can be envisaged that these composites will similarly provide performance benefits. Interestingly, PLA, in addition to other polyester-based materials, has been considered as a phase-change material for use in refrigeration (Perez-Masia *et al.* 2013).

5.2.4 WHY DO BLENDS OF PLA AND A SECOND POLYMER OFTEN YIELD SYNERGISTIC EFFECTS?

When product development engineers set out to formulate a film, they need to consider an assortment of different requirements that must be met simultaneously. It is unlikely that one pure plastic material will simultaneously satisfy all the requirements. Figure 5.4 presents an idealized concept in which two matrix components combined as a blend can achieve unique attributes relative to their recyclability, properties, and biodegradability. One of the inherent challenges faced by such engineers is the fact that polymers having different properties often fail to form good mixtures (Potschke and Paul 2003; Muthuraj *et al.* 2018; Bildik Dal and Hubbe 2021). According to Flory-Huggins theory (Flory 1942; Kuleznev *et al.* 1971), the free energy of typical polymer mixtures is most favorable when the unlike polymers are self-segregated into different phases. Because the segments of polymers have few degrees of freedom, there is relatively little gain in entropy upon mixing of the contrasting polymers. In this context, it can be difficult to explain how and why blends of PLA and various other polymers have been reported in the literature to provide various favorable combinations of properties, *e.g.* combinations of suitable modulus of elasticity with high enough stretch to breakage for various proposed applications. For example, a uniform blend of starch with poly(3-hyroxybutyrate), which is highly hydrophobic, has been reported.

One way to begin to explain the ability of certain combinations of polymers to achieve uniform blends, despite having different natures, is based on a phenomenon called molecular mimicry (Tolstoguzov 1999, 2003). This can be used to describe those combinations favorably developed within PLA blends. Certain polymers, including starch, are able to arrange their coiled conformations so as to adjust their solubility characteristics to suit differing environments. Such behavior appears to be a main mechanism by which biological organisms are able to biosynthesize a wide range of highly insoluble materials, such as cellulose, bone, and waxes, using processes that take place within aqueous environments of biological cells (Tolstoguzov 1999, 2003). Further evidence that starch, especially the amylose component of starch, can behave in this way comes from studies of inclusion complexes, which may involve PLA. The V-helix structures present in the amylose content of freshly

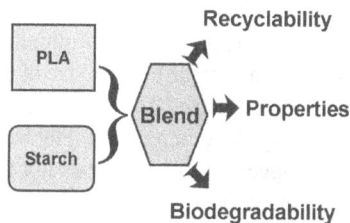

FIGURE 5.4 Blends of polymers, such as PLA with starches, offer prospects of achieving useful properties, high rates of biodegradation, and possible reprocessing routes.

gelatinized starch are so arranged that they can serve as hosts for hydrophobic substances such as lipids and for iodine oligomeric ions, which are associated with iodine staining procedures (Immel and Lichtenthaler 2000). Putseys *et al.* (2010) reviewed publications shedding light on starch inclusion complexes, including their ability to serve as hosts for fatty acids.

Another concept that may help to explain why useful blends can be prepared between PLA and other polymers having different average affinity properties is based on kinetics. The expected self-segregation of different polymers in a melt can be expected to take time. As a practical consideration, since both the PLA itself and any cellulosic reinforcing particles have only limited tolerances for the high temperatures required to melt PLA, the melt-extrusion, compounding, and molding processes are invariably set up so as to minimize the time during which the material is hot. Thermodynamic factors governing the onset of phase separation have been reviewed (Binder 1994). Based on the compositions, affinities, and molecular masses of a pair of polymers, as a function of temperature, it is possible to predict whether or not phase separation will occur, given sufficient time. Because PLA blends with other polymers are often prepared under high shear conditions, as in the case of melt-extrusion, there is a high likelihood of forming co-continuous phase structures (Potschke and Paul 2003). Although such structures eventually would disappear in a quiescent polymer melt, the relatively rapid cooling of most PLA-based polymer blends, in addition to their very high viscosities, would not provide sufficient time for this process to happen. There appears to be a great need for research to determine the existence and size of phase structures present in various blends of PLA and other polymers over a range of melt-processing conditions.

Based on the points made in the articles cited in this section, it is proposed that blends of PLA with other polymers can be effective when each of the polymeric materials can contribute useful attributes to the resulting blend, and that true and full mutual solubility is not required. Thus, if the PLA by itself it too hydrophobic and too brittle to meet all of the requirements, then a blend with a more hydrophilic and more compliant polymer can be considered. At the same time, it is clear from the cited studies that much remains unknown with respect to both theory and the nanostructural details of PLA-based blends. It follows that a lot of empirical testing will be needed in the formulation of polymer blends involving bioplastics such as PLA.

5.2.5 When Should Relatively Large Cellulose Particles Be Used in PLA-Cellulose Composites?

A review of cellulose particles incorporated into traditional olefin plastics found no advantages in mechanical performance regarding the size of the cellulosic reinforcement used in the composite (Hubbe and Grigsby 2020). It is reasonable to expect this also to be the case for cellulose-containing PLA composites (Hubbe *et al.* 2021). Based on this available research, it could be suggested that size does not matter, but this is only one aspect of PLA-cellulose composite performance. Practically, one must also consider the end-uses of PLA-cellulose composites. Such applications may include boxes and bins or as thin films. Beyond mechanical properties, there are aesthetic considerations, and the particle size may be critical in the visual appearance, surface texture, or water resistance

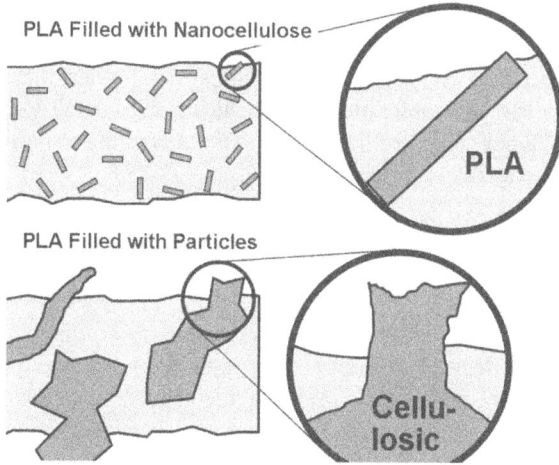

FIGURE 5.5 Relatively large cellulosic reinforcements (particles, fibers, etc.) are relatively low cost and can be surface-modified, but the composite surfaces will be generally rougher.

required in these applications. For example, Figure 5.5 considers a likely greater contribution to surface roughness when a polymer matrix such as PLA is filled with relatively large cellulosic fibers or particles, in contrast to using nanocellulose.

Researchers tend to produce smaller-scale composites and, as a result, the academic focus has been on mechanical testing and demonstrating the performance of PLA-cellulose composites with small, defined particles. Those considering PLA-cellulose composites will need to consider what is the maximum cellulose content used for each particle size category (Hubbe and Grigsby 2020). The available literature provides the expected mechanical performance for a range of composites comprising cellulose particles, size, content, and differing grades of PLA (Hubbe et al. 2021). However, consideration should be given to whether the cellulosic material is providing reinforcement or acting as a filler, interfacial adhesion with the PLA (Luedtke et al. 2019), aspect ratio (fines, particles, fibers), and whether particle attrition occurs on processing (Hubbe and Grigsby 2020). Beyond mechanical property enhancements, there has been research on the use of cellulosic inclusion to enhance the gas barrier properties of films and materials (Nazrin et al. 2020). Nanocellulose features predominantly in research for packaging applications compared to the use of larger-sized cellulose particles. Nanocellulose inclusion in PLA can reduce oxygen transmission up to 90%, with similar benefits to reducing moisture transmission (>75%) (Nazrin et al. 2020). High oxygen barrier efficacies have also been demonstrated in blown PLA films containing nanocelluloses (Karkhanis et al. 2018), and it can be expected this can be extended to PLA composites applied in laminated film form. Given the efficacy of nanoscale celluloses in gas barrier applications, this provides further direction and appropriateness of cellulose particle size for an intended functionality or use of the composite.

Overall, although research may suggest no disadvantages to mechanical properties of PLA composites varying in cellulose particle size, other factors will dictate which particle size is best used in the composite, including processing equipment, aesthetics, and practicalities. There will also be a need to match scales and processing equipment to opacity and visual appearance. Given that the food and packaging sectors are predominant users of the PLA-cellulose composites, these materials will also be a visual advertisement for products, so the inclusion of cellulose particles will need to maintain a physical appearance that is not detracting to performance, function, or visual appearance of products.

5.2.6 WHEN SHOULD ONE USE A SECOND TYPE OF FILLER IN ADDITION TO A CELLULOSIC IN PLA?

Substantially greater complexity is introduced into the formulation of a composite recipe whenever it is decided to employ two, rather than just one, types of reinforcing particles. In theory, it is conceivable that, due to some favorable combination of particle shapes or other attributes, a combination of two different kinds of particles might produce synergistic effects (Phillips 1976). However, an earlier search of the literature found only rare cases in which properties such as the strength or modulus of a hybrid composite exceeded what would have been predicted from a simple "rule of mixtures" linear calculation (Hubbe 2017). It follows that, when strength properties are of primary concern, it is probably a good idea to consider the use of just one type of reinforcing particle. By such an approach, one aims at making realistic modifications in properties to better meet and control the requirements for the product being manufactured. In particular, an analysis of the literature has shown that composites of PLA (and other plastics as well) with cellulosic particles are able to achieve higher tensile and flexural modulus values than the base polymer(s) alone, especially when suitable agents or treatments are used to enhance interfacial compatibility (Hubbe and Grigsby 2020).

Because of their predominantly fibrillar shapes, cellulosic reinforcing particles generally are not effective for increasing the tortuosity of the plastic matrix material to which they are added. In other words, the minimum diffusion path that must be taken by an average diffusing molecule (*e.g.* an oxygen molecule, water vapor, or grease, *etc.*) is not significantly lengthened by the presence of cellulosic particles (Hubbe *et al.* 2019). Nanosized reinforcing particles are typically considered for such barrier film applications. It has been argued that because nanoscale cellulosic reinforcing particles are often present at levels of 3% or less in plastic matrices (to avoid undesired agglomeration), and due to their fibrillar shape, their presence cannot account for the substantial reductions in permeability that sometimes are observed when they are incorporated into plastic composites (Khan *et al.* 2010).

A very extensive analysis of published data showed that only in the case of highly platy particles, such as montmorillonite clay, is there reliable evidence of increasing resistance to transport of permeant molecules through barrier films (Wolf *et al.* 2018). Even though the pattern was clear from overall averages, there was a severe scatter in data obtained from multiple studies. Such differences in results likely can be attributed to cases in which permeants were able to leak through defects, some of which might have been caused by the presence of the reinforcing particles. On the other hand, there appear to be sound reasons to employ combinations of cellulosic particles (aimed at strength properties) together with highly platy particles (aimed

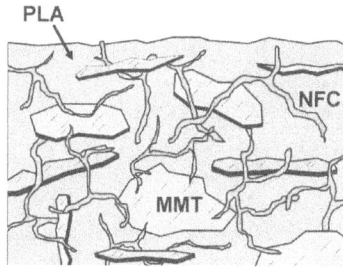

FIGURE 5.6 In a hybrid composite with nanofibrillated cellulose and montmorillonite (nanoclay), the fibrillar material provides reinforcement and the platy mineral slows down the diffusion of oxygen through the film.

at increasing the tortuosity), when the two kinds of goals are both important. Such combinations have been demonstrated at a fundamental level (Gamelas and Ferraz 2015; Trifol et al. 2016). This approach is illustrated in Figure 5.6.

Based on the literature cited in this section, it can be proposed that mixtures of reinforcing particles (or "filler" particles) might be justified when both strength issues and barrier issues need to be adjusted in the same composite system. However, the question of whether hybrid composite systems for PLA can be justified, from a practical standpoint, will require more research before it can be regarded as fully settled. That is because highly platy filler particles, especially if they have excellent interfacial compatibility with a matrix, also have the potential to contribute to increased modulus and strength values, at the same time that they are tending to enhance barrier properties (Jung et al. 2020; Chaiwutthinan et al. 2021).

5.2.7　SHOULD PILOT-SCALE PROCESSING BE USED WHEN DEVELOPING PLA COMPOSITES?

Researchers tend to work on smaller scales due to equipment and material requirements. PLA research literature is dominated by laboratory-scale composite processing and test specimen preparation in comparison to publications on large-volume PLA processing and product forms. In considering larger-scale processing and piloting materials, few publicly reported studies (Koppolu et al. 2019) are available on PLA and composite processing despite already large-scale uptakes in PLA packaging.

In industry, the advantages of processing at scale are well known, including benefits in design, prototyping, and testing. This can also extend to product evaluations and assessing the economics of processing and production (Ganeshan et al. 2001). Progression to larger-scale testing is well suited to the development of PLA-cellulose composites for applications identified in the challenging and demanding environments section. As illustrated in Figure 5.7, there often appears to be a "missing link" between academic research and the needs of industry, and this need may be addressed by increasing the amount of pilot-scale development and testing. For piloting PLA-cellulose processing, the production of large test specimens will be more uniform and less likely influenced by variations in tooling or molding conditions such as cooling. This is particularly the case for crystalline forms of PLA, for which

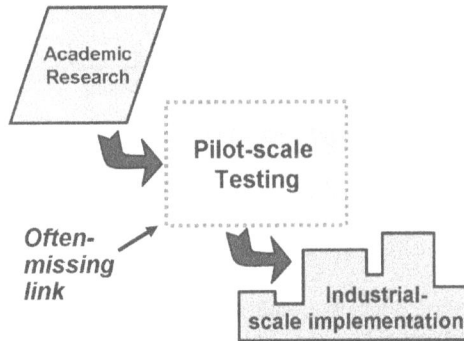

FIGURE 5.7   While there has been an acceleration of academic research related to bioplastics, there is a continuing need for pilot-scale testing, which can decrease risks associated with the implementation of commercial-scale production.

differences in processing conditions such as cooling rates can impact crystallinity and resulting polymer properties (Herrera *et al.* 2016). Production of larger samples will also provide opportunities to assess composites along a greater sample length as well as options for creating test specimens in suitable dimensions for more challenging mechanical property evaluations and testing of sample cross-sections. Pilot- and larger-scale composite manufacture and processing can allow for the testing of individual elements or fully fabricated materials such as composite construction materials. While industry uptake of PLA in packaging applications is widespread, it is reasonable to direct research and development into larger-scale PLA-cellulose composite processing evaluations. There are associated benefits and synergies to be gained by using equipment and approaches already pioneered in the more traditional processing of petrochemical plastics and composites.

5.2.8   WHERE OUGHT RESEARCHERS BE PLACING EMPHASIS WITH RESPECT TO APPLICATIONS OF PLA?

As has been documented in this chapter, substantial progress has been achieved in research and development related to PLA and PLA-cellulose composites and their introduction into some niche markets. It follows that risk avoidance or risk reduction should be a prominent focus of future research related to developing PLA and its composites. As noted by Leider and Rashid (2016), risks of technical failure can provide a strong disincentive for manufacturers to implement processes that involve new eco-friendly processes and materials. In principle, such risks can be reduced by carrying out experiments, especially when such tests are done at semi-commercial scales and speeds. Because research takes time and money, not every idea or concept will receive full attention nor be undertaken within desired timeframes. A general message that can be gained based on the findings considered in this chapter is that there are some quite diverse emerging and potential applications of PLA and its composites. It is reasonable to consider which applications fit best to the known properties of PLA and which applications are likely to benefit from additional research attention aimed at overcoming known deficiencies with PLA and mixtures and composites that contain PLA. Potential health risks would be another factor having the potential to slow down the adoption of a technology. However, in the case of PLA, safety assessments generally have been positive. Conn *et al.* (1995) reported only limited migration of PLA and its byproducts when used in houseware items and food packaging. Accordingly, the material is "generally recognized as safe" by the US Food and Drug Administration.

Further promising areas for the focus of future research follow from items already discussed in the preceding sections of this article. Selected items for research and further development are summarized in Table 5.1.

TABLE 5.1
Selected Areas for Future Research on PLA and Its Cellulosic Composites Based on Sources Considered in This Chapter

Area of Potential Research	Selected Citations
How to achieve interfacial compatibility between PLA and cellulosic surfaces in a cost-effective, eco-friendly manner at high production speeds	Yin *et al.* (2020); Hubbe and Grigsby (2020)
How to tailor the rate of biodegradation of PLA to meet different objectives, including the possible mitigation of harm that would result from littering or unplanned release into the ocean	Koh *et al.* (2018); Hubbe *et al.* (2021)
How to create durable PLA and PLA composites with tunable or on-demand biodegradation at the end of use or in the environment	Wadsworth *et al.* (2013); Wan *et al.* (2019)
How to optimize the usage of relatively large cellulosic reinforcing particles or fibers, minimizing the amount of surface area that would need to be treated for improved interfacial compatibility, while still achieving suitable surface smoothness and uniformity of PLA composites	Hubbe and Grigsby (2020); Choudhury *et al.* (2021)
How to enable PLA-based composites to be efficiently collected for reuse, recycling, or other beneficial end-fates, consistent with the goals of a circular economy	Wang *et al.* (2019b); Wojnowska-Baryla *et al.* (2020)
How to formulate PLA-based composites that are intended for extended use in the out-of-doors, refrigeration, or other challenging environments	Grigsby *et al.* (2013); Köhler-Hammer *et al.* (2016); González-López *et al.* (2020)
How to formulate PLA-based composites that are intended to achieve high barrier properties against the diffusion of oxygen	Karkhanis *et al.* (2018); Nazrin *et al.* (2020)
How to rapidly integrate and retain the dispersion of nanocellulose within PLA matrices while maintaining PLA film attributes including transparency	Hubbe and Grigsby (2020); Lee, Yun, and Park (2016); Rol *et al.* (2019)

5.3 CONCLUDING STATEMENTS

This review of the literature has revealed opportunities for researchers to develop some additional areas that are not traditionally served by plastics and composites research. This includes developing the economics of processing PLA-cellulose composites and new routes toward these exciting materials. In general, bioplastics are finding use in applications associated with high interactions with consumers. This includes 3D printing applications and boutique consumer goods. PLA and its composites can be a good fit in such applications due to their ability to be mass-produced, their film-forming ability, and their hydrophobic nature, which makes them suitable for the containment of liquids. Based on the discussion of the pivotal questions considered in this article, one can envision multiple paths of future research and future commercial implementation of PLA and its cellulose-based composites. In the near term, it is reasonable to expect a continuing emphasis on research related to relatively small-scale, specialty applications including the packaging sector. In the longer term, larger PLA composite products can be expected to play an increasing role, along with the development and increase of bioplastic production. Though composite formulation may contribute to some favorable properties of PLA, the widespread adoption of PLA composites for such applications as beverage containers will likely depend on the implementation of collection strategies and any regulations related to recycling and reuse, which represents a much more desirable end-fate compared to littering or landfilling. PLA-based items intended for outdoor use or other challenging environments are likely to be designed so as to maximize service life; possible end-of-life strategies can include incineration to recover the energy value. In cases where suitably rapid biodegradation is required, there is a need for more development and testing of PLA and its composites in blends with starch. Although nanocellulose is widely studied as a promising ingredient for PLA composites, especially after surface hydrophobization, its uptake and usage are likely to be limited to applications requiring a flow of the melt through small orifices (such as in 3D printing devices), formation of PLA films, or when the surface of a resulting composite needs to be smooth. For PLA-based barrier coatings, it seems likely that platy minerals such as nanoclay (montmorillonite) will gain increased usage, with or without simultaneous inclusion of cellulose or other fibrillar reinforcement particles.

REFERENCES

Abdel-Rahman, M. A., Tashiro, Y., and Sonomoto, K. 2013. Recent advances in lactic acid production by microbial fermentation processes, *Biotech. Advan.* 31(6):877–902.

Adom, F. K., and Dunn, J. B. 2017. Life cycle analysis of corn-stover-derived polymer-grade l-lactic acid and ethyl lactate: Greenhouse gas emissions and fossil energy consumption, *Biofuels Bioprod. Bioref.* 11(2):258–268.

Agarwal, M., Koelling, K. W., and Chalmers, J. J. 1998. Characterization of the degradation of polylactic acid polymer in a solid substrate environment, *Biotechnol. Prog.* 14(3):517–526.

Alves De Oliveira, R., Alexandri, M., Komesu, A., Venus, J., Rossell, C. E. V., and Maciel, R. 2020. Current advances in separation and purification of second-generation lactic acid, *Separ. Purif. Rev.* 49(2):159–175.

Arrieta, M. P., Samper, M. D., Aldas, M., and Lopez, J. 2017. On the use of PLA-PHB blends for sustainable food packaging applications, *Materials* 10(9): Article no. 1008.

Auras, R., Harte, B., and Selke, S. 2004. An overview of polylactides as packaging materials, *Macromol. Biosci.* 4:835–64.

Auras, R., Singh, S. P., and Singh, J. 2006. Performance evaluation of PLA against existing PET and PS containers, *J. Testing Eval.* 34(6):530–536.

Austin, H. P., Allen, M. D., Donohoe, B. S., *et al.* 2018 Characterization and engineering of a plastic-degrading aromatic polyesterase, *PNAS* 115(19):E4350–E4357.

Bildik Dal, A. E., and Hubbe, M. A. 2021. Hydrophobic copolymers added with starch at the size press of a paper machine: A review of findings and likely mechanisms, *BioResources* 16(1):2138–2180.

Binder, K. 1994. Phase transitions in polymer bends and block copolymer melts – Some recent developments, in: Theories and Mechanism of Phase Transitions, Heterophase Polymerizations, Homopolymerization, Addition Polymerization, Book Series, *Adv. Polym. Sci.* 112:181–299.

Bledzki, A. K., Reihmane, S., and Gassan, J. 1998. Thermoplastics reinforced with wood fillers: A literature review, *Polym. Plast. Technol. Eng.* 37:451–468.

Calvino, C., Macke, N., Kato, R., and Rowan, S. J. 2020. Development, processing and applications of bio-sourced cellulose nanocrystal composites, *Prog. Polym. Sci.* 103:101221.

Chaiwutthinan, P., Phutfak, N., and Larpkasemsuk, A. 2021. Effects of thermoplastic poly(ether-ester) elastomer and bentonite nanoclay on properties of poly(lactic acid), *J. Appl. Polym. Sci.* 138:Article no. e50443.

Chan, C. M., Vandi, L. J., Pratt, S., Halley, P., Richardson, D., Werker, A., and Laycock, B. 2018. Composites of wood and biodegradable thermoplastics: A review, *Polym. Rev.* 58: 444–494.

Chang, C. P., Wang, I. C., and Perng, Y. S. 2013. Enhanced thermal behavior, mechanical properties and UV shielding of polylactic acid (PLA) composites reinforced with nanocrystalline cellulose and filled with nanosericite, *Cellul. Chem. Technol.* 47:111–123.

Choudhury, M. R., Rao, G. S., Debnath, K., and Mahapatra, R. N. 2021. Analysis of force, temperature, and surface roughness during end milling of green composites, *J. Natural Fibers.* DOI: 10.1080/15440478.2021.1875350

Conn, R. E., Kostad, J. J., Borzelleca, J. F., Dixler, D. S., Filer, L. J., Ladu, B. N., and Pariza, M. W. 1995. Safety assessment of polylactide (PLA) for use as a food-contact polymer, *Food Chem. Toxicol.* 33(4):273–283.

Dahy, H. 2019. Efficient fabrication of sustainable building products from annually generated non-wood cellulosic fibres and bioplastics with improved flammability resistance, *Waste Biomass Valorization* 10:1167–1175.

Dong, Y., Milentis, J., and Pramanik, A. 2018. Additive manufacturing of mechanical testing samples based on virgin poly(lactic acid) (PLA) and PLA/wood fibre composites, *Adv. Manuf.* 6:71–82.

Doran-Peterson, J., Cook, D. M., and Brandon, S. K. (2008). Microbial conversion of sugars from plant biomass to lactic acid or ethanol, *Plant J.* 54(4):582–592.

Eiteman, M. A., and Ramalingam, S. (2015). Microbial production of lactic acid, *Biotech. Lett.* 37(5):955–972.

Farah, S., Anderson, D. G., and Langer, R. 2016. Physical and mechanical properties of PLA, and their functions in widespread applications: A comprehensive review, *Adv. Drug Delivery Rev.* 107:367–392.

Ferri, J. M., Garcia-Garcia, D., Sanchez-Nacher, L., Fenollar, O., and Balart, R. 2016. The effect of maleinized linseed oil (MLO) on mechanical performance of poly(lactic acid)-thermoplastic starch (PLA-TPS) blends, *Carbohydr. Polym.* 147:60–68.

Flory, P. J. 1942. Thermodynamics of high polymer solutions, *J. Chem. Phys.* 10: 51–61.

Gamelas, J. A. F., and Ferraz, E. 2015. Composite films based on nanocellulose and nanoclay minerals as high strength materials with gas barrier capabilities: Key points and challenges, *BioResources*. 10(4):6310–6313.

Gandini, A., and Lacerda, T. M. 2015. From monomers to polymers from renewable resources: Recent advances, *Prog. Polym. Sci.* 48:1–39.

Ganeshan, R., Kulkarni, S., and Boone, T. 2001. Production economics and process quality: A Taguchi perspective, *Int. J. Prod. Econ.* 71(1):343–350.

Gardner, D. J., Han, Y., and Wang, L. 2015. Wood–plastic composite technology, *Curr. For. Rep.* 1:139–150.

González-López, M. E., Martín del Campo, A. S., Robledo-Ortíz, J. R., Arellano, M., and Pérez-Fonseca, A. A. 2020. Accelerated weathering of poly(lactic acid) and its biocomposites: A review, *Polym. Degrad. Stab.* 179:109290.

Gorrasi G., and Pantani, R. 2013. Effect of PLA grades and morphologies on hydrolytic degradation at composting temperature: Assessment of structural modification and kinetic parameters, *Polym. Degrad. Stabil.* 98(5):1006–1014.

Grigsby, W. J., Bridson, J. H., Lomas, C., and Elliot, J. A. 2013. Esterification of condensed tannins and their impact on the properties of poly(lactic acid), *Polymers* 5:344–360.

Grzabka-Zasadzinska, A., Klapiszewski, L., Bula, K., Jesionowski, T., and Borysiak, S. 2016. Supermolecular structure and nucleation ability of polylactide-based composites with silica/lignin hybrid fillers, *J. Therm. Anal. Calorim.* 126(1):263–275.

Hakkarainen, M., Karlsson, S., and Albertsson, A. C. 2000. Rapid (bio)degradation of polylactide by mixed culture of compost microorganisms – Low molecular weight products and matrix changes, *Polymer* 41(7):2331–2338.

Hermansyah, H., Wijanarko, A., Dianursanti. *et al.* 2007. Kinetic model for triglyceride hydrolysis using lipase: Review, *Makara J. Technol.* 11(1):30–35.

Herrera, N., Salaberria, A. M., Mathew, A. P., and Oksman, K. 2016. Plasticized polylactic acid nanocomposite films with cellulose and chitin nanocrystals prepared using extrusion and compression molding with two cooling rates: Effects on mechanical, thermal and optical properties, *Compos. A Appl. Sci. Manuf.* 83:89–97.

Hubbe, M. A. 2017. Hybrid filler (cellulose/noncellulose) reinforced nanocomposites, in: *Handbook of Nanocellulose and Cellulose Nanocomposites*, Vol. 1, H. Kargarzadeh, I. Ahmad, S. Thomas, and A. Dufresne, eds., Wiley-VCH, Ch. 8, pp. 273–299.

Hubbe, M. A., and Grigsby, W. 2020. From nanocellulose to wood particles: A review of particle size vs. the properties of plastic composites reinforced with cellulose-based entities, *BioResources* 15(1):2030–2081.

Hubbe, M. A., Lavoine, N., Lucia, L. A., and Dou, C. 2021. Formulating bioplastic composites for biodegradability, recycling, and performance: A review, *BioResources* 16(1):2021–2083.

Hubbe, M. A., Rojas, O. J., Lucia, L. A., and Sain, M. 2008. Cellulosic nanocomposites, a review, *BioResources* 3(3):929–980.

Hubbe, M. A., Tyagi, P., and Pal, L. 2019. Nanopolysaccharides in barrier composites, in: *Advanced Functional Materials from Nanopolysaccharides*, N. Lin, J. T. Tang, A. Dufresne, and M. K. C. Tam, eds., Springer, Ch. 10.

Huda, M. S., Drzal, L. T., Mohanty, A. K., and Misra, M. 2007. The effect of silane treated- and untreated-talc on the mechanical and physico-mechanical properties of poly(lactic acid)/newspaper fibers/talc hybrid composites, *Compos. B Eng.* 38(3):367–379.

Immel, S., and Lichtenthaler, F. W. 2000. The hydrophobic topographies of amylose and its blue iodine complex, *Starch – Stärke* 52(1):1–8.

Jem, K. J., and Tan, B. 2020. The development and challenges of poly(lactic acid) and poly(glycolic acid), *Adv. Indust. Eng. Polym. Res.* 3(2):60–70.

Jung, B. N., Jung, H. W., Kang, D., Kim, G. H., and Shim, J. K. 2020. Synergistic effect of cellulose nanofiber and nanoclay as distributed phase in a polypropylene based nanocomposite system, *Polymers* 12(10): Article no. 2399.

Karkhanis, S. S., Stark, N. M., Sabo, R. C., and Matuana, L. M. 2018. Water vapor and oxygen barrier properties of extrusion-blown poly(lactic acid)/cellulose nanocrystals nanocomposite films, *Compos. A Appl. Sci. Manuf.* 114:204–211.

Khan, R. A., Salmieri, S., Dussault, D., *et al.* 2010. Production and properties of nanocellulose-reinforced methylcellulose-based biodegradable films, *J. Agric. Food Chem.* 58(13): 7878–7885.

Khosravi, A., Fereidoon, A., Khorasani, M. M., *et al.* 2020. Soft and hard sections from cellulose-reinforced poly(lactic acid)-based food packaging films: A critical review, *Food Packag. Shelf Life* 23: Article no. 100429.

Koh, J. J., Zhang, X. W., and He, C. B. 2018. Fully biodegradable poly(lactic acid)/starch blends: A review of toughening strategies, *Int. J. Biol. Macromol.* 109:99–113.

Köhler-Hammer, C., Knippers, J., and Hammer, M. R. 2016. Bio-based plastics for building facades, in: *Start-Up Creation: The Smart Eco-Efficient Built Environment*, Elsevier, Inc.

Koppolu, R., Lahti, J., Abitbol, T., Swerin, A., Kuusipalo, J., and Toivakka, M. 2019. Continuous processing of nanocellulose and polylactic acid into multilayer barrier coatings, *ACS Appl. Mater. Interfaces* 11:11920–11927.

Kuleznev, V. N., Krokhina, L. S., Oganesov, Y. G., and Zlatsen, L. M. 1971. Effect of molecular weight on mutual solubility of polymers, *Colloid J. USSR* 33(1):81.

Lagarón, J. M., Catala, R., and Gavara, R. 2004. Structural characteristics defining high barrier properties in polymeric materials, *Mater. Sci. Technol.* 20:1–7.

Lee, B. K., Yun, Y., and Park, K. 2016. PLA micro- and nano-particles, *Adv. Drug Deliv. Rev.* 107:176–191.

Lee, J. H., and Rhim, J. W. 2006. Polylactic acid coating affects the ring crush strength of linerboards, *Palpu Chongi Gisul/J. Korea Tech. Assoc. Pulp Pap. Ind.* 38:54–59.

Lee, S. H., Kim, I. Y., and Song, W. S. 2014. Biodegradation of polylactic acid (PLA) fibers using different enzymes, *Macromol. Res.* 22(6):657–663. DOI: 10.1007/s13233-014-2107-9

Leider, M., and Rashid, A. 2016. Towards circular economy implementation: A comprehensive review in context of manufacturing industry, *J. Cleaner Prod.* 115:36–51.

Liu, H. Z., and Zhang, J. W. 2011. Research progress in toughening modification of poly(lactic acid), *J. Polym. Sci. B Polym. Phys.* 49(15):1051–1083.

Luedtke, J., Gaugler, M., Grigsby, W. J., and Krause, A. 2019. Understanding the development of interfacial bonding within PLA/wood-based thermoplastic sandwich composites, *Ind. Crops Prod.* 127:129–134.

MacArthur, E., Waughray, D., and Stuchtey, M. R. 2021. https://www.mckinsey.com/~/media/McKinsey/dotcom/client_service/Sustainability/PDFs/The%20New%20Plastics%20Economy.ashx

Masek, A., and Latos-Brozio, M. 2018. The effect of substances of plant origin on the thermal and thermo-oxidative ageing of aliphatic polyesters (PLA, PHA), *Polymers* 10:1252.

Mazzoli, R. 2020. Metabolic engineering strategies for consolidated production of lactic acid from lignocellulosic biomass, *Biotech. Appl. Biochem.* 67(1):61–72.

Mertens, O., Gurr, J., and Krause, A. 2017. The utilization of thermomechanical pulp fibers in WPC: A review, *J. Appl. Polym. Sci.* 134:45161.

Muthuraj, R., Misra, M., and Mohanty, A. K. 2018. Biodegradable compatibilized polymer blends for packaging applications: A literature review, *J. Appl. Polym. Sci.* 135(24): Article no. 45726.

Nakajima, H., Dijkstra, P., and Loos, K. 2017. The recent developments in biobased polymers toward general and engineering applications: Polymers that are upgraded from biodegradable polymers, analogous to petroleum-derived polymers, and newly developed, *Polymers* 9:523.

Nazrin, A., Sapuan, S. M. Zuhri, M. Y. M., Ilyas, R. A., Syafiq, R., and Sherwani, S. F. K. 2020. Nanocellulose reinforced thermoplastic starch (TPS), polylactic acid (PLA), and polybutylene succinate (PBS) for food packaging applications, *Front. Chem.* 8:213.

Perez-Masia, R., Lopez-Rubio, A., Fabra, M. J., and Lagaron, J. M. 2013. Biodegradable polyester-based heat management materials of interest in refrigeration and smart packaging coatings, *J. Appl. Polym. Sci.* 130:3251–3262.

Phillips, L. N. 1976. The hybrid effect – Does it exist? *Composites.* 7(1):7–8.

Potschke P., and Paul, D. R. 2003. Formation of co-continuous structures in melt-mixed immiscible polymer blends, *J. Macromol. Sci. Polym. Rev.* C43(1):87–141.

Putseys, J. A., Lamberts, L., and Delcour, J. A. 2010. Amylose-inclusion complexes: Formation, identity and physico-chemical properties, *J. Cereal Sci.* 51(3):238–247.

Râpă, M., Mieluţ, A. C., Tănase, E. E., Grosu, E., Popescu, P., Popa, M. E., Rosnes, J. T., Sivertsvik, M., Darie-Niţă, R. N., and Vasile, C. 2016. Influence of chitosan on mechanical, thermal, barrier and antimicrobial properties of PLA-biocomposites for food packaging, *Compos. B Eng.* 102:112–121.

Raquez, J. M., Habibi, Y., Murariu, M., and Dubois, P. 2013. Polylactide (PLA)-based nanocomposites, *Prog. Polym. Sci.* 38:1504–1542.

Reddy, G., Altaf, M., Naveena, B. J., Venkateshwar, M., and Kumar, E. V. 2008. Amylolytic bacterial lactic acid fermentation – A review, *Biotech. Adv.* 26(1):22–34.

Rol, F., Belgacem, M. N., Gandini, A., and Bras, J. 2019. Recent advances in surface-modified cellulose nanofibrils, *Prog. Polym. Sci.* 88:241–264.

Saba, N., Jawaid, M., and Al-Othman, O. 2017. An overview on polylactic acid, its cellulosic composites and applications, *Curr. Organic Synth.* 14(2):156–170.

Scaffaro, R., Maio, A., Sutera, F., Gulino, E. F., and Morreale, M. 2019. Degradation and recycling of films based on biodegradable polymers: A short review, *Polymers* 11: 651.

Siemann, U. 1992. The solubility parameter of poly(DL-lactic acid), *Eur. Polym. J.* 28(3):293–297.

Stuchtey, M. R., and Dillion, T. 2021. https://www.pewtrusts.org/-/media/assets/2020/07/breakingtheplasticwave_report.pdf

Sun, J. Y., Shen, J. J., Chen, S. K., Cooper, M. A., Fu, H. B., Wu, D. M., and Yang, Z. G. 2018. Nanofiller reinforced biodegradable PLA/PHA composites: Current status and future trends, *Polymers* 10(5): Article no. 505.

Tolstoguzov, V. 1999. Origins of globular structure in proteins. *FEBS Lett.* 444:145–148.

Tolstoguzov, V. 2003. Thermodynamic considerations of starch functionality in foods. *Carbohydr. Polym.* 51(1):99–111.

Trifol, J., Plackett, D., Sillard, C., Szabo, P., Bras, J., and Daugaard, A. E. 2016. Hybrid poly(lactic acid)/nanocellulose/nanoclay composites with synergistically enhanced barrier properties and improved thermomechanical resistance, *Polym. Intl.* 65(8): 988–995.

Vaidya, A. N., Pandey, R. A., Mudliar, S., Kumar, M. S., Chakrabarti, T., and Devotta, S. 2005. Production and recovery of lactic acid for polylactide – An overview, *Crit. Rev. Environ. Sci. Technol.* 35(5):429–467.

Wadsworth, L. C., Hayes, D. G., Wszelaki, A. L., Washington, T. L., Martin, J., Lee, J., Raley, R., Pannell, C. T., Dharmalingam, S., Miles, C., Saxton, A. M., and Inglis, D. A. 2013. Evaluation of degradable spun-melt 100% polylactic acid nonwoven mulch materials in a greenhouse environment, *J. Eng. Fibers Fabr.* 8(4):50–59.

Wan, L., Li, C. X., Sun, C., Zhou, S., and Zhang, Y. H. 2019. Conceiving a feasible degradation model of polylactic acid-based composites through hydrolysis study to polylactic acid/wood flour/polymethyl methacrylate, *Compos. Sci. Technol.* 181: Article no. 107675.

Wang, Q., Yao, Q., Liu, J., Sun, J., Zhu, Q., and Chen, H. 2019a. Processing nanocellulose to bulk materials: A review, *Cellulose* 26:7585–7617.

Wang, X., Zhao, N. J., Yin, G. F., Meng, D. S., Ma, M. J., Yu, Z. M., Shi, C.-Y., Qin, Z. S., and Liu, J. G. 2019b. Classification and identification of plastic with laser-induced fluorescence spectroscopy based on back propagation neural network model, *Spectroc. Spec. Anal.* 39(10):3136–3141.

Wojnowska-Baryla, I., Kulikowska, D., and Bernat, K. 2020. Effect of bio-based products on waste management, *Sustainability* 12(5): Article no. 2088.

Wolf, C., Angellier-Coussy, H., Gontard, N., Doghieri, F., and Guillard, V. 2018. How the shape of fillers affects the barrier properties of polymer/non-porous particles nanocomposites: A review, *J. Membr. Sci.* 556: 393–418.

Yin, Y.-Y., Lucia, L. A., Pal, L., Jiang, X., and Hubbe, M. A. 2020. Lipase-catalyzed laurate esterification of cellulose nanocrystals and their use in reinforcement in PLA composites, *Cellulose* 27:6263–6273.

Zhang, Y. C., Wu, H. Y., and Qiu, Y. P. (2010). Morphology and properties of hybrid composites based on polypropylene/polylactic acid blend and bamboo fiber, *Bioresour. Technol.* 101(20):7944–7950.

6 Morphology of PLA/ Cellulose Composites

Sabarish Radoor, Aswathy Jayakumar, and Suchart Siengchin
King Mongkut's University of Technology North Bangkok
Bangkok, Thailand

Amritha Bemplassery
National Institute of Technology
Calicut, India

Jasila Karayil
Government Women's Polytechnic College
Calicut, India

Jyothi Mannekote Shivanna
Dayananda Sagar College of Engineering
Bengaluru, India

Jyotishkumar Parameswaranpillai
Alliance University
Bengaluru, India

CONTENTS

6.1 INTRODUCTION

The demand for green composite is increasing as the synthetic polymers are associated with serious environmental impact, high cost, and high energy requirements (Vroman and Tighzert 2009). Biocomposites are fabricated from natural polymers,

DOI: 10.1201/9781003160458-6

which is cheap and biodegradable material (Rogina 2014, DeFrates et al. 2017). PLA is one of the most explored biopolymers which is extracted from a renewable resource, corn starch (Shogren et al. 2003). The attractive features of PLA include good transparency, desirable mechanical properties, and good processability (Hu and Lim 2016, Tham et al. 2016, Geng et al. 2020). Owing to its excellent properties, PLA was considered as an alternative for petrochemically derived polymer (Cheng et al. 2009, Chen et al. 2016, Li et al. 2020). PLA-based systems have been commercialized and used in industries such as textiles, food packaging, medical field (Lopes e al. 2012, Tawakkal et al. 2014, Lunt and Shafer 2016, Radoor et al. 2021). PLA-based systems are bio-degradable and will degrade completely after 120 days in compost. The properties of PLA, especially barrier properties, thermal stability, and mechanical properties, could be improved by reinforcing it with nanoparticles (Carrasco et al. 2010, Fukushima et al. 2013). Cellulose nanofibers (CNFs) are one of the best choices to reinforce polymer matrix (Saito et al. 2007). Cellulose is an eco-friendly and inexpensive material extracted from plants (Somerville 2006). The additional features of CNFs are its desirable aspect ratio, high specific stiffness and strength, high sound attenuation, high recyclability, low energy consumption, easy processability, low density, and possess high stiffness (Abdollahi et al. 2013, Han et al. 2013, Fattahi Meyabadi et al. 2014, Li et al. 2015, Sabarish and Unnikrishnan 2018, Ramesh and Radhakrishnan 2019, Radoor et al. 2020b). The previous report shows that the incorporation of CNF in PLA improves its oxygen barrier and mechanical properties. However, the poor dispersion and high hydrophilic nature of CNF limits their reinforcing effect. The compatibility between hydrophilic CNF and hydrophobic PLA is generally improved by surface treatment such as sodium hydroxide, acetylation. The surface modification enhances the roughness and decreases the hydrophilicity of CNF fibers. These features make CNF a promising candidate as a reinforcement filler for polymers (Oksman et al. 2006, Sullivan et al. 2015).

Morphological analysis is done to understand the fiber characteristic (diameter and structure); it also gives an idea about the dispersion of fiber in the polymer matrix and the compatibility of fiber with the polymer matrix (Singha and Thakur 2009). An idea of fiber-matrix compatibility is essential to explain the mechanical properties and viscoelastic properties of composites. This chapter gives a brief discussion on the morphology of PLA/cellulose composites. The morphology of composites was explained using techniques such as SEM, TEM, and AFM.

6.2 SCANNING ELECTRON MICROSCOPY (SEM)

Scanning electron microscopy (SEM) is a topographic technique that uses a beam of electrons to provide the surface morphology and composition of specimens. In this technique, the sample is bombarded with a beam of electrons. The interaction of the electrons with the specimen surface emits high-energy backscattered electrons, low-energy secondary electrons, and Auger electrons *SEM uses backscattered electrons and secondary electrons* for developing the 2D image of the sample. An electroconductive surface is required for SEM analysis. Therefore, non-conducting samples are coated with a conducting layer to improve their conductivity (Zhou et al. 2006, Inkson 2016, Radoor et al. 2020a).

Shi et al. (2012) developed a biodegradable poly(lactic acid) (PLA)/cellulose nanofibrils composite through electrospinning process. They reported that cellulose content has a significant effect on the physical properties of the composite. The SEM analysis revealed the formation of smooth and homogenous fiber without any beads. Uniform distribution of cellulose on the PLA was noted. However, at 10 wt.%, cellulose particles started to aggregate. The fiber diameters of pure PLA, PLA/1 wt.% cellulose, PLA/2 wt.% cellulose, PLA/32 wt.% cellulose, PLA/5 wt.% cellulose, and PLA/10 wt.% cellulose are 1540, 1046, 826, 642, and 405 nm, respectively. The decrease in average fiber diameter with cellulose content was explained on the basis of conductivity and viscoelasticity of the electrospinning solution. The highly conductive solution when passed through the spinneret generates a jet with high surface charge density. Consequently, the repulsive force between polymer chains in the electric field increases. The high repulsive force split the jet into a small and thin one. The conductivity of PLA and PLA/10 wt.% cellulose solution are respectively 0.45 and 0.66 (s cm^{-1}). The SEM image also indicates the formation of pores in PLA/CNFs (Figure 6.1). This is attributed to the volatilization of solvent during the spinning process. A similar observation was noted by Yang et al. (2019). However, at high cellulose content, the fiber diameter increases. This is because at high CNF content, the particle gets entrapped more into the PLA fiber and therefore, the fiber diameter

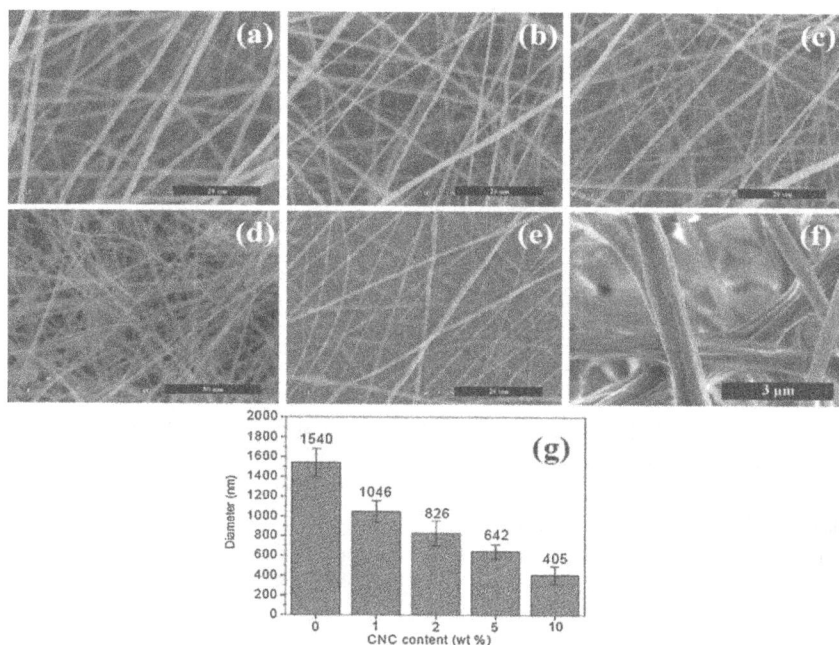

FIGURE 6.1 SEM images of PLA with different wt.% of CNC: (a) plain PLA, (b) PLA/1 wt.% CNC, (c) PLA/2 wt.% CNC, (d) PLA/5 wt.% CNC, (e) PLA/10 wt.% CNC, (f) PLA/5 wt.% CNC (high magnification), and (g) graph plotted against CNC content vs diameter (nm) for PLA/CNC fibers. (Reproduced with permission from Elsevier, License Number: 5118011384792; Shi et al. 2012.)

increases. Furthermore, the conductivity of the solution is also gets reduced at high cellulose content. The TEM observation was complementary to the SEM analysis. The CNF particles with a diameter around 500 nm were uniformly dispersed in the PLA fiber. The thermal and mechanical properties of microcrystalline cellulose (MCC)/polylactic acid (PLA) composite were significantly improved by the addition of a low cost and environment friendly filler; activated biochar (AB) (Zhang et al. 2020). In the absence of MCC, the PLA possess a smooth and flat structure. The SEM morphology of the PLA/MCC composite shows that the cellulose fiber was wrapped in the PLA matrix. However, a few cracks were seen on the composites; indicting weak interfacial interaction between PLA and cellulose fiber. The effect of filler on the morphology of PLA/MCC composites was also studied. The analysis shows that the addition of AB generates pores on the composites and the composites loaded with high AB, possess a large number of pores. Thus, it can be concluded that the biochar was an effective reinforcement for PLA/cellulose composites. The better physical/mechanical interlocking in AB incorporated system is responsible for its high mechanical properties (Figure 6.2).

Ying et al. (2018) investigated the effect of surface-treated CNF on PLA matrix. The surfaced modification was done through coupling and acetylation reaction. The TEM analysis indicates that acetylation decreases the transverse modules of CNF fiber. However, coupling treatment does not affect the morphology and transverse modulus of the fiber. The morphological analysis further confirmed that surface treatment improves the dispersion of particles in the polymer matrix. Thus, surface-treated fiber has good compatibility with the matrix. The formation of hydrogen bond between acetylated CNFs and PLA is responsible for increasing the affinity between fiber and matrix. Thus, surface-treated composites possess high mechanical and thermal properties. Furthermore, a high T_g value was obtained for surface-treated composites. High T_g value also indicated better compatibility between fiber and matrix.

PLA can exist in three forms, PLLA, PDLA, and poly(DL-lactide) PDLLA. In their study, Wang and Drzal (2012) used amorphous PDLLA. The compatibility of PDLLA with CNF was improved by converting it into PDLLA microparticle. The diameter of CNF and PDLLA was found to be in the order 20–30 nm and 150–1000 nm, respectively. PDLLA-CNF composite with 8 wt.%, 15 wt.%, and 32 wt.% were developed. The mechanical analysis shows that with increasing the wt.% of CNF, the flexural modulus and strength of the composite also increase. This indicates good dispersion of CNF in polymer and was later confirmed by SEM analysis. After flexural testing, composites displayed a rough fracture surface and hairy appearance through the surface of the composite. This indicates that CNF was well dispersed in the polymer matrix. However, fiber pull out and several cracks and voids were seen on the surface of fractured composites. This could be due to weak interfacial adhesion between fiber and matrix. Pracella et al. (2014) enhanced the adhesion and compatibility of PLA and cellulose by functionalizing cellulose with glycidyl methacrylate (GMA) or dispersing cellulose into PVAc (poly (vinyl acetate)). The morphology of cellulose nanocrystals (CNCs) was confirmed by TEM and AFM analysis. The rod-shaped CNC with a diameter in the order 200–400 nm were dispersed in the PLA matrix. Furthermore, SEM analysis revealed that at

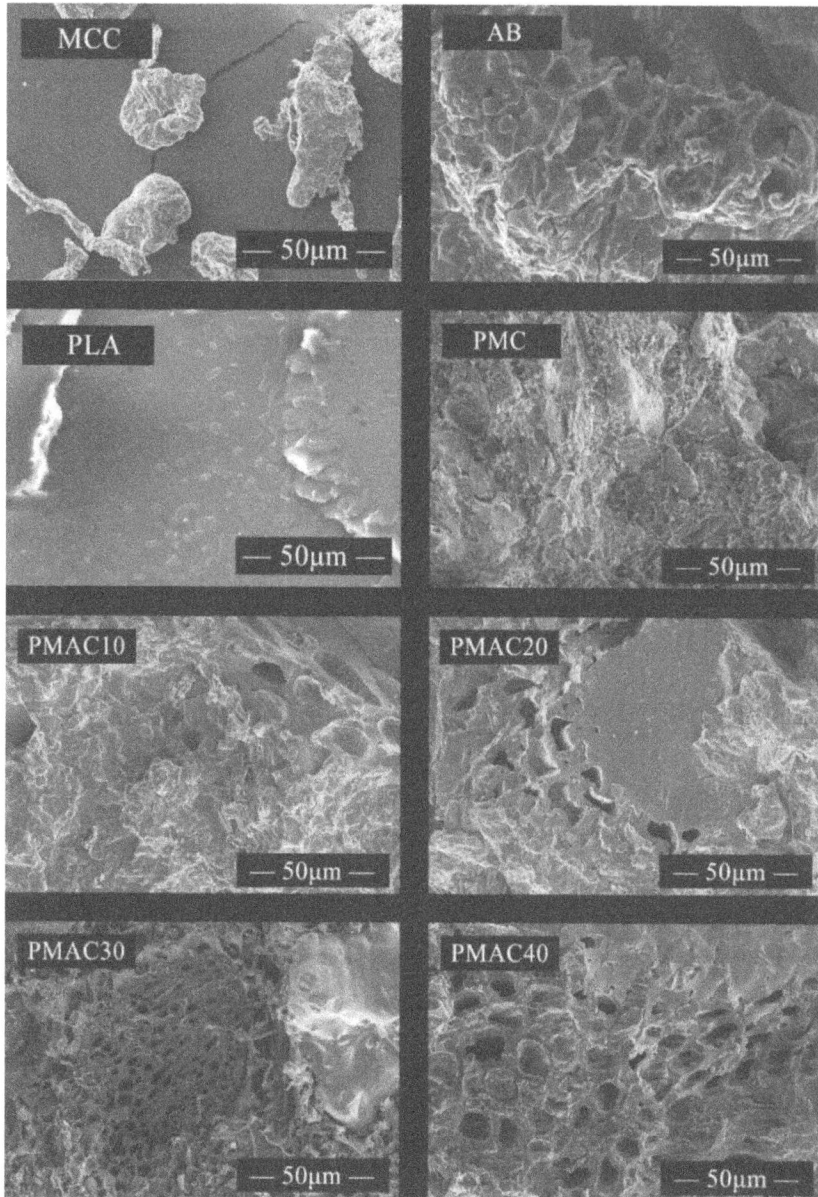

FIGURE 6.2 SEM micrographs of plain MCC, AB, PLA, and PMC with different blend ratio MCC/AB (%) composites. (Reproduced with permission from Elsevier, License Number: 5118020341141; Zhang et al. 2020.)

low concentrations, the nanocellulose was uniformly dispersed in the PLA matrix. However, the presence of aggregates was clearly visible in the composite with a high concentration of cellulose. Moreover, good dispersion was seen in composites that contained functionalized cellulose or cellulose dispersed in PVAc. The mechanical

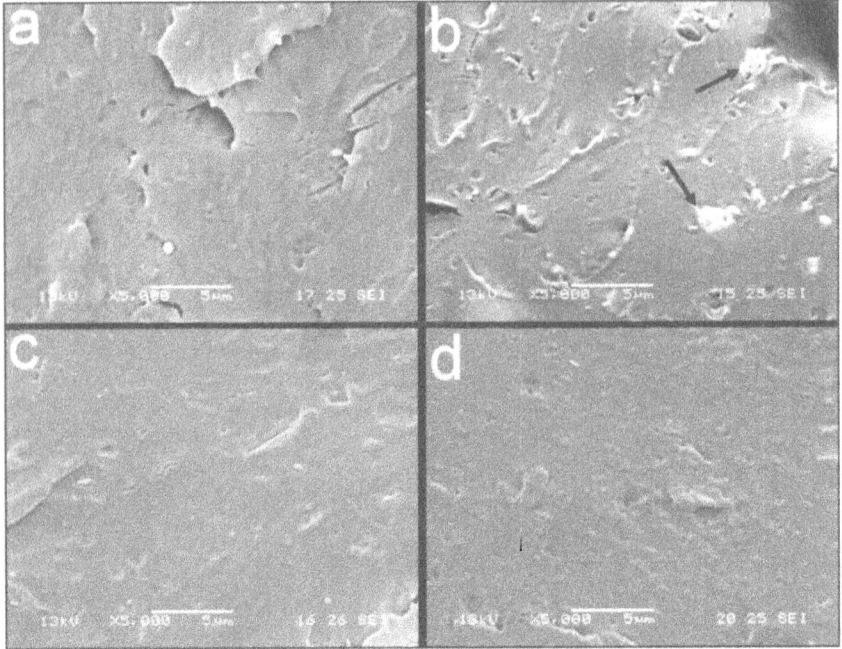

FIGURE 6.3   SEM images of PLA/CNC and PLA/PVAc/CNC nanocomposites: (a) CNC-1, (b) CNC-3, (c) CNC-GMA-3, and (d) CNC*-3. (Reproduced with permission from Elsevier, License Number: 5167741383892; Pracella et al. 2014.)

data well correlate with the morphological analysis. It was noted that at low CNC content, the composite possesses high modulus, whereas a decrease in modulus value was observed at high CNC content. This clearly indicates the presence of agglomeration of the nanocrystal into the polymer matrix (Figure 6.3). Niu et al. (2018) adopted the layer-by-layer method to develop a ternary composite film; CNF/PLA/chitosan. The compatibility of CNFs and their dispersion in polymer-matrix was improved by modifying it with rosin, a natural product obtained from the pine resin. Due to the synergistic effect of rosin modified cellulose nanofiber (R-CNF) and chitosan (CHT), the composite film displayed better mechanical, thermal, and antimicrobial properties. Owing to its brittle nature, pure PLA has a smooth and clear fracture surface. In the case of R-CNF (rosin modified PLA/CNT composite), the surface was transformed into rough and the roughness increases with an increase in the concentration of R-CNF (rosin modified). However, when the concentration of R-CNF reaches 10%, the particle began to aggregate and is clearly visible in the SEM image (Figure 6.4). Thus 10% R-CNF loaded film has a poor binding effect and thereby results in lower mechanical properties. A similar observation was noted by Jonoobi et al. (2012) in CNF reinforced PLA composite. The CNF was well dispersed in the polymer matrix at low concentration (1 wt.% and 3 wt.%). But agglomeration appears on nanocomposite with 5 wt.% of cellulose. The same group also compared the morphology of acetylated (PLA/ACNF) and non-acetylated PLA (PLA/CNF)

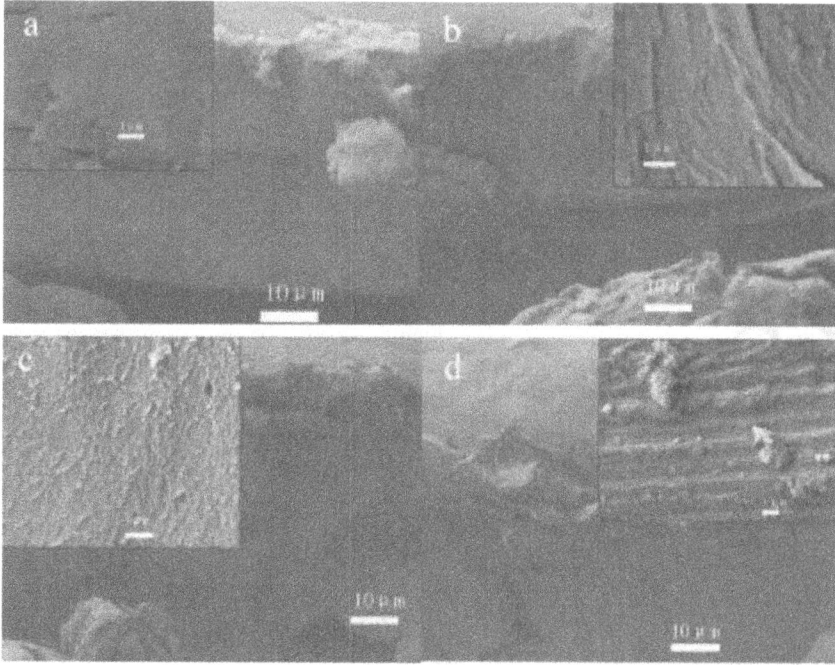

FIGURE 6.4 SEM micrographs of composite films with (a) 0 wt.%, (b) 8 wt.%, (c) 10 wt.% R-CNF loadings, and (d) CNF/PLA/CHT with 8 wt.% CNF loading. (Reproduced with permission from Elsevier, License Number: 5118020142374; Niu et al. 2018.)

composite. The analysis indicates that both PLA/CNF and PLA/ACNF composite have a rough surface. The particle tends to aggregate when the concentration of CNF reaches 5 wt.% (Jonoobi et al. 2010). Huda et al. (2005) developed green bio-composite from PLA and recycled cellulose fibers through the injection molding process. Three different cellulose fibers, namely TC 1004, R 0083, and R 0084 fiber, were used. The mechanical results show that the tensile and flexural moduli of the bio-composite were higher than the neat resin. This was due to the well dispersed fiber in the polymer matrix, which was confirmed by SEM analysis. Furthermore, prominent fiber pull out and large void between fiber and matrix is evident from SEM micrograph of PLA/TC 1004 composites. However, PLA/R 0083 and PLA/R 0084 displayed less gap between fiber and matrix and hence possessed better tensile properties than PLA/TC 1004 composites.

Fortunati et al. (2012) employed unmodified and surfactant-modified CNC to reinforce PLA. Modified composites have a superior barrier, thermal, and crystallinity when compared with unmodified ones. The TEM image of the unmodified composite showed a cellulose cluster. This is due to the formation of an intermolecular hydrogen bond between the CNCs. However, cluster formation was absent in the surfactant-modified sample. Well-defined single crystal formation was seen and it indicates good dispersion of cellulose crystal on the PLA matrix. The migration test was supported by FE-SEM. They observed that after incubating both modified and unmodified composites for 10 days in ethanol solution (10%), the clear surface erosion with some fractures and holes around 10 µm diameter in size were observed.

6.3   TRANSMISSION ELECTRON MICROSCOPY

Transmission electron microscopy (TEM) has evolved and developed into an inevitable tool in almost all fields of modern-day science. Invented by Ernst Ruska and Max Knoll in 1931, this microscopic technique is based on transmitting a beam of electrons through a specimen to produce an image (Tang and Yang 2017). These images of ultrathin samples provide valuable information regarding the fine structure and morphology of specimens. The technique is highly beneficial in characterizing soft matter and membrane-based materials as nanoscale images are obtained (Tang and Yang 2017). Further advancements like cryo-TEM and cryo-FIB milling extended the possibilities of TEM in a very positive way.

Haafiz et al. studied about the PLA reinforced cellulose nanowhiskers (CNWs) (Figure 6.5) (Haafiz et al. 2016). Cellulose nanoparticles have attracted great attention owing to their renewability, biocompatibility, ease of chemical and mechanical modification, etc. (Li et al. 2009, Fahma et al. 2010, Eichhorn 2011, Fernandes et al. 2011). The CNWs possess several superior qualities like excellent mechanical strength, biodegradability, optical, etc. (Azizi Samir et al. 2005, Goetz et al. 2009). Hence it would be highly advantageous to incorporate PLA with CNW-S (CNW isolated from oil palm empty fruit bunches) in order to enhance the properties of composites (Haafiz et al. 2016). TEM analysis of composites furnished a clear idea about the changes in mechanical properties of PLA when incorporated with CNW-S. For the analysis, they prepared 60 mm thick specimens using Leica Ultracut ultramicrotome with a diamond knife. An accelerating voltage of 120 kV was used for the examination of samples. The TEM images revealed that the PLA displayed smooth and clean surface features. When 3 phr CNW-S loading was done, the homogeneous dispersion of filler inside PLA matrices was clearly visible. This suggested the reason for good tensile strength. However, aggregation of CNW-S was revealed at higher filler loading of 5 phr, which could be the reason for the reduced mechanical properties of PLA-CNW-S5.

The synthesis of CNCs from MCC powder using acid hydrolysis was presented by Khoo et al. (2016). Nanocelluloses are used in polymer matrices as reinforcement material owing to their good thermal and mechanical properties (Lee et al. 2009). The morphology of CNC and MCC were examined using energy filtered transmission electron microscope (EFTEM). TEM micrographs of MCC powder before acid hydrolysis and CNC revealed striking morphological changes. While MCC powder was observed as irregularly shaped particles of up to 200 nm width and 400 nm length. CNC demonstrated a rod-like structure with about 250 nm length and 10 nm width. These images could assure that acid hydrolysis had taken place, which reduced fiber sizes to nanoscale. In addition to that, agglomeration of cellulose at some places due to the drying process pointed out the importance of drying of CNCs. This step is a major challenging step in order to use the material reliably in making products.

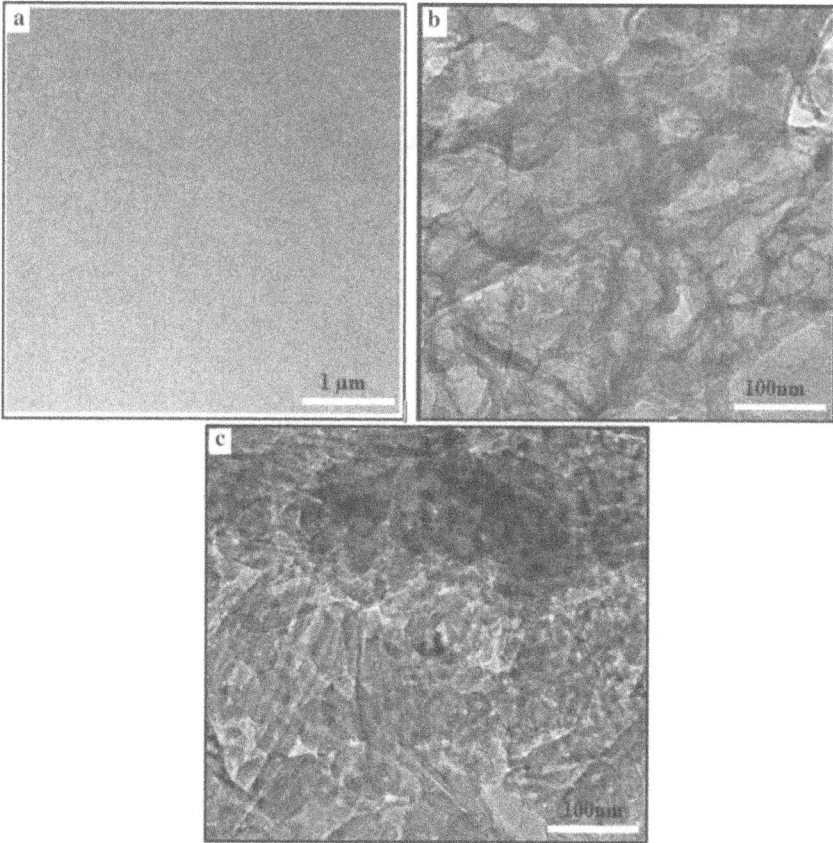

FIGURE 6.5 TEM images for (a) PLA, (b) PLA-CNW-S 3, and (c) PLA-CNW-S 5. (Reproduced with permission from Elsevier, License Number: 5118051467802; Haafiz et al. 2016.)

Liu et al. (2018) prepared PLA/CNW composite nanofibers by the electrospinning of uniform PLA/CNW mixtures. TEM was employed in detecting the morphological structures of composite nanofibers at an accelerating voltage of 100 kV. The micrographs revealed that sulfuric acid hydrolysis of cellulose produced rod-like fragments with a diameter of 20 nm and 300 nm length. Crystalline phases remained after hydrolysis of amorphous phases. Stained CNWs which are aligned along PLA nanofibers were observed to be uniformly distributed within matrix fibers.

There are numerous reports that are based on the preparation as well as morphological, thermal and thermomechanical properties of PLA/CNW nanocomposites. Hossain et al. (2011) used solvent casting method to prepare a range of cellulose nanocomposites from CNWs and PLA. They concentrated on the effect of different concentrations of nanowhiskers on the PLA matrix. The morphology of the nanocomposites was analyzed using TEM. The rod-like characteristics of nanowhiskers were revealed in the TEM images of CNWs. Here also, acid hydrolysis produced crystalline portions by hydrolyzing amorphous portions. Hence nanosized crystalline rod-like whiskers were obtained with an average length and width of 300 and 10 nm, respectively. CNWs were observed to be clearly dispersed in the polymer matrix, which was confirmed by the aggregated small bundles in TEM images. This established strong self-association of CNWs through hydrogen bonding of their hydroxyl groups and their high surface area (Thielemans et al. 2009). Arslan et al. (2020) prepared PLA/CNC nanocomposites using different concentrations of CNC, i.e., 1, 3, and 5 wt.%, by employing solution casting (SC) as well as dilution of solution cast PLA/CNC in a twin-screw extruder (TSE). To verify the importance of the preparation method of nanocomposites on the dispersion quality of CNC and hence the final properties of the resulting nanocomposites, they studied the morphological, rheological, and crystallization properties and compared. Morphological characterization was done using nanocomposites with 5 wt.% of CNC. For morphological analysis, they took microtomed slices of thickness 60–80 nm. TEM images exhibited better CNC dispersion in nanocomposites prepared through SC method. However, there is a possibility of re-agglomeration and hence the samples prepared via TSE still produced nanocomposites with better CNC dispersion quality. TEM could thus be concluded as an inevitable tool for analyzing the surface morphology of polymer nanocomposites.

6.4   ATOMIC FORCE MICROSCOPY (AFM)

Atomic force microscopy is the most widely used scanning probe microscopy technique. It was developed first by G. Binnig, Ch. Gerber, and C. F. Quate in 1985. This technique is used to produce a topographical image of a surface with atomic resolution. It can determine film's thickness, image non-conducting surfaces, imaging live, dead cells, etc. (Giessibl 2003). In short, AFM measures and manipulates matter at the nanoscale. AFM uses a tip on a cantilever that scans the surface of the sample. A position-sensitive photo-diode detects the deflection of the cantilever based on the raised and lowered features of the sample surface (Dixson et al. 2018). This helps to generate an accurate topographic map of the surface features of the sample. This probing technique is far better than other electrical and optical microscopy techniques. AFM finds applications in semiconductor science and technology, polymer chemistry, solid-state physics, molecular biology, etc.

Atomic force microscopy was made use of in analyzing the morphology of CNWs. A dilute drop of CNW-S suspension (0.1 mg/100 ml) was dispersed on the mica surface and dried before analysis (Haafiz et al. 2016). The AFM images of oil palm empty fruit bunches microcrystalline cellulose (OPEFB-MCC) were also analyzed using AFM and regular spherical particles were observed. The morphological changes on OPEFB-MCC after hydrolysis treatment were highly contrasting rod-like in appearance. Thus, the production of individual CNW-S from OPEFB-MCC with a good aspect ratio was confirmed using AFM analysis. Pracella et al. (2014) examined the morphology features of nanocellulose using AFM operating in contact mode in air. A diluted droplet of solution of CNC was cast on a microscopic slide and vacuum dried. AFM images clearly showed the presence of micrometric sized extended CNC particles and isolated nanofibers of diameter 34 nm. This observation was in close agreement with TEM analysis of the samples (Figure 6.6).

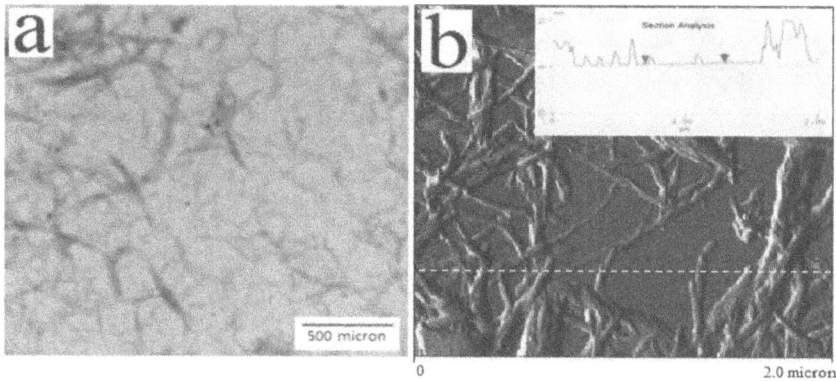

FIGURE 6.6 (a) TEM and (b) AFM images of cellulose nanocrystals (Insert: section analysis (along dashed line). (Reproduced with permission from Elsevier, License Number: 5167741383892; Pracella et al. 2014.)

In order to investigate the polymer composites containing different types of cellulose fillers, AFM was very effectively used (Frone et al. 2013). Owing to their valuable abilities to measure forces up to piconewton ranges, AFM by PeakForce QNM (Quantitative Mechanical Property Mapping at the nanoscale) mode was considered as the aptest method to estimate the nanomechanical properties of material surfaces. AFM images of both CNFs and PLA/CNF composites were studied (Frone et al. 2013). Samples used for analysis were composite films with uniform surface and thickness of about 30 µm. Scanning rates of 1.4 Hz and scan angle of 90° were used for recording images. The inferences from AFM images are that there was an excellent degree of crystallinity of PLA composites containing untreated nanofibers compared to silane-treated ones. QNM analysis revealed that the silane-treated CNFs have finer dispersion in PLA than untreated ones.

Fazeli et al. (2018) discussed the preparation and characterizations of polysaccharide-based biocomposite films produced from CNFs and starch. The thermoplastic starch matrix-based composites were characterized by AFM. AFM served the purpose of evaluating the surface quality of TPS/CNF films and characterizing the CNFs in TPS/CNF films. Surface topology can be used to measure the roughness parameters. The scale of the surface roughness of thermoplastic starch was obtained in the micrometer range, which suggested that TPS film surface is rough. Topography of TPS/CNF 1 wt.% of nanocomposite film was analyzed and was found to be in nanometer scale. Hence, the surface of TPS/CNF 1 wt.% film is so smooth and there existed the presence of nanoparticles even on the surface.

Employing twin-screw extrusion method to develop CNFs reinforced PLA was studied by Jonoobi et al. (2010). Masterbatch preparation of CNF in PLA and extrusion of diluted mixture were used to prepare nanocomposites. Morphological examination of composites using AFM images indicated that an entangled nanofiber network was obtained, which signified the completion of the fibrillation process.

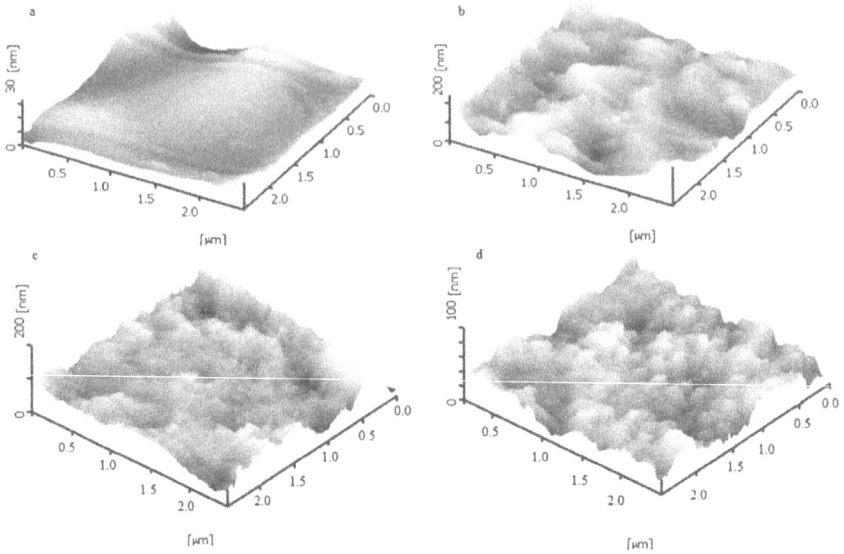

FIGURE 6.7   AFM images of (a) PLA, and PLA with different MCC loading: (b) 1%, (c) 3%, and (d) 5%. (Reproduced with permission from Elsevier, License Number: 5118040550218; Haafiz et al. 2013.)

The diameter of the fibers ranging from 40 to 70 nm was measured using Nanoscape V software and length was determined to be several micrometers. The SC technique was successfully employed in preparing PLA composites filled with MCC from oil palm biomass (Haafiz et al. 2013). AFM analysis was conducted to further investigate the nature of interactions between PLA and PLA/MCC composites. It was evident from the AFM images that the surface topography of pure PLA was altered significantly due to the inclusion of MCC into the PLA matrix. PLA surface is smooth. However, when MCC content was increased, surface features revealed clusters between PLA and MCC. The formation of clusters was elucidated as due to aggregation of MCC (Figure 6.7).

Spinella et al. (2015) attempted green solvent-less one-pot concurrent acid hydrolysis/Fischer esterification processes to prepare lactic acid and acetate modified CNCs and used them as a modifier for PLA. Surface topology studies by AFM explained the influence of CNC surface chemistry on CNC dispersion. AFM images unambiguously pointed out that lactic acid-modified CNCs possessed the highest CNC dispersion within the PLA matrix (Figure 6.8). The esterification process resulted in producing lactate oligomers from the surface of CNCs. These observations conclusively present that surface modifications of CNCs are highly significant in preparing CNC-polymer nanocomposites of desired properties. Atomic Force Microscopy is not only a morphological characterization technique of polymer composites but also presents important data regarding the role of surface features of composites. Hence, it could be considered as an important tool in understanding the surface as well as chemical features of polymer nanocomposites.

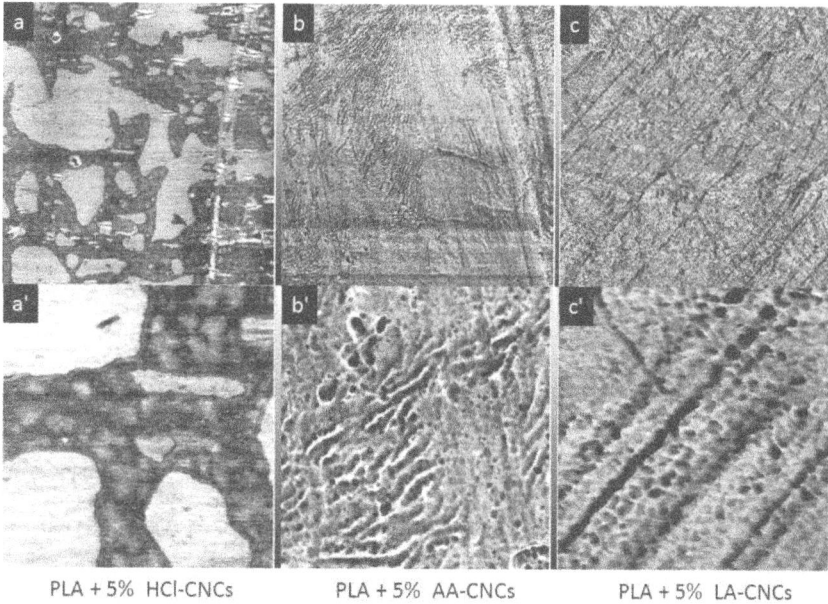

PLA + 5% HCl-CNCs   PLA + 5% AA-CNCs   PLA + 5% LA-CNCs

FIGURE 6.8 AFM images of PLA/CNC (5 wt.%) blends; (Different magnifications – 10 mm square (a–c) and 5 mm square (a′–c′). (Reproduced with permission from Elsevier, License Number: 5118031383542; Spinella et al. 2015.)

6.5 CONCLUSION

In this chapter, we summarized the morphological changes in PLA/cellulose composite by different morphological techniques such as SEM, TEM, and AFM. The weak mechanical and thermal properties of PLA-cellulose composite could be improved by the surface treatment of fiber. The morphological changes observed in the composites indicated that the surface treatment improves the compatibility between the cellulose and PLA matrix. The better physical/mechanical interlocking in the composite is responsible for the composite's superior performance.

ACKNOWLEDGMENTS

Authors gratefully thanks for financial support by the King Mongkut's University of Technology North Bangkok (KMUTNB), Thailand, through the Post-Doctoral Program (Grant No. KMUTNB-63-Post-03 and KMUTNB-64-Post-03 to SR) and (Grant No. KMUTNB-BasicR-64-16).

REFERENCES

Abdollahi, Mehdi, Mehdi Alboofetileh, Rabi Behrooz, Masoud Rezaei, and Reza Miraki. 2013. "Reducing Water Sensitivity of Alginate Bio-Nanocomposite Film Using Cellulose Nanoparticles." *International Journal of Biological Macromolecules* 54:166–173. doi: 10.1016/j.ijbiomac.2012.12.016.

Arslan, Dogan, Emre Vatansever, Deniz Sema Sarul, Yusuf Kahraman, Gurbuz Gunes, Ali Durmus, and Mohammadreza Nofar. 2020. "Effect of Preparation Method on the Properties of Polylactide/Cellulose Nanocrystal Nanocomposites." *Polymer Composites* 41 (10):4170–4180. doi: 10.1002/pc.25701.

Azizi Samir, My Ahmed Said, Fannie Alloin, and Alain Dufresne. 2005. "Review of Recent Research into Cellulosic Whiskers, Their Properties and Their Application in Nanocomposite Field." *Biomacromolecules* 6 (2):612–626. doi: 10.1021/bm0493685.

Carrasco, F., P. Pagès, J. Gámez-Pérez, O. O. Santana, and M. L. Maspoch. 2010. "Processing of Poly(lactic Acid): Characterization of Chemical Structure, Thermal Stability and Mechanical Properties." *Polymer Degradation and Stability* 95 (2):116–125. doi: 10.1016/j.polymdegradstab.2009.11.045.

Chen, Y., L. M. Geever, J. A. Killion, J. G. Lyons, C. L. Higginbotham, and D. M. Devine. 2016. "Review of Multifarious Applications of Poly (Lactic Acid)." *Polymer-Plastics Technology and Engineering* 55 (10):1057–1075. doi: 10.1080/03602559.2015.1132465.

Cheng, Yanling, Shaobo Deng, Paul Chen, and Roger Ruan. 2009. "Polylactic Acid (PLA) Synthesis and Modifications: A Review." *Frontiers of Chemistry in China* 4 (3):259–264. doi: 10.1007/s11458-009-0092-x.

DeFrates, Kelsey, Theodore Markiewicz, Kayla Callaway, Ye Xue, John Stanton, David Salas-de la Cruz, and Xiao Hu. 2017. "Structure–Property Relationships of Thai Silk–Microcrystalline Cellulose Biocomposite Materials Fabricated from Ionic Liquid." *International Journal of Biological Macromolecules* 104:919–928. doi: 10.1016/j.ijbiomac.2017.06.103.

Dixson, Ronald, Ndubuisi Orji, Ichiko Misumi, and Gaoliang Dai. 2018. "Spatial Dimensions in Atomic Force Microscopy: Instruments, Effects, and Measurements." *Ultramicroscopy* 194:199–214. doi: 10.1016/j.ultramic.2018.08.011.

Eichhorn, Stephen J. 2011. "Cellulose Nanowhiskers: Promising Materials for Advanced Applications." *Soft Matter* 7 (2):303–315. doi: 10.1039/c0sm00142b.

Fahma, Farah, Shinichiro Iwamoto, Naruhito Hori, Tadahisa Iwata, and Akio Takemura. 2010. "Isolation, Preparation, and Characterization of Nanofibers from Oil Palm Empty-Fruit-Bunch (OPEFB)." *Cellulose* 17 (5):977–985. doi: 10.1007/s10570-010-9436-4.

Fattahi Meyabadi, Tayebeh, Fatemeh Dadashian, Gity Mir Mohamad Sadeghi, and Hamid Ebrahimi Zanjani Asl. 2014. "Spherical Cellulose Nanoparticles Preparation from Waste Cotton Using a Green Method." *Powder Technology* 261:232–240. doi: 10.1016/j.powtec.2014.04.039.

Fazeli, Mahyar, Meysam Keley, and Esmaeil Biazar. 2018. "Preparation and Characterization of Starch-Based Composite Films Reinforced by Cellulose Nanofibers." *International Journal of Biological Macromolecules* 116:272–280. doi: 10.1016/j.ijbiomac.2018.04.186.

Fernandes, A. N., L. H. Thomas, C. M. Altaner, P. Callow, V. T. Forsyth, D. C. Apperley, C. J. Kennedy, and M. C. Jarvis. 2011. "Nanostructure of Cellulose Microfibrils in Spruce Wood." *Proceedings of the National Academy of Sciences* 108 (47):E1195–E1203. doi: 10.1073/pnas.1108942108.

Fortunati, E., M. Peltzer, I. Armentano, L. Torre, A. Jiménez, and J. M. Kenny. 2012. "Effects of Modified Cellulose Nanocrystals on the Barrier and Migration Properties of PLA Nano-Biocomposites." *Carbohydrate Polymers* 90 (2):948–956. doi: 10.1016/j.carbpol.2012.06.025.

Frone, Adriana N., Sophie Berlioz, Jean-François Chailan, and Denis M. Panaitescu. 2013. "Morphology and Thermal Properties of PLA–Cellulose Nanofibers Composites." *Carbohydrate Polymers* 91 (1):377–384. doi: 10.1016/j.carbpol.2012.08.054.

Fukushima, Kikku, Daniela Tabuani, Maria Arena, Mara Gennari, and Giovanni Camino. 2013. "Effect of Clay Type and Loading on Thermal, Mechanical Properties and Biodegradation of Poly(Lactic Acid) Nanocomposites." *Reactive and Functional Polymers* 73 (3):540–549. doi: 10.1016/j.reactfunctpolym.2013.01.003.

Geng, Yi, Hui He, Hao Liu, and Huaishuai Jing. 2020. "Preparation of Polycarbonate/Poly(Lactic Acid) with Improved Printability and Processability for Fused Deposition Modeling." *Polymers for Advanced Technologies* 31 (11):2848–2862. doi: 10.1002/pat.5013.

Giessibl, F. J. 2003. "Advances in Atomic Force Microscopy." *Reviews of Modern Physics* 75 (3):949–983. doi: 10.1103/RevModPhys.75.949.

Goetz, Lee, Aji Mathew, Kristiina Oksman, Paul Gatenholm, and Arthur J. Ragauskas. 2009. "A Novel Nanocomposite Film Prepared From Crosslinked Cellulosic Whiskers." *Carbohydrate Polymers* 75 (1):85–89. doi: 10.1016/j.carbpol.2008.06.017.

Haafiz, M. K. Mohamad, Azman Hassan, H. P. S. Abdul Khalil, M. R. Nurul Fazita, Md Saiful Islam, I. M. Inuwa, M. M. Marliana, and M. Hazwan Hussin. 2016. "Exploring the Effect of Cellulose Nanowhiskers Isolated from Oil Palm Biomass on Polylactic Acid Properties." *International Journal of Biological Macromolecules* 85:370–378. doi: 10.1016/j.ijbiomac.2016.01.004.

Haafiz, M. K. Mohamad, Azman Hassan, Zainoha Zakaria, I. M. Inuwa, M. S. Islam, and M. Jawaid. 2013. "Properties of Polylactic Acid Composites Reinforced with Oil Palm Biomass Microcrystalline Cellulose." *Carbohydrate Polymers* 98 (1):139–145. doi: 10.1016/j.carbpol.2013.05.069.

Han, Jingquan, Chengjun Zhou, Yiqiang Wu, Fangyang Liu, and Qinglin Wu. 2013. "Self-Assembling Behavior of Cellulose Nanoparticles during Freeze-Drying: Effect of Suspension Concentration, Particle Size, Crystal Structure, and Surface Charge." *Biomacromolecules* 14 (5):1529–1540. doi: 10.1021/bm4001734.

Hossain, Kazi M. Zakir, Ifty Ahmed, Andrew J. Parsons, Colin A. Scotchford, Gavin S. Walker, Wim Thielemans, and Chris D. Rudd. 2011. "Physico-Chemical and Mechanical Properties of Nanocomposites Prepared Using Cellulose Nanowhiskers and Poly(Lactic Acid)." *Journal of Materials Science* 47 (6):2675–2686. doi: 10.1007/s10853-011-6093-4.

Hu, Ruihua, and Jae-Kyoo Lim. 2016. "Fabrication and Mechanical Properties of Completely Biodegradable Hemp Fiber Reinforced Polylactic Acid Composites." *Journal of Composite Materials* 41 (13):1655–1669. doi: 10.1177/0021998306069878.

Huda, M. S., A. K. Mohanty, L. T. Drzal, E. Schut, and M. Misra. 2005. "'Green' Composites From Recycled Cellulose and Poly(lactic Acid): Physico-Mechanical and Morphological Properties Evaluation." *Journal of Materials Science* 40 (16):4221–4229. doi: 10.1007/s10853-005-1998-4.

Inkson, B. J. 2016. "Scanning Electron Microscopy (SEM) and Transmission Electron Microscopy (TEM) for Materials Characterization." In *Materials Characterization Using Nondestructive Evaluation (NDE) Methods*, Woodhead Publishing, 17–43.

Jonoobi, Mehdi, Jalaluddin Harun, Aji P. Mathew, and Kristiina Oksman. 2010. "Mechanical Properties of Cellulose Nanofiber (CNF) Reinforced Polylactic Acid (PLA) Prepared by Twin Screw Extrusion." *Composites Science and Technology* 70 (12):1742–1747. doi: 10.1016/j.compscitech.2010.07.005.

Jonoobi, Mehdi, Aji P. Mathew, Mahnaz M. Abdi, Majid Davoodi Makinejad, and Kristiina Oksman. 2012. "A Comparison of Modified and Unmodified Cellulose Nanofiber Reinforced Polylactic Acid (PLA) Prepared by Twin Screw Extrusion." *Journal of Polymers and the Environment* 20 (4):991–997. doi: 10.1007/s10924-012-0503-9.

Khoo, R. Z., H. Ismail, and W. S. Chow. 2016. "Thermal and Morphological Properties of Poly (Lactic Acid)/Nanocellulose Nanocomposites." *Procedia Chemistry* 19:788–794. doi: 10.1016/j.proche.2016.03.086.

Lee, Sun-Young, D. Jagan Mohan, In-Aeh Kang, Geum-Hyun Doh, Soo Lee, and Seong Ok Han. 2009. "Nanocellulose Reinforced PVA Composite Films: Effects of Acid Treatment and Filler Loading." *Fibers and Polymers* 10 (1):77–82. doi: 10.1007/s12221-009-0077-x.

Li, Ge, Menghui Zhao, Fei Xu, Bo Yang, Xiangyu Li, Xiangxue Meng, Lesheng Teng, Fengying Sun, and Youxin Li. 2020. "Synthesis and Biological Application of Polylactic Acid." *Molecules* 25 (21). doi: 10.3390/molecules25215023.

Li, Mei-Chun, Qinglin Wu, Kunlin Song, Sunyoung Lee, Yan Qing, and Yiqiang Wu. 2015. "Cellulose Nanoparticles: Structure–Morphology–Rheology Relationships." *ACS Sustainable Chemistry & Engineering* 3 (5):821–832. doi: 10.1021/acssuschemeng.5b00144.

Li, Rongji, Jianming Fei, Yurong Cai, Yufeng Li, Jianqin Feng, and Juming Yao. 2009. "Cellulose Whiskers Extracted from Mulberry: A Novel Biomass Production." *Carbohydrate Polymers* 76 (1):94–99. doi: 10.1016/j.carbpol.2008.09.034.

Liu, Wenqiang, Yu Dong, Dongyan Liu, Yuxia Bai, and Xiuzhen Lu. 2018. "Polylactic Acid (PLA)/Cellulose Nanowhiskers (CNWs) Composite Nanofibers: Microstructural and Properties Analysis." *Journal of Composites Science* 2 (1). doi: 10.3390/jcs2010004.

Lopes, M. S., A. L. Jardini, and R. M. Filho. 2012. "Poly (Lactic Acid) Production for Tissue Engineering Applications." *Procedia Engineering* 42:1402–1413. doi: 10.1016/j.proeng.2012.07.534.

Lunt, James, and Andrew L. Shafer. 2016. "Polylactic Acid Polymers from Com. Applications in the Textiles Industry." *Journal of Industrial Textiles* 29 (3):191–205. doi: 10.1177/152808370002900304.

Niu, Xun, Yating Liu, Yang Song, Jinquan Han, and Hui Pan. 2018. "Rosin Modified Cellulose Nanofiber as a Reinforcing and Co-Antimicrobial Agents in Polylactic Acid/ Chitosan Composite Film for Food Packaging." *Carbohydrate Polymers* 183:102–109. doi: 10.1016/j.carbpol.2017.11.079.

Oksman, K., A. P. Mathew, D. Bondeson, and I. Kvien. 2006. "Manufacturing Process of Cellulose Whiskers/Polylactic Acid Nanocomposites." *Composites Science and Technology* 66 (15):2776–2784. doi: 10.1016/j.compscitech.2006.03.002.

Pracella, Mariano, Md Minhaz-Ul Haque, and Debora Puglia. 2014. "Morphology and Properties Tuning of PLA/Cellulose Nanocrystals Bio-Nanocomposites by Means of Reactive Functionalization and Blending with PVAc." *Polymer* 55 (16):3720–3728. doi: 10.1016/j.polymer.2014.06.071.

Radoor, Sabarish, Jasila Karayil, Aswathy Jayakumar, Edayileveettil Krishnankutty Radhakrishnan, Jyotishkumar Parameswaranpillai, and Suchart Siengchin. 2021. "Alginate-Based Bionanocomposites in Wound Dressings." In *Bionanocomposites in Tissue Engineering and Regenerative Medicine*, Woodhead Publishing, 351–375.

Radoor, Sabarish, Jasila Karayil, Aswathy Jayakumar, E. K. Radhakrishnan, Lakshmanan Muthulakshmi, Sanjay Mavinkere Rangappa, Suchart Siengchin, and Jyotishkumar Parameswaranpillai. 2020a. "Structure and Surface Morphology Techniques for Biopolymers." In *Biofibers and Biopolymers for Biocomposites*, Springer, 35–70.

Radoor, Sabarish, Jasila Karayil, Jyotishkumar Parameswaranpillai, and Suchart Siengchin. 2020b. "Adsorption of Methylene Blue Dye from Aqueous Solution by a Novel PVA/ CMC/Halloysite Nanoclay Bio Composite: Characterization, Kinetics, Isotherm and Antibacterial Properties." *Journal of Environmental Health Science and Engineering* 18 (2):1311–1327. doi: 10.1007/s40201-020-00549-x.

Ramesh, Shruthy, and Preetha Radhakrishnan. 2019. "Cellulose Nanoparticles from Agro-Industrial Waste for the Development of Active Packaging." *Applied Surface Science* 484:1274–1281. doi: 10.1016/j.apsusc.2019.04.003.

Rogina, Anamarija. 2014. "Electrospinning Process: Versatile Preparation Method for Biodegradable and Natural Polymers and Biocomposite Systems Applied in Tissue Engineering and Drug Delivery." *Applied Surface Science* 296:221–230. doi: 10.1016/j. apsusc.2014.01.098.

Sabarish, R., and G. Unnikrishnan. 2018. "Polyvinyl Alcohol/Carboxymethyl Cellulose/ ZSM-5 Zeolite Biocomposite Membranes for Dye Adsorption Applications." *Carbohydrate Polymers* 199:129–140. doi: 10.1016/j.carbpol.2018.06.123.

Saito, Tsuguyuki, Satoshi Kimura, Yoshiharu Nishiyama, and Akira Isogai. 2007. "Cellulose Nanofibers Prepared by TEMPO-Mediated Oxidation of Native Cellulose." *Biomacromolecules* 8 (8):2485–2491. doi: 10.1021/bm0703970.

Shi, Qingfeng, Chengjun Zhou, Yiying Yue, Weihong Guo, Yiqiang Wu, and Qinglin Wu. 2012. "Mechanical Properties and In Vitro Degradation of Electrospun Bio-Nanocomposite Mats from PLA and Cellulose Nanocrystals." *Carbohydrate Polymers* 90 (1):301–308. doi: 10.1016/j.carbpol.2012.05.042.

Shogren, R. L., W. M. Doane, D. Garlotta, J. W. Lawton, and J. L. Willett. 2003. "Biodegradation of Starch/Polylactic Acid/Poly(Hydroxyester-Ether) Composite Bars in Soil." *Polymer Degradation and Stability* 79 (3):405–411. doi: 10.1016/s0141-3910(02)00356-7.

Singha, A. S., and Vijay Kumar Thakur. 2009. "Mechanical, Thermal and Morphological Properties of Grewia Optiva Fiber/Polymer Matrix Composites." *Polymer-Plastics Technology and Engineering* 48 (2):201–208. doi: 10.1080/03602550802634550.

Somerville, Chris. 2006. "Cellulose Synthesis in Higher Plants." *Annual Review of Cell and Developmental Biology* 22 (1):53–78. doi: 10.1146/annurev.cellbio.22.022206.160206.

Spinella, Stephen, Giada Lo Re, Bo Liu, John Dorgan, Youssef Habibi, Philippe Leclère, Jean-Marie Raquez, Philippe Dubois, and Richard A. Gross. 2015. "Polylactide/Cellulose Nanocrystal Nanocomposites: Efficient Routes for Nanofiber Modification and Effects of Nanofiber Chemistry on PLA Reinforcement." *Polymer* 65:9–17. doi: 10.1016/j.polymer.2015.02.048.

Sullivan, Erin, Robert Moon, and Kyriaki Kalaitzidou. 2015. "Processing and Characterization of Cellulose Nanocrystals/Polylactic Acid Nanocomposite Films." *Materials* 8 (12):8106–8116. doi: 10.3390/ma8125447.

Tang, C. Y., and Z. Yang. 2017. "Transmission Electron Microscopy (TEM)." In *Membrane Characterization*, Elsevier, 145–159.

Tawakkal, Intan S. M. A., Marlene J. Cran, Joseph Miltz, and Stephen W. Bigger. 2014. "A Review of Poly(lactic Acid)-Based Materials for Antimicrobial Packaging." *Journal of Food Science* 79 (8):R1477–R1490. doi: 10.1111/1750-3841.12534.

Tham, W. L., B. T. Poh, Z. A. Mohd Ishak, and W. S. Chow. 2016. "Transparent Poly(lactic Acid)/ Halloysite Nanotube Nanocomposites with Improved Oxygen Barrier and Antioxidant Properties." *Journal of Thermal Analysis and Calorimetry* 126 (3):1331–1337. doi: 10.1007/s10973-016-5834-7.

Thielemans, Wim, Catherine R. Warbey, and Darren A. Walsh. 2009. "Permselective Nanostructured Membranes Based on Cellulose Nanowhiskers." *Green Chemistry* 11 (4). doi: 10.1039/b818056c.

Vroman, Isabelle, and Lan Tighzert. 2009. "Biodegradable Polymers." *Materials* 2 (2):307–344. doi: 10.3390/ma2020307.

Wang, Tao, and Lawrence T. Drzal. 2012. "Cellulose-Nanofiber-Reinforced Poly(lactic Acid) Composites Prepared by a Water-Based Approach." *ACS Applied Materials & Interfaces* 4 (10):5079–5085. doi: 10.1021/am301438g.

Yang, Zhangqiang, Xiaojie Li, Junhui Si, Zhixiang Cui, and Kaiping Peng. 2019. "Morphological, Mechanical and Thermal Properties of Poly(Lactic Acid) (PLA)/Cellulose Nanofibrils (CNF) Composites Nanofiber for Tissue Engineering." *Journal of Wuhan University of Technology-Materials Science Edition.* 34 (1):207–215. doi: 10.1007/s11595-019-2037-7.

Ying, Zeren, Defeng Wu, Zhifeng Wang, Wenyuan Xie, Yaxin Qiu, and Xijun Wei. 2018. "Rheological and Mechanical Properties of Polylactide Nanocomposites Reinforced with the Cellulose Nanofibers with Various Surface Treatments." *Cellulose* 25 (7):3955–3971. doi: 10.1007/s10570-018-1862-8.

Zhang, Qingfa, Hanwu Lei, Hongzhen Cai, Xiangsheng Han, Xiaona Lin, Moriko Qian, Yunfeng Zhao, Erguang Huo, Elmar M. Villota, and Wendy Mateo. 2020. "Improvement on the Properties of Microcrystalline Cellulose/Polylactic Acid Composites by Using Activated Biochar." *Journal of Cleaner Production* 252. doi: 10.1016/j.jclepro.2019.119898.

Zhou, Weilie, Robert Apkarian, Zhong Lin Wang, and David Joy. 2006. "Fundamentals of Scanning Electron Microscopy (SEM)." In *Scanning Microscopy for Nanotechnology*, Springer, 1–40.

7 Thermogravimetric Analysis (TGA) and Differential Scanning Calorimetry (DSC) of PLA/Cellulose Composites

N. M. Nurazzi, N. Abdullah, and M. N. F. Norrrahim
Universiti Pertahanan Nasional Malaysia (UPNM)
Kuala Lumpur, Malaysia

S. H. Kamarudin and S. Ahmad
Universiti Teknologi MARA (UiTM)
Shah Alam, Malaysia

S. S. Shazleen, M. Rayung,
and M. R. M. Asyraf
Universiti Putra Malaysia (UPM)
Serdang, Malaysia

R. A. Ilyas
Universiti Teknologi Malaysia (UTM)
Johor, Malaysia

M. Kuzmin
Ogarev Mordovia State University
Saransk, Russia

CONTENTS

DOI: 10.1201/9781003160458-7

7.1   INTRODUCTION

Commercial production of eco-friendly and biodegradable polymers from a natural feedstock for the development of green composites is becoming extremely prevalent due to concerns of environmental pollution associated with plastic waste, depletion of petroleum and fossil resources, and environmental legislative pressure. A significant amount of research has been dedicated to the use of development of green composites due to recyclable, environmentally friendly, and compostable alternatives to petrochemical-based conventional plastics such as polypropylene (PP) and polyethylene (PE) [1, 2]. Among all green biopolymers available in the market, polylactic acid (PLA) has been extensively studied and most widely used owing to its good mechanical properties, which are similar to those of polystyrene (PS) and polyethylene terephthalate (PET), sustainability, biocompatibility, non-toxicity, and renewability [3, 4]. Despite the advantages listed, the inherent brittleness, stiffness, poor impact strength, and low thermal stability have limited its uses for several industrial applications [5, 6]. These drawbacks can be overcome by reinforcement, grafting, plasticizing, or blending PLA with flexible polymers [7].

In particular, the reinforcement of plant fibers and their components (e.g., cellulose, hemicellulose) with PLA has gained significant interest in developing fully biodegradable and sustainable composites. Biopolymers reinforced with cellulose fiber in several forms (macro, micro, or nano) have attracted significant interest to obtain more environmentally friendly life cycle products, owing to cellulose's excellent properties such as being economical, low density, high specific strength, wide availability, and promising sustainability [8–13]. It has become one of the top priorities to use composites that are non-petroleum based. Bio-nano composites are a great example as they are known to have tremendous benefits like high performing, low density, lightweight, recyclable, good compatibility between reinforcement and matrix phases, and have the potential to improve impact resistance and toughness of a brittle polymer like PLA [14–18].

PLA and cellulose biocomposites have been processed with the focus of using conventional compounding techniques used for thermoplastics, such as injection molding, extrusion, and blow molding. However, the applicability of such processes on a large scale is strongly limited due to the incompatibility between PLA and cellulose. Cellulose is an abundant polymeric material and consists of crystalline and amorphous structure parts [19, 20]. Cellulose can be extracted from cotton [21], flax [22], jute [23], hemp [24], sugar palm fibre [25], sugarcane bagasse [26], kenaf [27], etc. [28, 29]. Cellulose can be classified into three different types, namely cellulose nanocrystals (CNCs), cellulose nanofibrils (CNFs), and bacterial cellulose (BC), depending on their properties and isolation methodologies [30]. Cellulose is one of the most remarkable and is growing steadily in the CelluForce market each year, as can be seen in industrial production of nanocellulose such as CelluForce (CNC, 300 ton/year), American Process (CNF and CNC, each 130 ton/year), Nippon Paper (CNF, 560 ton/year), and University of Maine (CNF, 260 ton/year) [31].

The advancement in nanocellulose production has sparked significant interest in this context due to their intriguing mechanical properties and the abundance of cellulose material in biomass [11]. Furthermore, the excellent properties of cellulose, like low density, excellent biodegradability, high specific strength, affordability, and good sustainability, make it an important character and suitable for developing PLA/cellulose composites [32]. PLA/cellulose composites reinforced with CNCs, CNFs, and BCs have been fabricated and extensively studied [33]. Biocomposites from PLA/cellulose are desirable due to the future commercialization that would unlock the potential of these underutilized renewable materials. In addition, they provide advantages over unmanageable synthetic plastics in disposable applications. Numerous research works have reported the use of BC and pristine cellulose as reinforcement fillers without modification [34, 35]. In most cases, incorporating a small amount of cellulose had increased mechanical strength and thermal stability properties compared with the virgin PLA matrix. It was found that the optimal loadings of nanocellulose in the PLA/cellulose composites were between 0.5 wt.% and 2 wt.% [36, 37].

The potential of cellulose as reinforcement agents for PLA was studied in the experimental research conducted by De Paula et al., and the results showed improvement in mechanical and thermal properties [38]. In their study, the influence of cellulose nanowhiskers (CNWs) on the hydrolytic degradation behavior of PLA was examined. They have found that only a tiny amount of nanocellulose fillers (1%) had significantly affected PLA composites' thermal stability and biodegradability. In this context, when cellulose fillers were added to the PLA, the hydrolytic degradation of the PLA was substantially delayed. This phenomenon is caused by the properties of highly crystalline cellulose CNWs, which could inhibit water absorption, thus improving thermal stability and raising the initial temperature of mass loss.

The number of nanocellulose fillers being added to the PLA matrix is an important aspect that needs to be studied to determine good properties for PLA/cellulose composites. In another study conducted by Lee et al., they had fabricated PLA/CNWs composites with various compositions of cellulose content between 0.1 wt.% and 0.5 wt.% [39]. In this study, several aspects including thermal stability, thermal behaviors, and mechanical properties of PLA/CNWs were discussed. Tensile strength and tensile modulus of the PLA/CNWs composites increased significantly as CNWs content increased. This could be explained by the presence of highly crystalline behavior of CNWs that had successfully promoted the crystallization behavior of the PLA matrix. However, with the same amount of filler content, the thermal stability of PLA/CNWs composites is decreasing simultaneously with the CNWs loading. In this case, CNWs fillers did not affect the glass transition or melting temperatures of the PLA matrix.

As reported by Hossain et al., the thermal stability behavior of the CNWs and PLA/CNWs composites was found to be stable in the temperature region between 20°C and 210°C from the TGA study [40]. This finding was in line with the study conducted by Petersson, who reported that all materials from the PLA/CNWs composites in which the cellulose isolated from microcrystalline cellulose were found to be thermally stable in the temperature region between 25°C and 220°C [41]. From the results obtained, the CNWs started to lose their weight after temperature 210°C, while the PLA/CNWs composites showed decomposition above 210°C. According to the DSC analysis, the melting peak of PLA was around 170°C, and the addition of CNWs had increased the melting temperature of the nanocomposites.

Dispersion of nanocellulose in the PLA matrix is highly challenging due to the polarity difference. Cellulose fibers are challenging to disperse in the hydrophobic PLA matrix due to the highly hydrophobic behavior, thus resulting in the agglomeration and poor fiber-matrix interfacial adhesion that brings challenges for processing (melt flow and non-continuous feeding) as well as for the final biocomposites resulting in poor mechanical strength and non-homogeneous composites. Besides agglomeration, thermal degradation of nanocellulose fillers in the melt processing techniques, which can cause a reduction in the degree of adhesion in composites, is another urgent challenge that needs to be addressed. Solution casting was one of the methods used to process and produce PLA/cellulose composites for achieving good dispersibility for nanocellulose in the PLA matrix [42]. Other than solution casting, many strategies have been attempted to avoid poor dispersion of nanocellulose in such polymer matrix in polymer composites. The use of solvent-based processes [43], commercial surfactants [44], chemical grafting reactions [45], compatibilizers [46], and other types of surface modification techniques [47] had been applied to improve the dispersion of nanocellulose in PLA/cellulose composites.

7.2 THERMOGRAVIMETRIC ANALYSIS (TGA)

TGA has been extensively used as an analytical technique to study the thermal stability of materials used in various environmental, pharmaceutical, food, and petrochemical applications. TGA measures changes in the mass of a sample as a function of temperature and time. Usually, samples were weighed at around 5–10 mg and placed in a clean ceramic plate, where the thermal stability data will be recorded by heating the samples at a constant rate at a controlled temperature [48]. Simply put, TGA measures the sample's mass as it is heated or cooled in a furnace. A sample is continuously weighed while heating as an inert gas of the atmosphere is passed over it. The measured sample's mass loss curves will give information on three things: changes in sample composition, kinetic parameters for chemical reactions, and thermal stability in the sample. TGA instrument can quantify the mass loss and calculate the loss of water, plasticizer, solvent, oxidation, pyrolysis, decarboxylation, decomposition, and weight percentage (wt.%) of filler and ash.

TGA was used in various research works of PLA/cellulose composites to identify the thermal stability of materials, oxidative stability of materials, the composition of multi-component systems, estimated lifetime of a product, kinetic decompositions, and moisture and volatile content of materials. Furthermore, TGA was used to study cellulosic nanofibers as a reinforcing phase in PLA/CNF composites by Wang et al.. The thermal degradation behavior of membrane with and without cellulose nanofibers addition was investigated [49]. It was found that the initial degradation temperature of all PLA/CNF composites was higher than that of PLA, indicating the improvement of thermal stability of the PLA/CNF composites due to the introduction of cellulose nanofibers. From the derivative thermogram (DTG) analysis, there is only one decomposition peak for PLA at 331°C and two temperature peaks for cellulose nanofibers, at about 40°C, which is associated with the hydrophilic groups of nanofibers and the second peak at about 346°C, which was attributed to the dehydration reactions and the formation of volatile products by chain scission and decomposition.

Moreover, the initial degradation temperature and maximum degradation temperature of PLA/CNF composites were increased by 20°C and 10°C after the addition of

5.0 wt.% of CNF, respectively. These improvements are essential for the attribution of good dispersibility and, therefore, increased cellulose nanofibres' interfacial interaction in the PLA matrix. It can be concluded that the PLA/CNF composites have an improved thermal property compared to the PLA. Similar findings were reported from a study conducted by Rasheed et al. [50]. The thermal stability of PLA blends is enhanced upon the addition of cellulose up to 1 wt.% due to the uniform dispersion of fillers in the PLA matrix. Initial degradation temperature (Ti), the temperature at 50% weight loss (T50%), and final degradation temperature (Tf) all increased on the addition of up to 1 wt.% cellulose, and then the values decreased upon further addition of cellulose. This behavior was also reported by Cao et al., where the cellulose hindered the release of decomposition products of the polymers [51].

7.3 DIFFERENTIAL SCANNING CALORIMETRY (DSC)

DSC is a technique to measure the energy or the amount of heat absorbed or released when the sample is heated or cooled at a controlled rate. It is a direct assessment of the heat energy uptake, which usually occurs in a sample with a regulated decrease or increase in temperature. The calorimetry is being applied to monitor the phase transitions of polymers samples' behavioral changes [52]. Simply put, it is a thermal analysis technique that looks at how a material's heat capacity (Cp) is changed by temperature. The DSC could examine the thermal stability and crystallinity behavior of polymeric samples. It is used to study transition biochemical reactions, called a single molecular transition of a molecule from one conformation to another. Melting points of the samples are being identified in solid, solution, and mixed phases such as suspensions [53].

Thermal behavior and thermal stability analysis of PLA/cellulose composites are shown from DSC thermogram in CNF filled in PLA bio-composite film based on the studies conducted by the researchers [2, 49]. The thermal stability of PLA significantly increased with the addition of cellulose. From the results obtained, DSC analysis is illustrated and summarized in Table 7.1 [2]. The information on the phenomena of glass transition (T_g), crystallization temperature (T_c), and melting temperature (T_m), as well as the degree of crystallinity of PLA and its biocomposites, is presented.

TABLE 7.1
Summary of Thermal Characteristics and the Degree of Crystallinity of PLA and Biocomposites [2]

Samples	First Heating Scan			Second Heating Scan			Degree of Crystallinity (%)
	T_g (°C)	T_c (J/g)	T_m (°C)	T_g (°C)	T_c (J/g)	T_m (°C)	
PLA	60.2	–	150	62.2	–	150.3	7.8
PLA/cellulose (2 wt.%)	59.8	115.0	149.9	62.2	125.0	150.6	13.0
PLA/cellulose (6 wt.%)	59.6	112.9	149.6	62.2	116.4	149.4	11.4
PLA/cellulose (10 wt.%)	57.8	111.0	143.6	62.2	117.2	149.4	10.5

The double peaks were observed from the DSC curves for PLA and PLA/cellulose composites for the following reasons; firstly, it is due to melting, recrystallization, and re-melting during heating; secondly, polymorphism or isodimorphism that presents multiple crystal forms; thirdly, it is due to different lamellar thickness, distribution, and morphology; and lastly, it is a species with different molecular weights [54]. Furthermore, PLA exhibits T_g of 60°C and T_m of 150°C without the appearance of a T_c peak. The difference in T_g values during the first and second heating is expected as it directly depends on the thermal history [9]. This result is in agreement with the findings of previous studies carried out by Yang et al., who described the same trend of DSC curves for PLA plasticized nanocellulose composites [48]. For a better understanding, the results of DSC analysis for PLA and PLA/CNF composites are summarized and tabulated in Table 7.2. It can be observed that the T_g, T_c, and Tm of PLA/CNF biocomposites were lower than that of neat PLA film. With the addition of nanocellulose, PLA/CNF biocomposites easily crystallized as compared with PLA. This behavior is attributed to the alteration mobility of PLA chains by the incorporation of CNF [55]. The decreasing trend in Tc values for PLA/CNF composites proved that CNF acted as a nucleating agent to the PLA and further enhanced the crystallization behavior of PLA/CNF composites. Overall, the addition of CNF enhanced the degree of crystallinity and brought a positive impact in PLA/cellulose biocomposites film.

7.4 FACTOR AFFECTING THERMAL STABILITY OF PLA/CELLULOSE COMPOSITES

Thermal properties of PLA/cellulose include heat capacity, thermal transition, thermal decomposition, and crystallization. These can be measured through DSC, TGA, and dynamic mechanical analysis [56]. Sin and Tueen stated that the listed thermal stability parameters depend on the polymer/matrix molecular weight, polymerization condition, thermal history, and purity. Accordingly, the different isomers can suggestively affect the molecular number (Mn), T_g, T_m, enthalpy, and T_c of PLA/cellulose. However, a factor that strongly influences thermal stability is crystallization. Crystallization temperature in the crystallization process will affect the types of crystal modification, melting point, and glass transition, which will lead to a mobility chain in the polymer matrix and thermal stability. The three main phases of crystallization temperature that strongly influence the types of crystal modification are below 110°C, 110 to 130°C, and above 130°C, of which these three phases are higher than the T_g, 58°C.

The effect of crystallization temperature on crystal modifications should be well understood. With the increase in the crystallization temperature, the PLA molecular chain mobility, which causes the formation of an unstable crystal nucleus, is also affected. This can lead to prolong duration in completing the crystallization process as well as decreasing thermal stability. In addition, the effect of nucleating agents on the crystallization of PLA/cellulose also needs to be understood because it affects the rate of crystallization and crystallization behavior (degree of crystallization), which is interrelated to thermal stability. Many nucleating agents affect the rate of crystallization and crystallization behavior. However, the most influential in PLA/cellulose are TMC-328, amide compound, PET-C, modified montmorillonite clay, dibenzoyladipohydrazide (TMC-306), and $N^1,N^{1'}$-(ethane-1,2-diyl)bis(N^2-phenyloxalamide)

TABLE 7.2

Recent Studies on the Reinforcement Effect of Cellulose on the DSC Properties of PLA Composites

Cellulose Source	Type of Cellulose	Additive/ Modification	Fabrication Method	Thermal Properties	References
Komagateibacter xylinus, hydrated softwood	BNC, CNF	–	Solution casting	• The T_g of PLA/BNC nanocomposites increased as the BNC wt.% loading increased. The T_g of the 2.0 wt.% PLA/BNC film composite was comparable to that of the PLA/CNF composites. • The T_c of PLA/BNC films is lower than that of neat PLA and PLA/CNF films at a lower temperature. • The T_m and overall crystallinity of PLA increased after the addition of both CNF and BNC.	[62]
Cassava pulp	CNF	–	Melt blending	• The T_g, T_c, and T_m of PLA/CNF biocomposites slightly increased in comparison to neat PLA. • The degree of crystallinity (%.X_c) of PLA/CNF biocomposites is higher than neat PLA.	[63]
Hardwood kraft pulp	CNF	–	In situ polymerization	• From PLA to NC1 and NC2 samples, there is a decrease in crystallinity, which then increases for NC3 and reaches the highest value among the materials studied for NC4. • NC4 appears to have two clear melting peaks at 167.3°C and 171.8°C, while PLA, NC1, and NC2 appear to have a single melting peak at 174.2 to 177.8°C, which may be due to the formation of two different crystalline phases. • The NC4 sample had the lowest cold crystallization temperature of all the samples, suggesting a higher crystallization tendency.	[64]
Bleached softwood sulfate fluff pulp	Cellulose	Surface modified using poly(ethylene oxide) (PEO)	Extrusion	• During the second heating scan, the thermograms of PLA and PLA/CF show the presence of T_g, T_c, and T_m while the T_g and T_c are almost absent in the PLA/CF/PEO composites, indicating that the materials were highly crystalline after the cooling scan.	[65]

(Continued)

TABLE 7.2 (Continued)
Recent Studies on the Reinforcement Effect of Cellulose on the DSC Properties of PLA Composites

Cellulose Source	Type of Cellulose	Additive/ Modification	Fabrication Method	Thermal Properties	References
Not stated	CNF	Freeze-drying	Melt blending	• The T_g increases slightly from 56.9°C (neat PLA) to 60.9°C (4 wt.% CNF). • The crystallinity of PLA increased significantly from 6.8% to 34.5% and 37.7% with the addition of 2 wt.% and 4 wt.% CNF.	[66]
Not stated	CNF	–	Melt blending	• The addition of CNF into PLA reduced the Tg of nanocomposites. • The cold crystallization peak for PLA/CNF3 nanocomposite was almost unseen, indicating the complete crystallization process occurred during the cooling process, owing to the CNF nucleation effect. • The PLA/CNF3 nanocomposite had the highest crystallinity of 44.2%, with an almost 95% increment than neat PLA.	[67]
Sisal fibers	CNF	Incorporated with C30B (commercial clay)	Solvent casting	• PLA/CNF nanocomposites crystallize faster than neat PLA at the investigated temperatures. • The Avrami indices of the hybrid nanocomposite PLA/CNF 1%/C30B 1% showed moderate variance with increasing temperature between CNF and C30B, implying the simultaneous growth of both clay and nanocellulose-nucleated spherulites. • The Tm values of PLA/CNF (151.1°C) and PLA/CNF/C30B (153.7°C) were slightly below the values of neat PLA (155.0°C) and PLA/C30B (154.5°C), suggesting different crystalline morphologies for these two groups of samples.	[68]
Eucalyptus globulus Cellulose, unbleached kraft pulp	MCC	Surface modified using kraft black liquor	Emulsion-solvent evaporation	• Composite films have a slightly lower Tg than neat PLA. • The cold peaks shift to lower temperatures for the PLA/UP, PLA/GUP, and PLA/MCC composites, indicating that cold crystallization occurs earlier induced by UP (unbleached pulp), GUP (ground unbleached pulp), and microcrystalline cellulose (MCC), which act as nucleating agents for PLA crystallization. • The crystallinity of PLA increased from 29.2% to 54.6% after reinforcing cellulose fibers.	[69]

(OXA). It can be observed that 0.5% PET-C and TMC-328 would increase more than 35% of crystal size, which promotes more crystallinity of PLA/cellulose. In detail, adding 0.2 wt.% to 0.6 wt.% of TMC-328 would gradually induce the crystallinity of PLA/cellulose. Besides, TMC-306 and OXA also significantly affect the rate of crystallization and degree of crystallinity as they provide a nucleation efficiency of up to >49%. Furthermore, both nucleation agents, TMC-306 and OXA, significantly reduce the crystallization time in PLA/cellulose. In addition, as the loading level of TMC-306 increased from 0.05% to 0.5 wt.%, it promoted a significant increase in height and sharpened the deflection peak on the x-ray diffraction curve of PLA/cellulose and accelerated the overall isothermal crystallization process of PLA/cellulose. Furthermore, this could induce better crystallization behavior as well as promote the best thermal stability characteristics [57, 58]. A summary of the factors that affect the thermal stability of PLA/cellulose is presented in Figure 7.1.

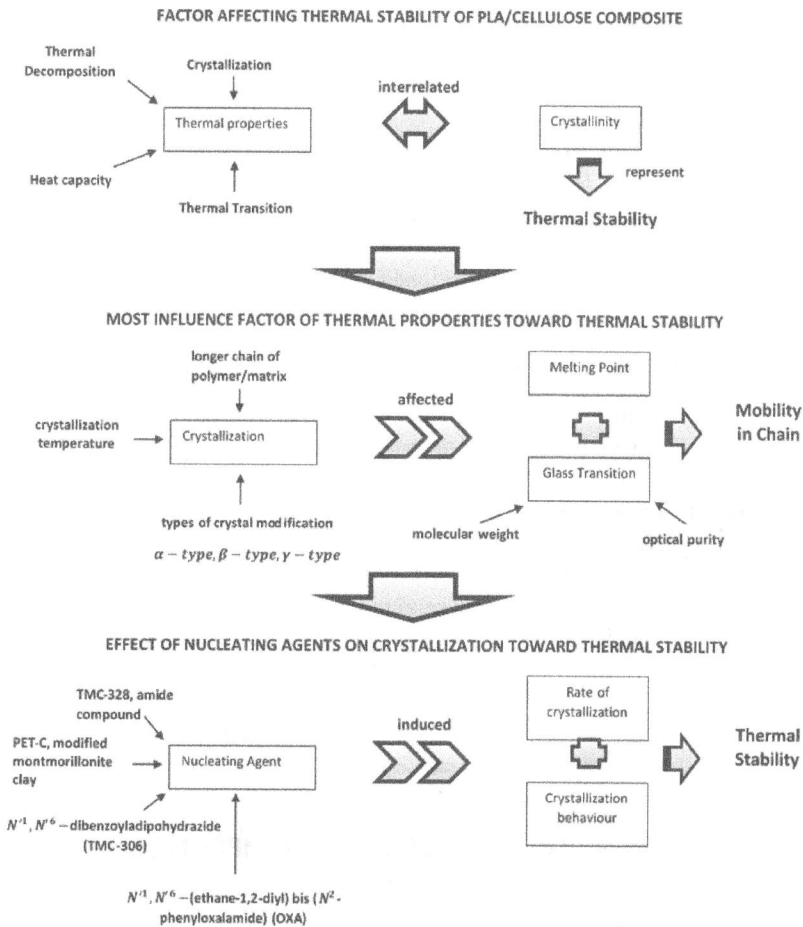

FIGURE 7.1 Factors affecting the thermal stability of PLA/cellulose composite.

7.5   TGA PERFORMANCE OF PLA/CELLULOSE COMPOSITE

One of the most important aspects of materials is their thermal stability, particularly when considering the heat involved in their preparation when using thermoplastic processing techniques. The evaluation of degradation performance encouraged researchers to optimize composite design and processing and develop high-performance polymer with better thermal stability [59, 60]. Generally, the thermal stability of the composite is determined by the polymer matrix and reinforcing fiber used. The use of cellulosic materials in PLA composites as fillers provides an effective and attractive way to enhance the thermal stability of PLA while sustaining the biological and environmental benefits associated with PLA matrices such as biodegradability, cytocompatibility, compostability, and renewability. The surface properties of cellulosic materials are known to play a significant role in fiber–fiber bonding within the cellulose network and interfacial adhesion between the fiber and the PLA matrix, which determines the thermal stability of composites [61].

Regarding this matter, the surface of cellulose materials can be modified to enhance the thermal stability of the resulting composites due to the strong interaction that enables the polymer to be insulated from heat. Several researchers have reported on the functionalization of cellulose materials to improve their dispersion and interaction with PLA matrix, including acetylation, silanization, polyethylene glycol (PEG), and poly(ethylene oxide) (PEO). Other factors that influenced the thermal stability of PLA/cellulose composites included cellulose types and size and the processing techniques used for composites fabrication. Table 7.3 summarizes the recent studies on the effect of cellulose reinforcement on the thermal stability of PLA composites using different types of cellulose, surface modification, and composites fabrication methods.

7.6   DSC PERFORMANCE OF PLA/CELLULOSE COMPOSITE

DSC is a popular thermoanalytical technique used to study the thermal properties of composites such as Tg, Tc, Tm, enthalpy of crystallization (ΔHc), enthalpy of fusion (ΔHm), and degree of crystallinity (Xc). DSC can also illustrate solid-state transitions in composites, including eutectic points, melting, and conversions of different crystalline phases such as polymorphic forms, crystallization, and re-crystallizations and degradation. Focusing on PLA/cellulose composites, the incorporation of cellulose improved the thermal properties of PLA composites, especially their crystallinity and Tm. Meanwhile, cellulose materials can act as a nucleating agent in composites, increasing their crystallization rate and overall crystallinity and thus improving their thermal properties. Table 7.2 summarizes recent studies regarding DSC properties of PLA/cellulose composites prepared from different cellulose types, surface modifications, and fabrication techniques.

7.7   APPLICATIONS OF PLA/CELLULOSE COMPOSITES

PLA has a wide range of applications, including food, personal care, telecommunications, fibers, and even automotive [75, 76]. However, it should be noted that the diversity of PLA application areas is due to their functionality [56]. This diversity

TABLE 7.3

Recent Studies on the Reinforcement Effect of Cellulose on the TGA of PLA Composites

Cellulose Source	Type of Cellulose	Additive/ Modification	Fabrication Method	Thermal Properties	References
Komagateibacter xylinus, hydrated softwood	BNC, CNF	–	Solution casting	• PLA/BNC films were more thermally stable than neat PLA and PLA/CNF films below 250°C. Above 330°C, PLA/CNF exhibited better thermal stability than PLA/BNC.	[62]
Cassava pulp	CNF	–	Melt blending	• As compared to neat PLA, the thermal stability of PLA/CNF biocomposites did not improve significantly.	[63]
Hardwood kraft pulp	CNF	–	In situ polymerization	• No significant differences were detectable among all the nanocomposites nor in comparison with neat PLA sample at degradation temperatures relative to 5% (T5%), 50% (T50%), and 95% (T95%) weight loss.	[64]
Bleached softwood sulfate fluff pulp	Cellulose	Surface modified using PEO	Extrusion	• PLA and PLA/CF composites degrade in a single step, while PLA/PEO/CF composites degrade in two steps. • The addition of CF to composites increases overall thermal stability, which may be attributed to the synergistic effect of PEO and CF on the PLA, which leads to better interfacial adhesion.	[65]
Not stated	CNF	Freeze drying	Melt blending	• Single degradation steps are observed for neat PLA and PLA/CNF composites with almost no char residue apparent at temperatures above 450°C. • The degradation temperature of PLA/CNF composites decreased with the increase in CNF material, which can be due to certain portions of the amorphous PLA being occupied by the less thermally stable cellulose fibers, reducing the composite's overall thermal stability.	[66]
Not stated	Cellulose	Surface modification via silanization	Compression molding	• The PLA/PHB/S (S-silanized cellulose paper) sample exhibited a significant shift in the primary degradation step, resulting in a decrease in thermal stability due to the presence of a cellulose interlayer.	[70]

(Continued)

TABLE 7.3 (Continued)
Recent Studies on the Reinforcement Effect of Cellulose on the TGA of PLA Composites

Cellulose Source	Type of Cellulose	Additive/ Modification	Fabrication Method	Thermal Properties	References
Textile fabric waste	CNF	–	Extrusion	• The T_{onset} reduced with the addition of CNF, while the percentage decomposition of the composite increased and less residue was found with the addition of CNF. • The DTG showed that biocomposites thermal stability is lowered with the addition of CNF	[71]
Not stated	CNF	–	Electrospinning	• All of the PLA/CNF composite nanofibers had a higher initial degradation temperature than the PLA fiber, suggesting that the composite nanofiber thermal stability had improved due to the addition of CNF.	[48]
MCC	CNC	–	Solvent casting	• The degradation of CNC-based composite was shifted to a lower range as compared to the MCC-based composite, owing to a reduction in cellulose particle size, resulting in higher surface area per unit volume and a higher heat transfer rate.	[72]
Not stated	MCC	Incorporated with activated biochar (AC)	Extrusion	• The thermal stability of PLA was found to be negatively affected by the addition of MCC, but the addition of AB overcame this drawback.	[73]
MCC	CNF	Freeze drying of CNF Addition PEG600 as a smoothing agent	Extrusion	• The thermal stability of the PLA/PEG600/CNF composite increases at low CNF additions and decreases as the CNF wt.% increases, implying that increasing the CNF proportion causes a significant shift in the composite's thermal degradation behavior.	[74]

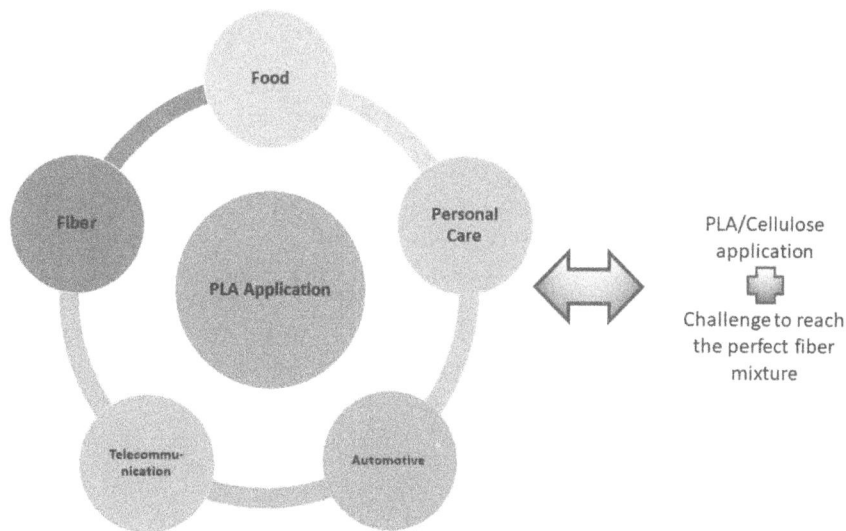

FIGURE 7.2 Correlation between PLA and PLA/cellulose application.

of functions is due to the unique features of PLA, namely, low weight, recyclability, higher strength, and stiffness [77]. However, the drawbacks of PLA are the high processing cost, brittleness, and low thermal conductivity, which somehow hinder its more comprehensive application [78, 79]. A reasonable approach to tackling this issue could be to reinforce the PLA with fiber/cellulose to enhance the aspect ratio and improve polymer crystallinity. This could provide better thermal stability with exceptional quality in terms of sustainability, renewability, recyclability, low density, biodegradability, and lower production cost [80, 81]. In addition, cellulose is known as the most abundant renewable material resource globally, at which its addition is expected to lower the price for PLA/cellulose composites. An implication of this is the possibility that a change of polymer crystallinity, aspect ratio, dispersion, network formation, content, and properties of reinforcement could occur [81–83]. Therefore, the main challenge is to produce PLA/cellulose that can recreate the variety of functions of the composites described in Table 7.4 at a low price. Previous researchers demonstrated a correlation between PLA and PLA/cellulose application, where a similar application will be served (Figure 7.2).

7.8 CONCLUSIONS AND FUTURE RECOMMENDATIONS

PLA/cellulose fiber composites have a vast potential to be developed as an ecologically friendly alternative to the petroleum-based polymer that functions in various applications. The challenges and concerns posed by the PLA and incompatibility between the PLA and cellulose fiber have been and are being addressed by many researchers. These limitations could be alleviated through chemical modifications, reinforcement, plasticizing, the addition of coupling agents, and others. Both techniques, TGA and DSC, are helpful to gain information on thermal behavior related to

TABLE 7.4

Potential Application with Market Product and Company Based on PLA Grade and Production Type from NatureWorks, the Largest PLA Producer Globally

Production Type	PLA Grade	Application	Company	Market Product	References
Thermoform/ injection molding	2003D	Food packaging for beverages, food containers, food service ware	– Naturally Iowa – Sant' Anna, Swangold, Cool Change, Good Water, Primo Water	– Bottles for debuted Yogurt 7.0 – Bottles for juice and still water	[84]
	3001D	Food service wear with heat exposure <55°C		– SunChips – ECO OctaView and ECO Expressions	
	3051D	Any application for heat exposure temperature, <55°C	– Frito-Lay – InnoWare Plastics	– Dyne-A-Pak Nature tray	
	3251D	Application for higher melting flow capability, higher gloss, UV resistance, and stiffness	– Dyne-A-Pak – Foam trays – CDS srl	– Trays for meat, fish, and cheese – Cutlery	
	3801X	Any application for heat exposure ranging from 65°C to 140°C without food contact			
Film and bottle grade	4043D	Biaxial oriented film application with specifications such as superior optics, twist, and dead fold, block to flavor, grease, and superior oil resistance	– Natures Organics – Priori – Cargo Cosmetics – Carrefour Belgium	– Beauty care/shampoo bottles in Australia – Cosmetic packaging/ CoffeeBerry – Casings for cosmetics – Clear film overwrap for trays	–

the thermal stability, composition, and kinetic decompositions of materials. Based on the findings of the existing literature, several factors are known to affect the thermal performance of the composites include polymer/matrix molecular weight, polymerization condition, thermal history, and purity of the material used. In essence, it is essential to have good cellulose fiber dispersion in the PLA matrix, increasing the PLA/cellulose interfacial interaction that will increase thermal stability.

Furthermore, cellulose materials act as a nucleating agent in composites, increasing their crystallization rate and overall crystallinity, thereby improving their thermal properties. Overall, the addition of cellulose fiber had a positive impact on the thermal properties of the composites. It is interesting to investigate the fundamental theory that could dictate the thermal characteristics of the composites. Therefore, a theoretical study based on simulation and modeling would give a clear picture of this matter. Such tools will provide valuable information to design a hybrid system with intended features.

ACKNOWLEDGMENT

We would like to express our gratitude to Universiti Pertahanan Nasional Malaysia (UPNM) for the opportunity and financial support.

REFERENCES

1. Kamarudin, S.H., et al., *Thermal and structural analysis of epoxidized jatropha oil and alkaline treated kenaf fiber reinforced poly (Lactic acid) biocomposites. Polymers*, 2020. **12**(11): p. 2604.
2. Ishak, W.H.W., N.A. Rosli, and I. Ahmad, *Influence of amorphous cellulose on mechanical, thermal, and hydrolytic degradation of poly (lactic acid) biocomposites. Scientific Reports*, 2020. **10**(1): pp. 1–13.
3. Sangeetha, V.H., et al., *State of the art and future prospectives of poly (lactic acid) based blends and composites. Polymer Composites*, 2018. **39**(1): pp. 81–101.
4. Kamarudin, S.H., et al., *Mechanical and physical properties of kenaf-reinforced poly (lactic acid) plasticized with epoxidized jatropha oil. BioResources*, 2019. **14**(4): pp. 9001–9020.
5. González-López, M.E., et al., *Polylactic acid functionalization with maleic anhydride and its use as coupling agent in natural fiber biocomposites: a review. Composite Interfaces*, 2018. **25**(5–7): pp. 515–538.
6. Kamarudin, S.H., et al., *A study of mechanical and morphological properties of PLA based biocomposites prepared with EJO vegetable oil based plasticiser and kenaf fibres. Materials Research Express*, 2018. **5**(8): p. 085314.
7. Mukherjee, T. and N. Kao, *PLA based biopolymer reinforced with natural fibre: a review. Journal of Polymers and the Environment*, 2011. **19**(3): p. 714.
8. Ayu, R.S., et al., Bioplastics: The Future of Sustainable Biodegradable Food Packaging, in *Bio-based packaging: material, environmental and economic aspects*. 2021, Wiley, pp. 335–351.
9. Norizan, M.N., et al., Treatments of natural fibre as reinforcement in polymer composites-short review. *Functional Composites and Structures*, 2021. **3**, p. 024002.
10. Baihaqi, N.M.Z.N., et al., *Effect of fiber content and their hybridization on bending and torsional strength of hybrid epoxy composites reinforced with carbon and sugar palm fibers. Polimery*, 2021. **66**(1): pp. 36–43.

11. Norrrahim, M.N.F., et al., *Emerging development on nanocellulose as antimicrobial material: an overview. Materials Advances*, 2021. **2**, pp. 3538–3551.

12. Nurazzi, N.M., et al., *A review on natural fiber reinforced polymer composite for bullet proof and ballistic applications. Polymers*, 2021. **13**(4): p. 646.

13. Norizan, N.M., et al., Green Materials in Hybrid Composites for automotive applications: Green materials, in *Implementation and evaluation of green materials in technology development: emerging research and opportunities*. 2020, IGI Global: pp. 56–76.

14. Ilyas, R.A., et al., Properties and characterization of PLA, PHA, and other types of bio-polymer composites, in *Advanced processing, properties, and applications of starch and other bio-based polymers*. 2020, Elsevier: pp. 111–138.

15. Ilyas, R.A., et al., *Polylactic acid (PLA) biocomposite: processing*, additive manufacturing and advanced applications. *Polymers*, 2021. **13**(8): p. 1326.

16. Mohd Nurazzi, N., et al., *Curing behaviour of unsaturated polyester resin and interfacial shear stress of sugar palm fibre. Journal of Mechanical Engineering and Sciences*, 2017. **11**(2): pp. 2650–2664.

17. Zin, M.H., et al., *Automated spray up process for pineapple leaf fibre hybrid biocomposites. Composites Part B: Engineering*, 2019. **177**: p. 107306.

18. Aisyah, H.A., et al., *A Comprehensive review on advanced sustainable woven natural fibre polymer composites. Polymers*, 2021. **13**(3): p. 471.

19. Zhang, B.X., J.I. Azuma, and H. Uyama, Preparation and *characterization of a transparent amorphous cellulose film. RSC Advances*, 2015. **5**(4): pp. 2900–2907.

20. Ilyas, R.A., et al., *Sugar palm (Arenga pinnata (Wurmb.) Merr) cellulosic fibre hierarchy: a comprehensive approach from macro to nano scale. Journal of Materials Research and Technology*, 2019. **8**(3): pp. 2753–2766.

21. Xu, C.A., et al., *Effects of polysiloxanes with different molecular weights on in vitro cytotoxicity and properties of polyurethane/cotton–cellulose nanofiber nanocomposite films. Polymer Chemistry*, 2020. **11**(32): pp. 5225–5237.

22. Hu, J., et al. Extract nano cellulose from flax as thermoelectric enhancement material. *Journal of Physics: Conference Series*. 2021. **1790**, p. 012087.

23. Shah, S.S., et al., *Present status and future prospects of jute in nanotechnology: a review. The Chemical Record*, 2021. **21**, pp. 1631-1665.

24. Beluns, S., et al., *From wood and hemp biomass wastes to sustainable nanocellulose foams. Industrial Crops and Products*, 2021. **170**: p. 113780.

25. Ilyas, R.A., et al., *Sugar palm (Arenga pinnata [Wurmb.] Merr) starch films containing sugar palm nanofibrillated cellulose as reinforcement: Water barrier properties. Polymer Composites*, 2020. **41**(2): pp. 459–467.

26. Yang, Y., et al., *Bio-based antimicrobial packaging from sugarcane bagasse nanocellulose/nisin hybrid films. International Journal of Biological Macromolecules*, 2020. **161**: pp. 627–635.

27. Sabaruddin, F.A., et al., *The effects of unbleached and bleached nanocellulose on the thermal and flammability of polypropylene-reinforced kenaf core hybrid polymer bionanocomposites. Polymers*, 2021. **13**(1): p. 116.

28. Sharip, N.S., et al., A review on nanocellulose composites in biomedical application, in *Composites in biomedical applications*. 2020, CRC Press: pp. 161–190.

29. Ilyas, R.A., et al., *Effect of hydrolysis time on the morphological, physical, chemical, and thermal behavior of sugar palm nanocrystalline cellulose (Arenga pinnata (Wurmb.) Merr). Textile Research Journal*, 2021. **91**(1–2): pp. 152–167.

30. Zhu, Q., et al., *Stimuli-responsive cellulose nanomaterials for smart applications. Carbohydrate Polymers*, 2020. **235**: p. 115933.

31. Miller, J., *Nanocellulose: producers, products, and applications: a guide for end users.* 2017, TAPPI Press.

32. Rosli, N.A., et al., *Effectiveness of cellulosic Agave angustifolia fibres on the performance of compatibilised poly (lactic acid)-natural rubber blends. Cellulose*, 2019. **26**(5): pp. 3205–3218.

33. Haafiz, M.K.M., et al., *Isolation and characterization of cellulose nanowhiskers from oil palm biomass microcrystalline cellulose. Carbohydrate Polymers*, 2014. **103**: pp. 119–125.

34. Kargarzadeh, H., et al., *Recent developments in nanocellulose-based biodegradable polymers, thermoplastic polymers, and porous nanocomposites. Progress in Polymer Science*, 2018. **87**: pp. 197–227.

35. Norrrahim, M.N.F., et al., *Emerging development of nanocellulose as an antimicrobial material: an overview. Materials Advances*, 2021. **2**(11): pp. 3538–3551.

36. Arjmandi, R., et al., *Partial replacement effect of montmorillonite with cellulose nanowhiskers on polylactic acid nanocomposites. International Journal of Biological Macromolecules*, 2015. **81**: pp. 91–99.

37. Khoo, R.Z., H. Ismail, and W.S. Chow, *Thermal and morphological properties of poly (lactic acid)/nanocellulose nanocomposites. Procedia Chemistry*, 2016. **19**: pp. 788–794.

38. De Paula, E.L., V. Mano, and F.V. Pereira, *Influence of cellulose nanowhiskers on the hydrolytic degradation behavior of poly (D, L-lactide). Polymer Degradation and Stability*, 2011. **96**(9): pp. 1631–1638.

39. Lee, J.H., S.H. Park, and S.H. Kim, *Preparation of cellulose nanowhiskers and their reinforcing effect in polylactide. Macromolecular Research*, 2013. **21**(11): pp. 1218–1225.

40. Hossain, K.M.Z., et al., *Physico-chemical and mechanical properties of nanocomposites prepared using cellulose nanowhiskers and poly (lactic acid). Journal of Materials Science*, 2012. **47**(6): pp. 2675–2686.

41. Petersson, L., I. Kvien, and K. Oksman, *Structure and thermal properties of poly (lactic acid)/cellulose whiskers nanocomposite materials. Composites Science and Technology*, 2007. **67**(11–12): pp. 2535–2544.

42. Scaffaro, R., et al., *Polysaccharide nanocrystals as fillers for PLA based nanocomposites. Cellulose*, 2017. **24**(2): pp. 447–478.

43. Sanchez-Garcia, M.D. and J.M. Lagaron, *On the use of plant cellulose nanowhiskers to enhance the barrier properties of polylactic acid. Cellulose*, 2010. **17**(5): pp. 987–1004.

44. Fortunati, E., et al., *Processing of PLA nanocomposites with cellulose nanocrystals extracted from Posidonia oceanica waste: innovative reuse of coastal plant. Industrial Crops and Products*, 2015. **67**: pp. 439–447.

45. Martínez-Sanz, M., et al., *Incorporation of poly (glycidylmethacrylate) grafted bacterial cellulose nanowhiskers in poly (lactic acid) nanocomposites: Improved barrier and mechanical properties. European Polymer Journal*, 2013. **49**(8): pp. 2062–2072.

46. Pandey, J.K., et al., *Bio-nano reinforcement of environmentally degradable polymer matrix by cellulose whiskers from grass. Composites Part B: Engineering*, 2009. **40**(7): pp. 676–680.

47. Abdulkhani, A., et al., *Preparation and characterization of modified cellulose nanofibers reinforced polylactic acid nanocomposite. Polymer Testing*, 2014. **35**: pp. 73–79.

48. Yang, Z., et al., *Morphological, mechanical and thermal properties of poly (lactic acid) (PLA)/cellulose nanofibrils (CNF) composites nanofiber for tissue engineering. Journal of Wuhan University of Technology-Materials Science Edition*, 2019. **34**(1): pp. 207–215.

49. Wang, Q., et al., *Structure and properties of polylactic acid biocomposite films reinforced with cellulose nanofibrils. Molecules*, 2020. **25**(14): p. 3306.

50. Rasheed, M., et al., *Morphology, structural, thermal, and tensile properties of bamboo microcrystalline cellulose/poly (lactic acid)/poly (butylene succinate) composites. Polymers*, 2021. **13**(3): p. 465.

51. Cao, Z., et al., *Effects of the chain-extender content on the structure and performance of poly (lactic acid)–poly (butylene succinate)–microcrystalline cellulose composites. Journal of Applied Polymer Science*, 2017. **134**(22): p. 44895.

52. Van Holde, K.E., W.C. Johnson, and P.S. Ho, *Principles of physical biochemistry*. 2006, Pearson Prentice Hall.

53. Cooper, A., M.A. Nutley, and A. Wadood, *Protein-ligand interactions: hydrodynamics and calorimetry*. 2000, Oxford University Press.

54. Zhao, H., et al., *Morphology and properties of injection molded solid and microcellular polylactic acid/polyhydroxybutyrate-valerate (PLA/PHBV) blends. Industrial & Engineering Chemistry Research*, 2013. **52**(7): pp. 2569–2581.

55. Wang, Q., et al., *Kinetic thermal behavior of nanocellulose filled polylactic acid filament for fused filament fabrication 3D printing. Journal of Applied Polymer Science*, 2020. **137**(7): p. 48374.

56. Sin, L.T. and B.S. Tueen, *Polylactic acid: a practical guide for the processing, manufacturing, and applications of PLA*. 2019, William Andrew.

57. Wang, L., et al., *Heat resistance, crystallization behavior, and mechanical properties of polylactide/nucleating agent composites. Materials & Design (1980–2015)*, 2015. **66**: pp. 7–15.

58. Xu, T., et al., *Crystallization kinetics and morphology of biodegradable poly (lactic acid) with a hydrazide nucleating agent. Polymer Testing*, 2015. **45**: pp. 101–106.

59. Gan, P.G., et al., *Thermal properties of nanocellulose-reinforced composites: a review. Journal of Applied Polymer Science*, 2020. **137**(11): p. 48544.

60. Nurazzi, N.M., et al., *Thermal properties of treated sugar palm yarn/glass fiber reinforced unsaturated polyester hybrid composites. Journal of Materials Research and Technology*, 2020. **9**(2): pp. 1606–1618.

61. Mokhena, T.C., et al., *Thermoplastic processing of PLA/cellulose nanomaterials composites. Polymers*, 2018. **10**(12): p. 1363.

62. Gitari, B., et al., *A comparative study on the mechanical, thermal, and water barrier properties of PLA nanocomposite films prepared with bacterial nanocellulose and cellulose nanofibrils. BioResources*, 2019. **14**(1): pp. 1867–1889.

63. Nguyen, T.C., C. Ruksakulpiwat, and Y. Ruksakulpiwat, *Effect of cellulose nanofibers from cassava pulp on physical properties of poly (lactic acid) biocomposites. Journal of Thermoplastic Composite Materials*, 2020. **33**(8): pp. 1094–1108.

64. Gazzotti, S., et al., *Cellulose nanofibrils as reinforcing agents for PLA-based nanocomposites: An in situ approach. Composites Science and Technology*, 2019. **171**: pp. 94–102.

65. Singh, A.A., et al., *Green processing route for polylactic acid–cellulose fiber biocomposites. ACS Sustainable Chemistry & Engineering*, 2020. **8**(10): pp. 4128–4136.

66. Xu, L., et al., *Green-plasticized poly (lactic acid)/nanofibrillated cellulose biocomposites with high strength, good toughness and excellent heat resistance. Composites Science and Technology*, 2021. **203**: p. 108613.

67. Shazleen, S.S., et al., *Functionality of cellulose nanofiber as bio-based nucleating agent and nano-reinforcement material to enhance crystallization and mechanical properties of Polylactic Acid Nanocomposite. Polymers*, 2021. **13**(3): p. 389.

68. Trifol, J., et al., *Impact of thermal processing or solvent casting upon crystallization of PLA nanocellulose and/or nanoclay composites. Journal of Applied Polymer Science*, 2019. **136**(20): p. 47486.

69. Sousa, S., et al., *Poly (lactic acid)/cellulose films produced from composite spheres prepared by emulsion-solvent evaporation method. Polymers*, 2019. **11**(1): p. 66.

70. Radu, E.R., et al., *The soil biodegradability of structured composites based on cellulose cardboard and blends of polylactic acid and polyhydroxybutyrate. Journal of Polymers and the Environment*, 2021. **29**, pp. 2310–2320.

71. Rizal, S., et al., *Isolation of textile waste cellulose nanofibrillated fibre reinforced in polylactic acid-chitin biodegradable composite for green packaging application. Polymers*, 2021. **13**(3): p. 325.

72. Bhiogade, A. and M. Kannan, Studies on thermal and degradation kinetics of cellulose micro/nanoparticle filled polylactic acid (PLA) based nanocomposites. *Polymers and Polymer Composites*, 2021. **29**, pp. S85–S98. https://doi.org/10.1177/0967391120987170

73. Zhang, Q., et al., *Improvement on the properties of microcrystalline cellulose/polylactic acid composites by using activated biochar. Journal of Cleaner Production*, 2020. **252**: p. 119898.

74. Wang, Q., et al., *Cellulose nanofibrils filled poly (Lactic Acid) biocomposite filament for FDM 3D printing. Molecules*, 2020. **25**(10): p. 2319.

75. Arjmandi, R., A. Hassan, and Z. Zakaria, Polylactic acid green nanocomposites for automotive applications, in *Green biocomposites*. 2017, Springer: pp. 193–208.

76. Notta-Cuvier, D., et al., *Tailoring polylactide properties for automotive applications: effects of co-addition of halloysite nanotubes and selected plasticizer. Macromolecular Materials and Engineering*, 2015. **300**(7): pp. 684–698.

77. Oksman, K., M. Skrifvars, and J.F. Selin, *Natural fibres as reinforcement in polylactic acid (PLA) composites. Composites Science and Technology*, 2003. **63**(9): pp. 1317–1324.

78. Frone, A.N., et al., *Morphology and thermal properties of PLA–cellulose nanofibers composites. Carbohydrate Polymers*, 2013. **91**(1): pp. 377–384.

79. Suryanegara, L., A.N. Nakagaito, and H. Yano, *The effect of crystallization of PLA on the thermal and mechanical properties of microfibrillated cellulose-reinforced PLA composites. Composites Science and Technology*, 2009. **69**(7–8): pp. 1187–1192.

80. Frone, A.N., et al., *Cellulose fiber-reinforced polylactic acid. Polymer Composites*, 2011. **32**(6): pp. 976–985.

81. Lu, J., P. Askeland, and L.T. Drzal, *Surface modification of microfibrillated cellulose for epoxy composite applications. Polymer*, 2008. **49**(5): pp. 1285–1296.

82. Huda, M.S., et al., *A study on biocomposites from recycled newspaper fiber and poly (lactic acid). Industrial & Engineering Chemistry Research*, 2005. **44**(15): pp. 5593–5601.

83. Iwatake, A., M. Nogi, and H. Yano, *Cellulose nanofiber-reinforced polylactic acid. Composites Science and Technology*, 2008. **68**(9): pp. 2103–2106.

84. Jamshidian, M., et al., *Poly-lactic acid: production, applications, nanocomposites, and release studies. Comprehensive Reviews in Food Science and Food Safety*, 2010. **9**(5): pp. 552–571.

8 Dynamic Mechanical Thermal Analysis (DMTA) of Polylactic Acid (PLA)/ Cellulose Composite

Sandeep Kumar
University of Warwick
Coventry, UK

Georg Graninger
Queen's University Belfast
Belfast, UK

Brian G. Falzon
RMIT University
Melbourne, Australia

and

Queen's University Belfast
Belfast, UK

CONTENTS

8.1 INTRODUCTION

Polylactic acid (PLA) is currently considered one of the most promising, commercially scalable, recyclable bioplastics. More accurately referred to as poly(lactic acid), it possesses excellent thermal processability, biocompatibility, biodegradability, high Young's modulus, high transparency, and properties comparable to conventional

DOI: 10.1201/9781003160458-8

petrochemical-based polymers (Mukherjee, 2013; Oksman, 2003). The poly(lactic) acid (PLA) basic monomer is lactic acid, which is derived from starch by a fermentation process and then used to synthesize PLA through different polymerization techniques such as ring-opening polymerization, polycondensation, and other methods (i.e., azeotropic dehydration and enzymatic polymerization) (Mokhena, 2018; Scaffaro, 2017). However, drawbacks such as poor thermo-mechanical properties, slow crystallization, brittleness, and poor barrier properties have limited its wider utilization (Bledzki, 2015). Expanding its scope of application is a primary objective of considerable current scientific research. There is enormous emphasis on property improvement by the inclusion of bio-based additives. If the reinforcement agent is also bio-degradable and can be extracted from renewable resources, it can potentially enhance the exploitation and biodegradability profile of PLA. For instance, there have been studies available in literature that show that microcrystalline cellulose (MCC) (Li, 2016; Paul, 2021), cellulose nanocrystals (CNCs) (Petersson, 2007), and cellulose nanofibrils (CNFs) (Jonoobi, 2012) can significantly enhance the thermo-mechanical performance of PLA composites. Thermo-mechanical properties of PLA/cellulose are extensively studied by dynamic mechanical thermal analysis (DMTA).

8.2 DYNAMIC MECHANICAL THERMAL ANALYSIS (DMTA)

DMTA is a thermal analysis technique that measures the viscoelastic response of materials over a broad temperature range, as they are deformed under periodic stress (Menard, 2008). In DMTA, a sinusoidal deformation (stress(σ)) is applied, which results in a sinusoidal deformation (strain (ϵ)) under control temperature/or frequency program, as shown in Figure 8.1. DMTA measures stiffness and damping, which are reported as *modulus* and *tan δ*. Because we apply a sinusoidal stress, the response signal strain (ϵ) is split into two components. We express the storage modulus, E', as an in-phase component and loss modulus, E'', as an out of phase component (Menard, 2008). The storage modulus provides a measure of elastic response/energy absorption ability of materials and as well as the molecular relaxation taking place as a function of temperature. The loss modulus represents the viscous response of a material and the amount of energy dissipated in a sample during one cyclic load. Tan δ is the ratio of loss modulus to storage modulus, E''/E', and is often called

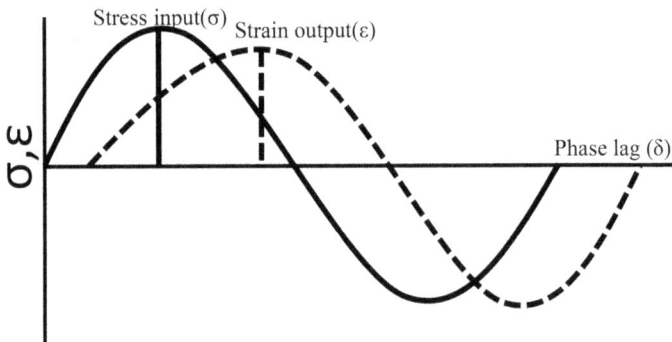

FIGURE 8.1 Relationship between applied sinusoidal stress, strain, and phase lag. (Redrawn and modified from Jones (2012).)

damping. It is a measure of the energy dissipation of a material. A higher area under the tan δ peak suggests higher energy dissipation in the system. It is worth mentioning that the maximum tan δ peak occurs at the glass transition temperature of the polymeric material.

8.3 DMTA OF PLA/CELLULOSE COMPOSITES

DMTA is extensively used to study the thermo-mechanical response of PLA/cellulose composites and is highly sensitive to microstructural changes. It is found that storage modulus, E', glass transition temperature (T_g), and damping (tan δ) properties of PLA composites are significantly affected by the types of cellulose reinforcements, i.e., CNC, cellulose nanofibers (CNF) or MCC, and filler/matrix interfacial interactions.

8.3.1 PLA/CELLULOSE NANOCRYSTALS (CNC) COMPOSITES

Among bio-based fillers, CNCs are extensively used to enhance the thermo-mechanical performance of PLA. Several studies have shown that loading level, processing methods, and surface modification of CNCs, can significantly affect the storage modulus, E' and glass transition temperature (T_g) of the PLA as summarized in Table 8.1.

Sullivan et al. used CNC obtained from wood pulp to reinforce PLA. The PLA/CNC films were fabricated by a unique two-step process: (1) melt compounding using direct liquid feeding, followed by melt fiber spinning and (2) compression molding. Compared to the neat PLA, PLA/CNC composites showed remarkable enhancement in E', below and above T_g. In the glassy state region at 35°C, incorporation of 3.0 wt.% CNC, increased E' by 47.0% with respect to the neat PLA. Above T_g (rubbery state) at 70°C, E' of the PLA/CNC composite significantly increased by 280%. This remarkable jump in E', below and above T_g, is attributed to the uniform dispersion of CNCs within the PLA matrix and increased polymer matrix crystallinity coupled with the hindered polymer chain mobility caused by the addition of CNCs. Moreover, the addition of 3.0 wt.% CNCs to PLA composites increased the T_g to 82.2°C from 79.3°C for the neat PLA (Sullivan, 2015). Several research papers deal with the optimization of chemical interfacial interactions between CNCs and PLA at a molecular level. With surface functionalization of CNCs, the strong interfacial interactions between CNCs and PLA leads to an effective stress transfer from matrix to CNCs and consequently strengthening of the resulting composites. In particular, the optimum improvement in the thermo-mechanical performance of PLA/CNC composites can be achieved through the uniform dispersion of CNCs in PLA matrix. To this end, to overcome the major issues of non-uniform dispersion of CNCs in hydrophobic PLA matrix, Shojaeiarani et al. studied the influence of unmodified and modified CNCs on thermo-mechanical properties of PLA/CNC composites. Modified CNCs were prepared by introducing benzoic acid as a grafting agent, as shown in Figure 8.2.

The PLA/CNC composites were prepared using a masterbatch approach: firstly, PLA/CNC masterbatch films with 15 wt.% of CNC (un-modified and modified) were prepared using a solvent casting approach. These films were chopped into small pieces and diluted to a final concentration of 1.0 wt.% and 3.0 wt.% of CNC in the PLA/CNC composites through melt processing followed by injection molding. It was observed that E' of composites gradually increased with the addition of CNCs and benzoic

TABLE 8.1

Dynamic Mechanical Thermal Analysis (DMTA) Data for PLA/CNC Composites

Composite System	$\Delta E'$ (%) Glassy State at 35°C	$\Delta E'$ (%) Rubbery State at 70°C	Glass Transition Temp. (ΔT_g) °C	Processing Procedure	References
PLA/CNC (3.0 wt.%)	47%	280%	+3	Melt compounding + compression molding	Sullivan (2015)
PLA/lignin coated-CNC (0.5 wt.%)	60%	5900%	–	Melt mixed master batch further diluted in melt mixing + compression molding	Gupta (2017)
PLA/acetylated-CNC (0.5 wt.%)	8%	500%	+2	Solution mixing + hot pressing	Cho (2013)
PLA/maleic anhydride grafted-CNC (5.0 wt.%)	25%	218%	+9	Solution mixing	Zhou (2018)
PLA/lactate groups modified CNC (5.0 wt.%)	31%	450%	+5	Melt compounding + injection molding	Spinella (2015)
PLA/ methacryloxy-based silane modified CNC (3.0 wt.%)	235%	–	–	Melt compounding + injection molding	Raquez (2012)
PLA/CNC (15%)	15%	378%	+8	Pickering emulsion approach + compression molding	Zhang (2019)

acid-CNCs as compared with neat PLA, which indicates the reinforcing effect of CNCs. In particular, E′ of the benzoic acid-modified PLA/CNCs composites was higher when compared to the unmodified PLA/CNC composites in the glassy state, which is attributed to the enhanced interfacial bonding between PLA and benzoic acid-modified-CNCs, which resulted in a better stress transfer from PLA matrix to CNCs. The maximum improvement in storage modulus was observed in benzoic acid-modified CNC/PLA composite at 3.0 wt.% loading of CNCs. The storage modulus, E′, tan δ peak value, and glass transition temperature (T_g) are tabulated in Table 8.2. The temperature at the peak value of tan δ curve is ascribed as the glass transition temperature (T_g) of the PLA matrix and PLA/CNC composites.

DMTA results confirm that the incorporation of either unmodified or modified CNCs into PLA matrix exhibits a very slight increase in the glass transition

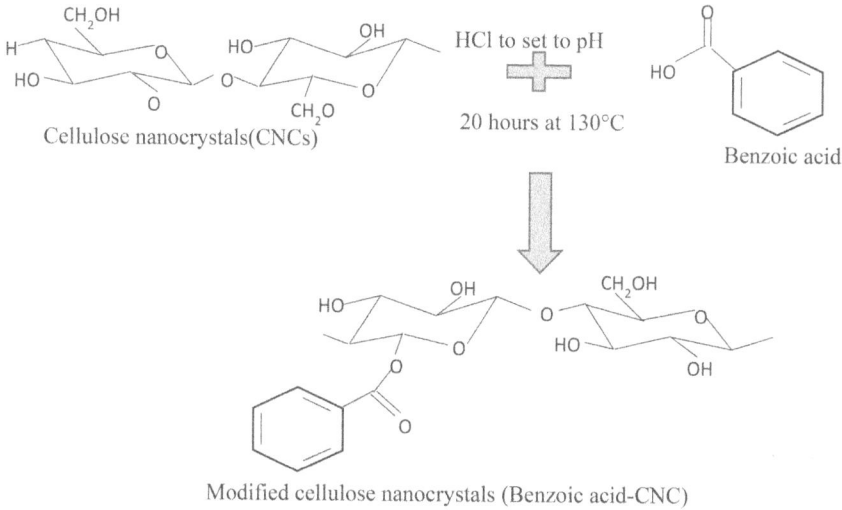

FIGURE 8.2 Esterification process to modify CNCs. (Redrawn from Shojaeiarani (2018).)

temperature of composites. The results imply that the CNCs played little role in the immobilization of the polymer chains. In another report, Shojaeiarani et al. used a time-efficient esterification technique using valeric acid to enhance the compatibility between CNCs and PLA (Shojaeiarani, 2019). Figure 8.3 shows the proposed process for the esterification of CNCs (e-CNCs).

DMTA reveals that the E′ of the PLA composites increases with increasing the loading of CNCs from 1.0 wt.% to 3.0 wt.%. This increase is rather small for the PLA composites containing unmodified CNCs and it becomes significant for the esterified (e-CNC)/PLA composites at same loading levels. For comparison, the storage modulus of all PLA composites samples at two different temperatures (above and below the glass transition temperature) are summarized in Table 8.3. Interestingly, in the glassy state, E′ increased (at 30°C) by 58% for the e-CNC/PLA composite at

TABLE 8.2
Dynamic Mechanical Properties of PLA and PLA/CNC Composites Obtained from DMTA

Composites	E′ (MPa) at 30°C	E′ (MPa) at 70°C	Tan δ Peak Values	T_g (°C)
PLA	1944	15.62	2.72	64.05
PLA-1.0 wt.% CNC	2524	8.69	2.62	66.77
PLA-3.0 wt.% CNC	2708	15.99	2.15	65.12
PLA-1.0 wt.% benzoic acid-CNC	2740	11.54	1.90	64.23
PLA-3.0 wt.% benzoic acid-CNC	3041	11.48	1.81	64.05

Source: Data from Shojaeiarani (2018).

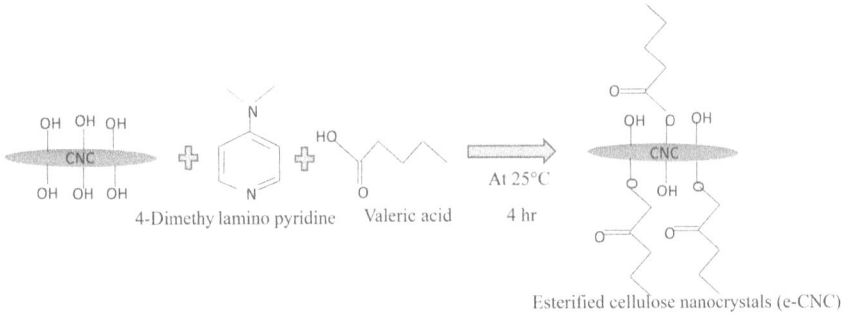

FIGURE 8.3 Proposed process for esterification of CNCs (e-CNCs). (Redrawn from Shojaeiarani (2019).)

3.0 wt.% CNCs compared to neat PLA and by 40% compared to that of unmodified CNC/PLA composite. It is expected due to improved compatibility between e-CNC and PLA, the applied stresses easily transferred from the matrix the CNCs, resulting in an enhancement of the mechanical properties.

In a rubbery state at 120°C, the e-CNCs/PLA composites showed a substantial jump (by 1296%) in storage modulus compared to that of neat PLA. This increase in E′ with an increase in temperature suggests that CNCs can help to retain modulus at higher temperatures. This dramatic increase in storage modulus at higher temperatures indicates that the interaction between functional groups present on the surface of e-CNCs and the PLA are strong enough to restrict polymer chain mobility. Additionally, the height of the tan δ peak decreased with an increase in the CNCs loading in the PLA composites. The height of the tan δ peak is a measure of the energy dissipation characteristics of a material. Incorporation of CNCs into PLA induced a reduction in the tan δ peak value, suggesting a stiffening effect of nanofillers in composites. The lower tan δ peak values for e-CNC/PLA composites as compared to CNC/PLA composites, simply suggest that the strong interactions and compact packing of the e-CNC in the matrix resulted in a decrease in damping properties. The better filler-matrix interaction strongly restricts the polymer chains and result in less energy dissipation at the interface (Shojaeiarani, 2019).

Gupta and co-workers investigated the effect of lignin-coated cellulose nanocrystals (L-CNCs) on the thermo-mechanical properties of PLA composites (Gupta, 2017).

TABLE 8.3
Dynamic Mechanical Properties of PLA and PLA/CNC Composites Obtained from DMTA

Composites	E′ (MPa) at 30°C	E′ (MPa) at 120°C	Tan δ Peak Values
PLA	2413	36.18	2.78
PLA-1.0 wt.% CNC	2688	65.91	2.54
PLA-3.0 wt.% CNC	2715	85.39	2.23
PLA-1.0 wt.% esterified-CNC	3927	167.75	1.98
PLA-3.0 wt.% esterified-CNC	3812	503.15	1.73

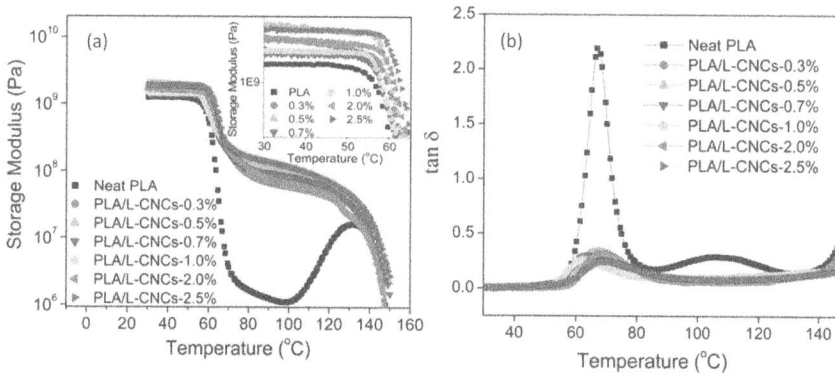

FIGURE 8.4 (a) DMTA chart shows a change in storage modulus as a function of temperature (the larger image shows the increase in storage modulus in the glassy region) and (b) Plot of tan δ vs temperature for neat PLA and PLA/L-CNC composites at 1 Hz. (Reprinted with permission from Gupta (2017). Copyright © 2017 American Chemical Society.)

Lignin is the second most abundant biopolymer on earth, which is basically amorphous macromolecules consisting of a combination of hydrocarbon aliphatic chains and aromatic hydroxyl and carboxyl groups. It has been found to improve interactions with hydrophobic polymers like PLA. Figure 8.4 clearly shows that the incorporation of only 0.5 wt.% of L-CNCs improved the storage modulus, E′, by 60% compared to the neat PLA. This shift in E′ indicates that lignin not only improved the dispersion of CNCs within the PLA matrix, but also enhanced the interfacial interactions with PLA chains.

The observed improvement in E′ of PLA composites is also attributed to the increase in the degree of crystallinity because of the excellent dispersion and strong compatibility of L-CNCs with PLA promoting a high density of nucleating sites. The significant decrease in the tan δ peak height of L-CNC/PLA composites compared to the neat PLA indicates that uniformly dispersed CNC significantly affected the chain mobility due to the confinement effect (Gupta, 2017). Zhou et al. prepared CNCs from cottonseed hulls, available as an abundant and inexpensive agricultural waste (Zhou, 2018). They designed a very novel route to enhance the dispersion and compatibility of CNCs with PLA matrix by maleic anhydride (MA) modification of CNCs and further cross-linking between PLA and MA-CNCs using an initiator (DCP: Dicumyl peroxide) during melt processing. The solution casting approach was used to prepare PLA/CNC (designated as PCX) and PLA/MA-CNC composites (designated as: PCmX), but to prepare PLA/MA-CNC/DCP cross-linked composites (designated as PDCmX), the solution casted films were hot pressed at melting temperature. X represents the weight ratio of CNCs (MA-CNCs) in PLA. For example, X = 1, 3, 5 represents 1 wt.%, 3 wt.%, and 5 wt.% CNCs/(MA-CNCs) in PLA matrix, respectively. The results show that addition of unmodified CNCs at 5 wt.% loadings (PC5) led to a slight decrease in E′ (<5%) in the glass region, possibly due to the poor interfacial interactions and aggregation of unmodified CNCs in PLA matrix, as shown in Figure 8.5. However, the incorporation of MA-CNCs resulted in an obvious increase of E′ by 15.6% for PCm5 in the glassy region as compared to neat PLA, which clearly reflects that improved dispersion of MA-CNCs facilitated the effective stress transfer from PLA matrix to CNCs. Interestingly for

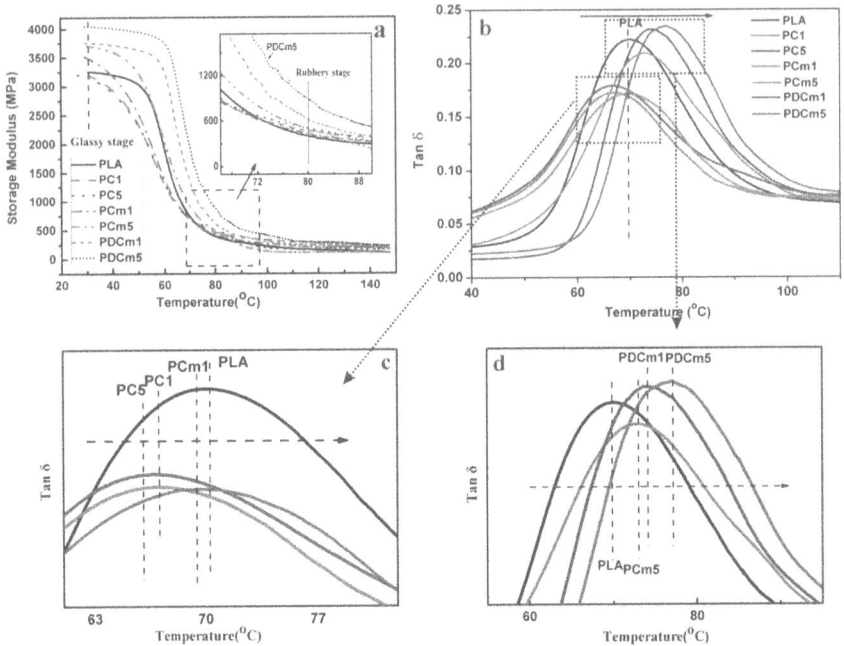

FIGURE 8.5 Storage modulus curves (a) and tan δ (b, c, d) peaks from DMTA analysis of PLA, PC, PCm, and PDCm films at CNCs loadings of 1 wt.% and 5 wt.%. (Reprinted with permission from Zhou (2018). Copyright© 2017 Elsevier B.V. All rights reserved.)

the cross-linked PLA/MA-CNCs (PDCm) composite films, the E′ showed a remarkable increase by 24.3% for PDCm5 (24.3%), which is higher than PLA, PC, and PCm composite films. It is assumed that the formation of cross-linked PLA-MA-CNCs structure largely enhanced the interfacial adhesion between PLA and CNCs. Moreover, for PDCm5 composite film, the tan δ peaks (T_g) showed a significant shift to a higher temperature to 78°C from 69°C for neat PLA, which further served as compelling evidence for enhanced intermolecular interaction between PLA matrix and MA-CNCs after cross-linking. Scanning electron microscopic (SEM) examination of the fractured surfaces of PDCm5 films further validates the presence of individual strongly bonded MA-CNCs to PLA molecular chains and therefore effective stress transfer from PLA to CNCs (Figure 8.6).

Similarly, Raquez et al. studied the influence of different silane coupling agents modified CNCs on the properties of PLA/CNC composites prepared using the melt-extrusion technique. The storage modulus, E′, results are tabulated in Table 8.4.

The results showed that the storage modulus, E′, of PLA/CNC composite films increased with the addition of CNCs over the range of temperatures tested. The improvement in modulus is attributed to the increase of crystallinity together with the good dispersion of CNCs. All the silane-modified CNCs, except amino-modified CNCs, showed similar trends in improvement. However, lower improvement for amino-based silane-modified CNC/PLA composites, was obtained, possibly due to the occurrence of aminolysis reactions during the extrusion implementation.

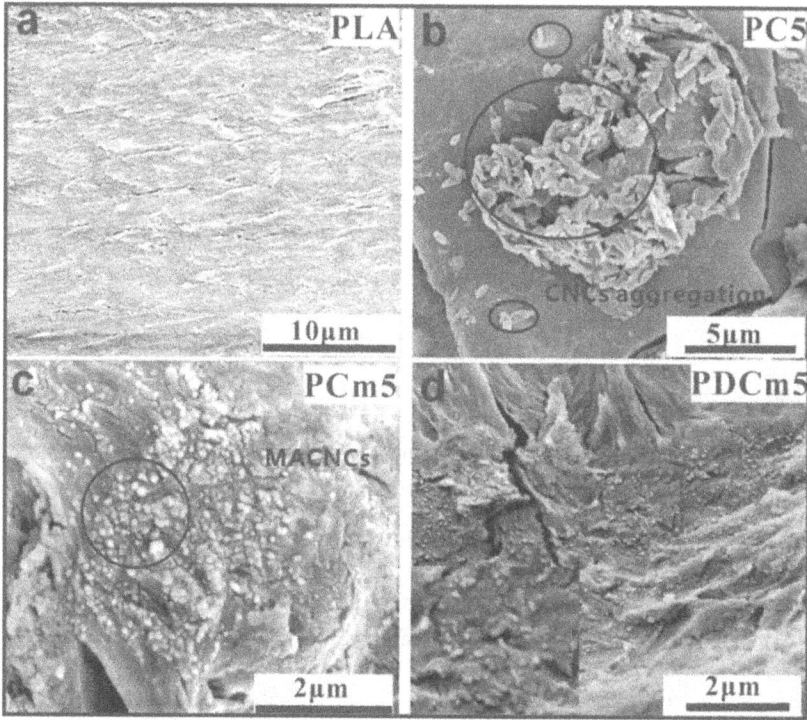

FIGURE 8.6   FE-SEM microscopic images of the cryogenic fractured surfaces of (a) neat PLA, (b) PC5, (c) PCm5, and (d) PDCm5 composite films. (Reprinted with permission from Zhou (2018). Copyright© 2017 Elsevier B.V. All rights reserved.)

To gain a comprehensive understanding of the surface topochemistry of CNCs on the viscoelastic properties of the PLA/CNC composites, Spinella et al. melt blended three different CNCs with PLA (Spinella, 2015). To explore the effect of interactions of unmodified CNC (HCl-CNC) and esterified-CNCs (acetate esterified (AA-CNC) and lactic acid esterified (LA-CNCs)) with PLA on the degree of reinforcement, DMTA was performed over a broad temperature range. Figure 8.7 displays the variation of storage modulus (E′) for the PLA, unmodified CNC/PLA, AA-CNC/PLA, and LA-CNC/PLA composites, and data is tabulated in Table 8.5.

TABLE 8.4

Storage Modulus, E′ and Glass Transition Temperature (T_g) of PLA-Based Composites Compared with Neat PLA

PLA/CNC Composites	E′ (MPa) at 25°C	Glass Transition Temp. (T_g) °C
PLA	1300	71.0
Amino-based silane-CNC	3740	70.5
n-Propyl-based silane-CNC	4250	71.0
Methacrylic-based silane-CNC	4360	70.9
Acrylic-based silane-CNC	4130	69.9

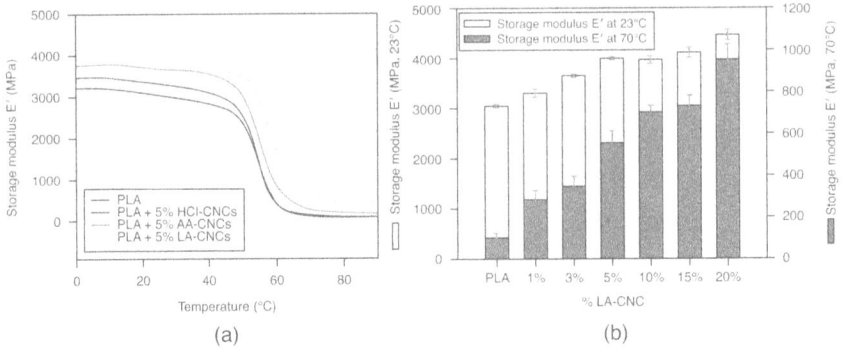

FIGURE 8.7 (a) Storage modulus, E', vs temperature, which shows the effect of surface chemistry of different CNCs on the E' of CNC/PLA composites over a broad temperature range and (b) E' for PLA/LA-CNC nanocomposites at 23°C and 70°C. (Reprinted with permission from Spinella (2015). Copyright © 2015 Elsevier Ltd. All rights reserved.)

It was found that the addition of modified CNCs led to a marginal increase in E' in the glassy state (23°C) and a significant increase in the rubbery state (70°C). In the glassy state, the presence of unmodified CNC (HCl-CNCs) and esterified CNC (by acetate (AA-CNCs) and lactic acid (LA-CNCs)) increased E' from 3048 to 3300±30 MPa (8.2%), 3676±40 (20.6%) and 4000±25 MPa (31.2%), respectively. E' increased from 100±20 MPa to 146±30 MPa (46%), 265±35 MPa (165%) to 550±65 MPa (450%) in the rubbery state, respectively. These findings indicate that the inherent high modulus of CNCs and enhanced interfacial interactions, enable the efficient stress transfer and improved energy storage mechanism at the interphase (Graninger, 2020). Hence, these results reveal that affinity between PLA and the CNCs vary as a function of the surface chemistry of CNCs in the following order: LA-CNC > AA-CNC > HCL-CNC.

Notably, LA-CNC was found to significantly enhance E' in the rubbery state as compared to the PLA, unmodified CNC (HCl-CNC) and AA-CNCs/PLA composites, which indicates an increased stability of stiffness at higher temperatures. As shown in Figure 8.6b, the shift in E' in the rubbery state with further increase in the loading of LA-CNCs, provides compelling evidence of the formation of a highly interactive network among LA-CNCs and strong hydrogen bonding between LA-CNC and PLA matrix.

TABLE 8.5

Storage Modulus (E') of PLA-Based Composites Compared with Neat PLA

PLA/CNC Composites	E' (MPa) at 23°C	E' (MPa) at 70°C
PLA	3048±30	100±20
5.0 wt.% HCl-CNC	3300±30	146±30
5.0 wt.% AA-CNC	3676±40	265±35
5.0 wt.% LA-CNC	4000±25	550±65

8.3.2 PLA/Cellulose Nanofiber (CNF) Composites

The incorporation of CNFs as reinforcement in PLA has been reported by several researchers and the composites have been prepared by melt compounding and solution processing. Table 8.6 summarizes the DMTA data for PLA/CNF composites. The incorporation of CNFs was found to enhance the dynamic mechanical performance of the PLA composites, which can be related to the high aspect ratio, high intrinsic stiffness, their ability to form strong percolating network structures, and the interfacial interaction between CNFs and the PLA matrix (Qu, 2010).

Kowalczyk et al. reported that 2.0 wt.% of CNFs markedly enhanced E′ of CNF/PLA composites (by 14%) compared to the neat PLA. They have attributed this enhancement to the uniform dispersion of CNFs within the PLA matrix, as shown in the SEM image, Figure 8.8. The intensity and position of the tan δ peaks remain unchanged, which confirm that the glass transition temperature remained largely unaffected by the presence of CNFs (Kowalczyk, 2011).

Jonoobi et al. produced PLA/CNFs composites using CNFs extracted from kenaf pulp. To fabricate uniformly dispersed PLA/CNFs composites, a two-step process was used. Firstly, solution mixing was used to make highly concentrated PLA/CNFs films, which were then diluted in a twin-screw extruder to a final CNFs content of (1.0 wt.%, 3.0 wt.%, and 5.0 wt.%) in the PLA/CNFs composites. As shown in Figure 8.9, DMTA studies reflect an increasing trend in E′ with increasing content of CNFs in the PLA/CNFs composites. In the glass state, at 35°C, E′ shows a remarkable increase by 69%

TABLE 8.6
DMTA Data for PLA/CNF Composites

Composite System	ΔE′ (%) Glassy Stage 35°C	ΔE′ (%) Rubbery Stage 70°C	Glass Transition Temp. (ΔT_g) °C	Processing Procedure	References
PLA/CNF (5.0 wt.%)	8.5%	131%	+2	Melt mixing + compression molding	Panaitescu (2017)
PLA/CNFs (2.0 wt.%)	14%	–	–	Solution mixing + melt mixing	Kowalczyk (2011)
PLA/CNF (32 wt.%)	63%	–	+3	Solution + compression molding	Wang (2012)
PLA/CNF (5 wt.%)	69.2%	2400%	+6	Master batch solution mixing + extrusion + injection molding	Jonoobi (2010)
PLA/acetylated-CNFs (5.0 wt.%)	80%	2800%	+7	Master batch solution mixing + extrusion mixing	Jonoobi (2012)

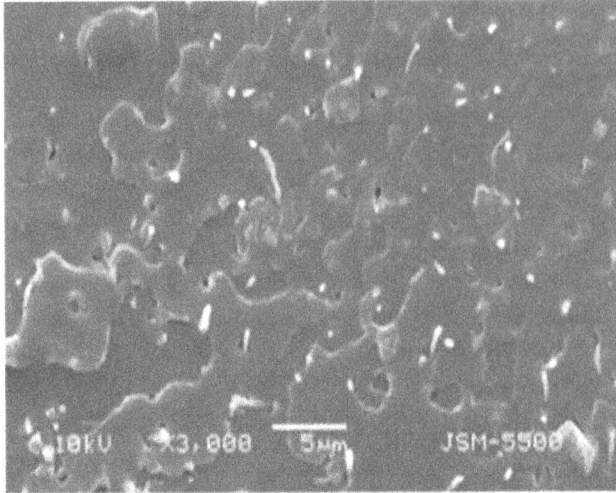

FIGURE 8.8 Scanning electron microscope image of the cryo-fractured surface of PLA/ CNFs composites. (Reprinted with permission from Kowalczyk (2011). Copyright © 2011 Elsevier Ltd. All rights reserved.)

at 5.0 wt.% of CNF, due to the networking forming tendency of CNFs. The improvements were more pronounced in the rubbery state at 70°C, where at 5.0 wt.% CNFs, the E' increased by 2400% compared to that of neat PLA. The intensity of tan δ peaks decreased and the peak position shifted to a higher temperature for PLA/CNFs composites with increasing CNFs content due to the confinement of the chain mobility of the PLA molecular chains by the percolating network of CNFs (Jonoobi, 2010).

However, using hydrophilic CNFs with hydrophobic PLA can hinder the full reinforcement potential of CNFs. To overcome this challenge, in another report, Jonoobi et al. produced PLA/CNFs composites using acetylated-CNC as a reinforcing agent. The addition of surface-modified CNCs improved E' by 80% in the glassy state

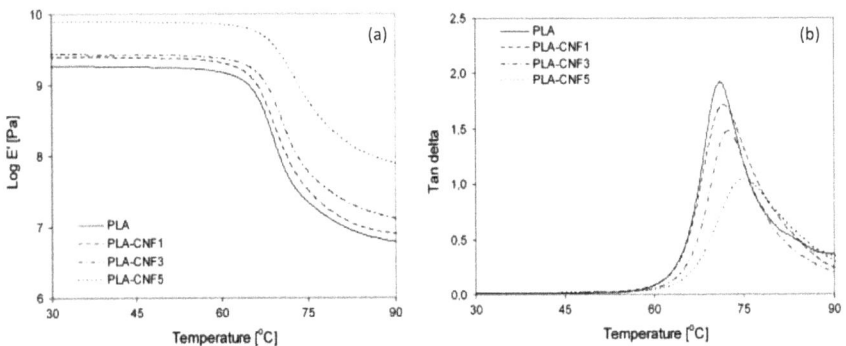

FIGURE 8.9 DMTA analysis of PLA and PLA/CNF nanocomposites; (a) storage modulus vs temperature and (b) tan δ vs temperature. (Reprint with permission from reference Jonoobi (2010). Copyright © 2010 Elsevier Ltd. All rights reserved.)

(at 35°C) and 2800% in the rubbery state (at 70°C) with 5 wt.% modified CNFs, which is higher compared to the PLA composites using unmodified CNFs made by the same process (Jonoobi, 2010, 2012). It is expected that the reaction between the functional of acetylated-CNC and the PLA matrix facilitated the percolating network formation and increased interfacial interactions, which enhanced the thermomechanical properties.

8.3.3 PLA/Microcrystalline (MCC) Composites

MCC, owing to read availability, strong rigidity, high crystallinity and as a cheaper alternative to CNCs/CNFs, has recently received attention as a reinforcement additive to improve the mechanical performance of thermoplastics. MCC is commercially available in many different grades for applications such as a thickening agent and drug shaping agent in pharmaceuticals, emulsifier for cosmetic products, and bulking agent in food processing (Trache, 2016; Kiziltas, 2011; Graninger, 2020). Several recent research studies on fabricating PLA/MCC composites using different processes, as summarized in Table 8.7, show the effectiveness of MCC in strengthening PLA.

However, due to the low aspect ratio and low surface area of MCC as compared to CNC and CNF, on average, high concentration of MCC is required to achieve considerable improvement in E' and T_g. Mathew et al. prepared the PLA/MCC composites

TABLE 8.7

Summarize the Dynamic Mechanical Properties Data for PLA/MCC Composites

Composite System	$\Delta E'$ (%) Glassy Stage 35°C	$\Delta E'$ (%) Rubbery Stage 70°C	Glass Transition Temp. (ΔT_g) °C	Processing Procedure	References
PLA/MCC	62.5%	293%	+2	Melt mixing + injection molding	Mathew (2005)
PLA/MCC	7.3%	38%	–	Melt mixing + 3D printing	Murphy (2018)
PLA/MCC	28.9%	100%	–	Melt mixing + Injection molding	Zhang (2020)
PLA/MCC	23%	–	+2	Melt mixing and injection molding	Johari (2016)
PLA/MCC	39.3%			Melt mixed extruded filaments	Wang (2019)
PLA/iron-surface modified-MCC	17.8%	–	+1	Melt mixing + Injection molding	Sundar (2011)

by direct melt blending with up to 25 wt.% MCC content. DMTA showed an increase in E' with increasing addition of MCC. Addition of MCC, also influenced the tan δ, which shifted slightly to higher temperatures with increasing MCC content, indicating restricted segmental motion of PLA molecules due to interactions between filler and polymer matrix (Mathew, 2005).

8.4 CONCLUSION AND OUTLOOK

Poly(lactic acid), or PLA, is an environmentally friendly material due to its excellent biocompatibility and degradability profile and has the potential to play a key role in the development of sustainable commercial polymers for a circular economy. PLA properties are highly tailorable and numerous researchers are exploring their use in transportation, food, agriculture, and medicine. However, known shortcomings associated with relatively poor thermo-mechanical properties have hindered PLA's development for structural applications. This chapter has systematically presented the use of different forms of cellulose reinforcement such as CNCs, CNFs, and MCC to improve the thermo-mechanical properties of PLA composites while preserving the environmental profile. The incorporation of CNC, CNF, and MCC in PLA has been shown to increase the storage modulus of the PLA composite materials. Higher reinforcing effect is seen for CNC and CNFs, due to their nano size and higher tendency to form a rigid network. Furthermore, the improvement of interfacial interactions through functionalization of CNCs/CNFs was found to have a significant effect on storage modulus in both glassy and rubbery states.

However, based on the current research status and development trend in cellulose-based PLA composites, a significant in-depth research is still required to form a proper and detailed understanding of the role of surface chemistry of cellulose particles at filler-polymer and filler-filler interface. Attention should also be paid toward the development of single-step manufacturing processes, which will have further appeal to the industry.

REFERENCES

Abdulkhani, A., Hosseinzadeh, J., Ashori, A., Dadashi, S., Takzare, Z., "Preparation and characterization of modified cellulose nanofibers reinforced polylactic acid nanocomposite," *Polymer Testing*, 2014, 35, 73–79.

Bledzki, A.K., Franciszczak, P., Meljon, A., "High performance hybrid PP and PLA biocomposites reinforced with short man-made cellulose fibres and softwood flour," *Composites Part A: Applied Science and Manufacturing*, 2015, 74, 132–139.

Cho, S.Y., Park, H.H., Yun, Y.S., Jin, H.J., "Cellulose nanowhisker-incorporated poly(lactic) acid composites for high thermal stability," *Fibers and Polymers*, 2013, 14, 1001–1005.

Graninger, G., Kumar, S., Garrett, G., Falzon, B.G., "Effect of shear forces on dispersion-related properties of microcrystalline cellulose-reinforced EVOH composites for advanced applications," *Composites Part A: Applied Science and Manufacturing*, 2020, 139, 106103.

Gupta, A., Simmons, W., Schueneman, G.T., Hylton, D., Mintz, E.A., "Rheological and thermo-mechanical properties of poly(lactic acid)/lignin-coated cellulose nanocrystal composites," *ACS Sustainable Chemistry & Engineering* 2017, 5, 1711–1720.

Johari, A.P., Kurmvanshi, S.K., Mohanty, S., Nayak, S.K., "Influence of surface modified cellulose microfibrils on the improved mechanical properties of poly(lactic) acid," *International Journal of Biological Macromolecules*, 2016, 84, 329–339.

Jones, D.S., Tian, Y., Diak, O.A., Andrews, G.P., "Pharmaceutical applications of dynamic mechanical thermal analysis," *Advanced Drug Delivery Reviews*, 2012, 64, 440–448.

Jonoobi, M., Harun, J., Mathew, A.P., Oksman, K., "Mechanical properties of cellulose nanofiber (CNF) reinforced polylactic acid (PLA) prepared by twin screw extrusion," *Composites Science and Technology*, 2010, 70,1742–1747.

Jonoobi, M., Mathew, A., Abdi, M.M., Makinejad, M.D., Oksman, K., "A comparison of modified and unmodified cellulose nanofiber reinforced polylactic acid (PLA) prepared by twin screw extrusion," *Journal of Polymers and the Environment*, 2012, 20, 991–997.

Kiziltas, A., Gardner, D.J., Han, Y., Yang, H.S., "Dynamic mechanical behavior and thermal properties of microcrystalline cellulose (MCC)-filled nylon 6 composites," *Thermochim Acta*, 2011, 519, 38–43.

Kowalczyk, M., Piorkowska, E., Kulpinski, P., Pracella, M., "Mechanical and thermal properties of PLA composites with cellulose nanofibers and standard size fibers," *Composites Part A: Applied Science and Manufacturing*, 2011, 42, 1509–1514.

Li, H., Cao, Z., Wu, D., Tao, G., Zhong, W., Zhu, H., Qiu, P., Liu, C., "Crystallisation, mechanical properties, and rheological behaviour of PLA composites reinforced by surface modified microcrystalline cellulose," *Plastics, Rubber and Composites*, 2016, 45, 181–187.

Mathew, A.P., Oksman, K., Sain, M., "Mechanical properties of biodegradable composites from poly lactic acid (PLA) and microcrystalline cellulose (MCC)," *Journal of Applied Polymer Science*, 2005, 45, 2014–2025.

Menard, K.P., *Dynamic Mechanical Analysis: A Practical Introduction*, 2nd ed., CRC Press, New York (2008).

Mokhena, T.C., Sefadi, J.S., Sadiku, E.R., John, M.J., Mochane, M.J., Mtibe, A., "Thermoplastic processing of PLA/cellulose nanomaterials composites," *Polymers*, 2018, 10(12), 1363.

Mukherjee, T., Sani, M., Kao, N., Gupta, R.K., Quazi, N., Bhattacharya, S., "Improved dispersion of cellulose microcrystals in polylactic acid (PLA) based composites applying surface acetylation," *Chemical Engineering Science*, 2013, 101, 655–662.

Murphy, C.A., Collins, M.N., "Microcrystalline cellulose reinforced polylactic acid biocomposite filaments for 3D printing," *Polymer Composites*, 2018, 39, 1311–1320.

Oksman, K., Skrifvars, M., Selin, J.F., "Natural fibres as reinforcement in polylactic acid (PLA) composites," *Composites Science and Technology*, 2003, 63, 1317–1324.

Panaitescu, D.M., Frone, A.N., Chiulan, I., Gabor, R.A., Spataru, I.C., and Căşărică, A., "Biocomposites from polylactic acid and bacterial cellulose nanofibers obtained by mechanical treatment," *BioResources*, 2017, 12, 662–672.

Paul, U.C., Fragouli, D., Bayer, I.S., Zych, A., Athanassiou, A., "Effect of green plasticizer on the performance of microcrystalline cellulose/polylactic acid biocomposites," *ACS Applied Polymer Materials*, 2021, 3, 3071–3081.

Petersson, L., Kvien, I., Oksman, K., "Structure and thermal properties of poly(lactic) acid/cellulose whiskers nanocomposite materials," *Composites Science and Technology*, 2007, 67, 11, 2535–2544.

Qu, P., Gao, Y., Wu, G.F., Zhang, L.P., "Nanocomposites of polylactic acid reinforced with cellulose nanofibrils," *BioResources*, 2010, 5, 1811–1823.

Raquez, J.M., Murena, Y., Goffin, A.L., Habibi, Y., Ruelle, B., DeBuyl, F., Dubois, P., "Surface-modification of cellulose nanowhiskers and their use as nanoreinforcers into polylactide: A sustainably-integrated approach," *Composites Science and Technology*, 2012, 72, 544–549.

Scaffaro, R., Botta, L., Lopresti, F., Maio, A., Sutera, F., "Polysaccharide nanocrystals as fillers for PLA based nanocomposites," *Cellulose* 2017, 24, 447–478.

Shojaeiarani, J., Bajwa, D.S., Hartman, K., "Esterified cellulose nanocrystals as reinforcement in poly(lactic acid) nanocomposites," *Cellulose*, 2019, 26, 2349–2362.

Shojaeiarani, J., Bajwa, D.S., Stark, N.M., "Green esterification – a new approach to improve thermal and mechanical properties of polylactic acid composites reinforced by cellulose nanocrystals," *Journal of Applied Polymer Science*, 2018, 135, 46468.

Spinella, S., Re, G.L., Liu, B., Dorgan, J., Habibi, Y., Leclere, P.L., Raquez, J.M., Dubois, P., Gross, R.A., "Polylactide/cellulose nanocrystal nanocomposites: Efficient routes for nanofiber modification and effects of nanofiber chemistry on PLA reinforcement," *Polymer*, 2015, 65, 9–17.

Sullivan, E.M., Robert, M., Kyriaki, K., "Processing and characterization of cellulose nanocrystals/polylactic acid nanocomposite films," *Materials*, 2015, 8, 8106–8116.

Sundar, S., Sain, M., Oksman, K., Thermal characterization and electrical properties of Fe-modified cellulose long fibers and micro crystalline cellulose," *Journal of Thermal Analysis and Calorimetry*, 2011, 104, 841–847.

Trache, D., Hussin, M.H., Chuin, C.T.H., Sabar, S., Fazita, M.R.N., Taiwo, O.F.A., "Microcrystalline cellulose: Isolation, characterization and bio-composites application- A review," *International Journal of Biological Macromolecules*, 2016, 93, 789–804.

Wang, T., Drzal, L.T., "Cellulose-nanofiber-reinforced poly(lactic acid) composites prepared by a water-based approach," *ACS Applied Materials Interfaces*, 2012, 4, 10, 5079–5085.

Wang, C., Smith, L.M., Zhang, W., Li, M., Wang, G., Shi, S.Q., Cheng, H., Zhang, S., "Reinforcement of polylactic acid for fused deposition modeling process with nano particles treated bamboo powder," *Polymers*, 2019, 11, 1146.

Zhang, Q., Lei, H., Cai, H., Mateo, W., "Improvement on the properties of microcrystalline cellulose/polylactic acid composites by using activated biochar," *Journal of Cleaner Production*, 2020, 252, 119898.

Zhang, Y., Cui, L., Xu, H., Feng, X., Wang, B., Pukánszky, B., Mao, Z., Sui, X., "Poly(lactic acid)/cellulose nanocrystal composites via the Pickering emulsion approach: Rheological, thermal and mechanical properties," *International Journal of Biological Macromolecules*, 2019, 137, 197–204.

Zhou, L., He, H., Li, M.C., Huang, S., Mei, C., Wu, Q., "Enhancing mechanical properties of poly(lactic acid) through its in-situ crosslinking with maleic anhydride-modified cellulose nanocrystals from cottonseed hulls," *Industrial Crops and Products*, 2018, 112, 449–459.

9 Mechanical Properties of PLA/Nanocellulose Composites

Premkumar Anil Kothavade
and Kadhiravan Shanmuganathan
CSIR-National Chemical Laboratory
Pune, India

and

Academy of Scientific and Innovative Research
Ghaziabad, India

CONTENTS

9.1   INTRODUCTION

Global warming, increased greenhouse gas emissions, and other environmental changes are currently affecting the world. Plastic pollution is one of the major causes of such significant environmental change. Most plastics made from petroleum and fossil fuels can take hundreds of years to break down (Webb et al. 2013; Chamas et al. 2020). Furthermore, fossil fuels are non-renewable, finite, and quickly depleting

DOI: 10.1201/9781003160458-9

resources. As a result, it is necessary to switch to renewable bio-derived plastics and reduce the consumption of non-renewable petroleum-based plastics. The development of bio-based sustainable polymeric materials is at the forefront of efforts to address these environmental challenges (Wasti et al. 2021). Biopolymers such as cellulose, starch, lignin, chitosan, silk fibroin, Poly(lactic acid) (PLA), Poly(hydroxy butyrate) (PHB), Poly(hydroxyalkanoate) (PHA), and others are currently attracting industrial and academic attention as a potential alternative to petroleum-based plastics (Aaliya, Sunooj, and Lackner 2021).

Among many biopolymers, PLA has been widely explored as a replacement for petroleum-based polymers. PLA is a biobased, biocompatible, and biodegradable linear aliphatic thermoplastic polyester made from renewable resources such as starch derived from biological sources such as corn, sugarcane, sugar beet, and wheat (Alam et al. 2020). The monomer lactic acid (LA) is obtained by the bacterial fermentation process of starch using the strain of *Lactobacillus* bacteria, and this LA monomer is then polymerized to get PLA. The LA produced via this route contains 99.5% of L-isomer, which gives high mechanical strength to PLA after polymerization (El-Hadi 2017). The LA monomer can also be commercially manufactured using petroleum feedstocks; however, this method is neither cost-effective nor environmentally benign.

Furthermore, LA derived from petroleum feedstocks produces a racemic mixture of D- and L-enantiomers, which is optically inactive and unsuitable for industrial applications. The lactic acid can be polymerized to get PLA via two common routes. The first one is the direct melt polycondensation of lactic acid in the presence of a solvent under a high vacuum. The second route is catalytic ring-opening polymerization of lactide, a cyclic dimer of lactic acid, to get high-molecular-weight PLA. This route does not require solvent, and hence it is commonly followed to synthesize the commercial high molecular weight PLA (Nam, Sinha Ray, and Okamoto 2003; Vatansever, Arslan, and Nofar 2019). The PLA repeating unit consists of a stereocenter, which results in optically active Poly(L-lactic acid) (PLLA) and Poly(D-lactic acid) (PDLA) enantiomers, or optically inactive Poly(D, L-lactic acid) (PDLLA) enantiomer, as well as a racemic mixture (Vestena et al. 2016). Commercially available PLA is produced from the L-lactide and D, L-lactide copolymers. The properties of PLA can be controlled by changing the catalyst, the content of D- or L-isomer present, the concentration of catalyst, residence time, and tuning the temperature during the polymerization process. PLA shows good mechanical properties such as high modulus (~3 GPa), high strength (~60 MPa), excellent optical and barrier properties (Vatansever, Arslan, and Nofar 2019). Owing to its biocompatibility, biodegradability, bioresorbality, nontoxicity, non-immunogenicity, non-inflammatory properties toward the human body, PLA has been a favorable polymer for biomedical applications as drug delivery, tissue engineering, scaffolding, etc. (Ruan and Feng 2003; Xi et al. 2017; Paolini et al. 2018). PLA is an eco-friendly biopolymer and does not release any toxic components during its degradation. Moreover, PLA requires 50% less energy during its production as compared to petroleum-derived thermoplastics like poly(propylene) (PP), poly(ethylene) (PE), poly(styrene) (PS), and poly(amides) (PA) (Vink et al. 2003; Rasal, Janorkar, and Hirt 2010).

Despite having all these excellent characteristics, applications of PLA are restricted to some extent due to slow crystallization rate, poor melt strength, poor processability and formability, inherent brittleness, low service temperature, moisture sensitivity, slow degradation rate, etc. (Saeidlou et al. 2012; Nofar, Salehiyan, and Sinha Ray 2019; Nofar et al. 2019; Naser, Deiab, and Darras 2021). These drawbacks of PLA have been addressed to some extent by copolymerization or blending with other polymers, addition of plasticizers, addition of bio-fillers to PLA, etc. Among these strategies, the addition of cellulose fibers to PLA has been widely explored to improve the properties of PLA (Cheng et al. 2009; Farah, Anderson, and Langer 2016; Su et al. 2019).

Cellulose is the most abundant renewable biomaterial available on the earth. Due to increasing environmental awareness, cellulose has gained tremendous attention as a reinforcement for biopolymeric composites. Besides its natural abundance and environmental friendliness, cellulose has excellent chemical and physical features, including outstanding mechanical properties, low cost, low specific density, chemical inertness, etc. Furthermore, the presence of many hydroxyl groups on its surface allows for surface modification in various ways (Vazquez et al. 2015; Trache et al. 2016; Foster et al. 2018; Mokhena and John 2019; Moohan et al. 2019; Trache et al. 2020b). Surface modification of cellulose fibers plays a vital role in bio-composites of hydrophilic cellulose and hydrophobic PLA matrix (Phanthong et al. 2018; Rajinipriya et al. 2018; Naz, Ali, and Zia 2019; Vineeth et al. 2019; Köse, Mavlan, and Youngblood 2020).

9.2 CELLULOSE

9.2.1 GENERAL DISCUSSION

Cellulose is a semicrystalline bio-polymer, abundantly available on the earth's surface. It is present in wood, hemp, cotton, and other plants and is a natural reinforcement in plant structures (Kargarzadeh et al. 2018). Cellulose and its derivatives have been used in diverse applications for more than 150 years. In recent years, the focus has been mainly on nanocellulose owing to its attractive properties (Salimi et al. 2019).

9.2.2 NANOCELLULOSE AND ITS SOURCES

One dimensional isolated cellulose having at least one dimension in nanosize is generally referred to as nanocellulose (Yang et al. 2017). It is derived from various sources of cellulose via different production routes. Sources of cellulose comprise plants as well as animals, bacteria, and algae. According to the type of cellulose fibers and their dimensions, the cellulosic family is classified into three forms, cellulose nanofibers (CNFs), crystalline nanocellulose (CNC), and bacterial cellulose (BC). Plant-based cellulosic nanofillers such as CNC and CNF could be produced using a top-down approach. The BC nanofiller could be obtained by a specific strain of bacteria using a bottom-up approach (Trache et al. 2017; Hazwan Hussin et al. 2019; Pennells et al. 2019).

CNCs are rod-like nanoparticles that are extracted from bulk cellulose using controlled acid hydrolysis. The acid removes the amorphous portion of the cellulose to get highly crystalline rod-like cellulose nanocrystals (Moberg et al. 2017). In various literature, CNCs are also referred to as nanocrystalline cellulose (NCC) or cellulose nanowhiskers (CNWs). CNF is also named nanofibrillated cellulose (NFC). They could be extracted from various cellulosic sources through mechanical and chemical treatments. The structure of CNFs is interconnected, having an aspect ratio of 50–100 (Iwatake, Nogi, and Yano 2008).

On the other hand, BC could be obtained from the strains of bacteria like *Acetobacter xylenium*, *Gluconacetobacter xylinus*, *Acetobacter hasneii*, or *Acetobacter pasterianus* (Trache et al. 2020a). In plant-based nanocellulose and bacterial nanocellulose, the chemical structure is similar. However, BC has good mechanical properties because of higher crystallinity and fewer defects in its crystalline structure. BC nanofiller also has a higher aspect ratio. Still, its applications are limited because of the limitation in the production of BC (Oun, Shankar, and Rhim 2019).

The most common plant sources of cellulose are cotton linter, bamboo, sugarcane bagasse, and eucalyptus. In Table 9.1, various common sources of cellulose for the production of nanocellulose are mentioned.

9.2.3 STRUCTURAL ORGANIZATION, CLASSIFICATION, AND ISOLATION PROCESSES OF NANOCELLULOSE

Cellulose consists of repeating β (1,4)-bound D-glucopyranosyl units in the $^4C_1$-chain configuration. Each monomer unit is twisted at 180° relatives to its neighbored monomeric unit (Moon et al. 2011; Gopi et al. 2019). The hierarchical structure of cellulose nanomaterials is shown in Figure 9.1.

Conversion of cellulose into nanocellulose involves two stages. Pre-treatment of cellulose is the first stage in which hemicellulose, lignin, fatty acids, tannins, and other impurities are removed partially or totally to obtain pure cellulose. Alkali treatment and bleaching are the most common pre-treatment processes for cellulose. Native cellulose contains both crystalline (ordered) as well as amorphous

---

TABLE 9.1

Various Common Sources of Cellulose for the Production of Nanocellulose

Source Group	Source
Softwood, hardwood, and agricultural residues	Cotton linter, sugarcane bagasse, hemp, jute, coconut husk, eucalyptus, pine, oak, balsa, aspen, maple, sisal, kenaf, wheat, rice, bamboo, corn leaf, potato peel, peanut shells, etc.
Animal, bacteria, and algae	Tunicates, Chordata, Drasche, *Sarcina*, *Rhodobacter*, *Acetobacter*, *Salmonella*, *Cystoseria myrica*, *Azotobacter*, *Aerobacter*, *Cladophora*, *Posidonia oceania*, etc.

Source: Adapted from Trache et al. (2020a). Under Creative Commons Attribution License (CC BY) Copyright (2020), Trache et al. published by Frontiers in chemistry.

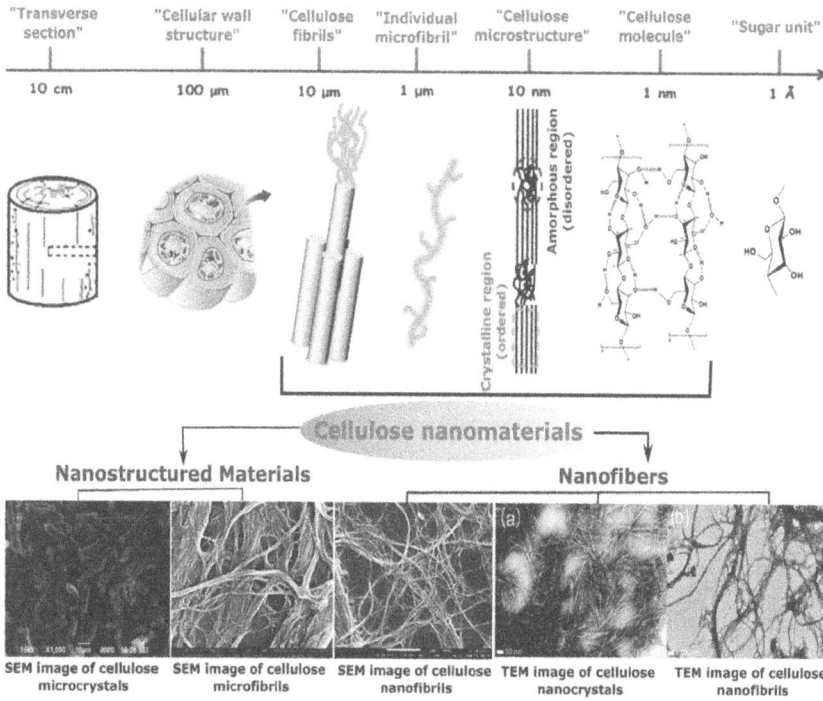

FIGURE 9.1 The hierarchical structure of cellulose nanomaterials. (Republished with permission from The Royal Society of Chemistry, from Trache et al. (2017). Copyright (2017); permission conveyed through Copyright Clearance Center, Inc.)

(disordered) regions. The amorphous region of cellulose is weak, less dense, and less resistant to chemical, mechanical and enzymatic treatments than the crystalline region. Thus, in the second stage of preparing nanocellulose, removing the amorphous domains from native cellulose is an important step to get crystalline cellulose in nanoform. The removal of the amorphous domains can be achieved by various chemical, physical, and biological treatments. Acid hydrolysis is the common chemical treatment used to remove the amorphous domains from native cellulose. The final properties, morphology, and crystallinity of nanocellulose are dependent on the source of cellulose and its production method (Lin and Dufresne 2014; Foster et al. 2018; Kargarzadeh et al. 2018; Vatansever, Arslan, and Nofar 2019; Trache et al. 2020a; Zheng and Pilla 2020). As shown in Figure 9.2, the nanoscale cellulose material can be classified into nanostructured materials and nanofibers depending on the isolation process used. Nanostructured materials are further classified as microcrystalline cellulose (MCC) and cellulose microfibrils. Nanofibers are classified as cellulose nanocrystals (CNC), CNF, and BC.

As mentioned above, the CNCs are cylindrical, elongated, rod-like nanocrystals with diameters in the range of 3–50 nm, lengths in the range of 100–500 nm, and 54%–88% crystallinity index (Naz, Ali, and Zia 2019). These are not flexible as CNFs. CNCs can be extracted from cellulose sources via different acid hydrolysis

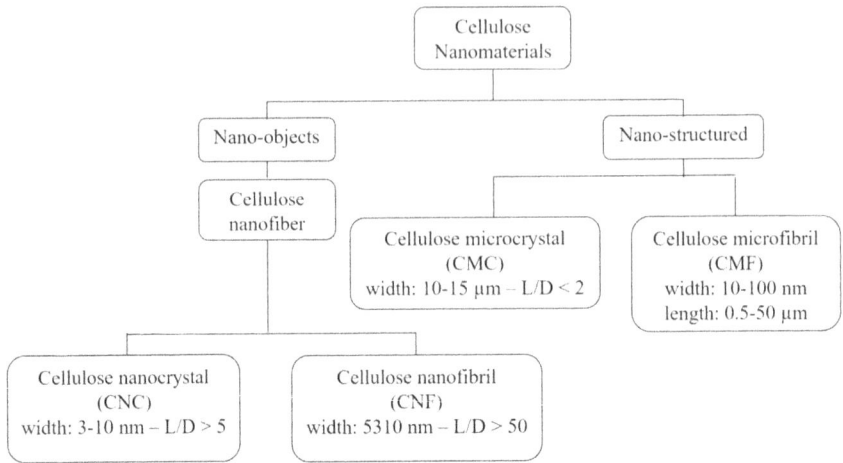

FIGURE 9.2 Classification of nanoscale cellulose according to Technical Association of Pulp and Paper Industries (TAPPI). (Reproduced from Mariano et al. (2014); with permission from John Wiley and Sons.)

processes, out of which acid hydrolysis using sulfuric acid is the most common process. Generally, before converting into CNC, cellulosic biomass has to pass through four different stages. The first stage is the disintegration of biomass or mechanical size reduction. The second stage involves purification by alkali and bleaching treatments. In the third stage, the amorphous portions have to be removed by controlled acid hydrolysis to get highly crystalline CNC's. In the last and fourth stage, mechanical or ultrasound treatment helps to disintegrate the CNC's. Different acids can be used for the acid hydrolysis process. The selection of strong or mild acid will decide the final properties of CNCs (Kargarzadeh et al. 2018). As opposed to CNCs, CNFs are longer and flexible, having diameters in the range of 20–50 nm, lengths in the range of 500–2000 nm, and lower crystallinity index compared to CNC's. CNF's can be extracted from different lignocellulosic sources via a combination of mechanical and chemical routes. Most commonly, CNF's are extracted from plant cell walls by simple mechanical shearing. Because of high shear force, a highly entangled network of nanofibrils with crystalline and amorphous domains is produced. Depending on the processing conditions, this highly entangled network of nanofibrils is disintegrated into CNFs. High-pressure homogenization, grinding, cryocrushing, high-intensity ultrasonication, steam explosion process, extrusion, and aqueous counter collision are some of the mechanical processes to prepare CNFs (Kentaro Abe, Shinichiro Iwamoto, and Yano 2007; Deepa et al. 2011; Chirayil et al. 2014; He et al. 2014; Li et al. 2014). Bacterial nanocellulose (BC), also known as microbial cellulose, is ultra-fine, pure, and ribbon-shaped nanofibers having a diameter in the range of 20–100 nm and length spanning a few micrometers. This type of cellulose possesses a high crystallinity index. BC, unlike plant-derived NC, does not require any pre-treatments since it lacks extractives such as hemicellulose and lignin. Despite possessing desirable properties such as a high crystallinity index and excellent purity, BC uses are still limited to biomedical applications. The reason for this is BC's high cost, which is due to its expensive synthesis procedure and synthetic media. (Trache and Trache 2018; Carvalho et al. 2019; Sharma and Bhardwaj 2019).

9.2.4 SURFACE MODIFICATION OF NANOCELLULOSE AND ITS IMPORTANCE

Due to the specific properties of nanocellulose such as high surface area (150 m^2/g), high Young's modulus (~170 GPa), high crystallinity (~50%–90%), good optical properties, non-toxic nature, abundant availability, eco-friendliness, and the presence of a large amount of primary reactive sites (hydroxyl groups) on its surface and high surface area to volume ratio, nanocellulose has become the choice material among other reinforcing fillers like carbon nanotubes, mica, silica, graphite, graphene oxide, graphene, etc. (Trache et al. 2020a). Nanocomposites of thermoplastics and nanocellulose are getting tremendous attention. Common thermoplastics like PP, PE, PLA, ABS, and nylon are hydrophobic. On the other hand, nanocellulose is hydrophilic. Poor compatibility, poor dispersion, and thus aggregation of nanocellulose nanomaterial in the thermoplastic matrix are some of the challenges in processing these nanocomposites. This ultimately affects the overall thermomechanical properties of nanocomposites (Lin, Huang, and Dufresne 2012). Surface hydroxyl groups can be functionalized with suitable molecules to achieve compatible and well-dispersed nanocellulose nanocomposite. The schematic of the most common surface modification routes for nanocellulose is shown in Figure 9.3. The nanocellulose surface can be modified either by chemical interactions, physical interactions or biological interactions. Covalent surface modification of nanocellulose via chemical interactions is the most common way, and generally, it includes sulfonation, silylation, oxidation, polymer grafting, esterification, nucleophilic substitution, etherification, carbamation (George and Sabapathi 2015; Afrin and Karim 2017; Daud and Lee 2017; Huang et al. 2019). Physical surface modification techniques include plasma treatment, ultrasonic treatment, surface fibrillation, and irradiation (Islam, Alam, and Zoccola 2013).

Similarly, the biological surface modification includes the enzymatic approach (Afrin and Karim 2017). The main reason behind the surface modification of nanocellulose is to impart stable electrostatic charges (either positive or negative) on its surface for the uniform dispersion of particles and thus to enhance their compatibility (Kaboorani and Riedl 2015). Enhanced compatibility ultimately leads to better thermo-mechanical properties of nanocomposites. The surface modification should be carried out in a controlled way so that the morphology and crystal structure remains unchanged (Islam, Alam, and Zoccola 2013).

9.3 PLA/CELLULOSE COMPOSITES

9.3.1 GENERAL DISCUSSION

The nanocomposite of thermoplastic PLA biopolymer and cellulose has received much attention from the scientific community because of its environmentally benign nature and its remarkable qualities. PLA/cellulose nanocomposites have superior properties than PLA/cellulose composites. Various processing techniques, such as

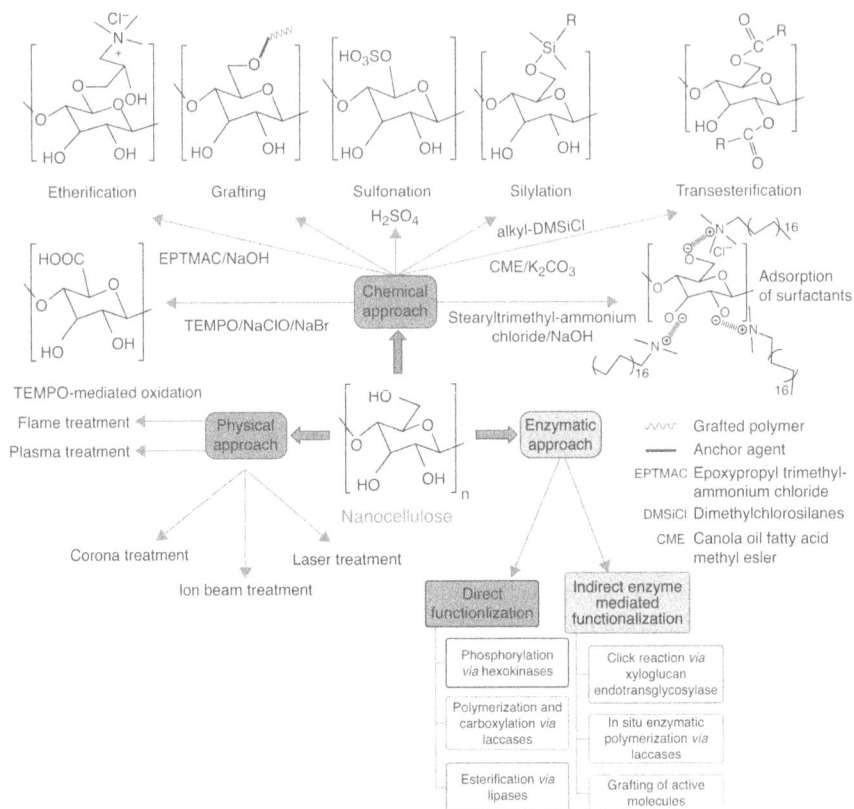

FIGURE 9.3 The most widely used techniques for surface functionalization/modification of nanocellulose. (Reproduced from Trache et al. (2020). Under Creative Commons Attribution License (CC BY) Copyright (2020), Trache et al. published by Frontiers in Chemistry.)

melt blending, solution casting, solution casting followed by melt blending, have been explored to tune the mechanical properties of nanocomposites. In this context, PLA/CNC nanocomposites, PLA/CNF nanocomposites, different processing routes of PLA/cellulose nanocomposites, their effect on mechanical properties, and the surface modification of nanocellulose and its impact on final mechanical properties of PLA/cellulose nanocomposites are discussed thoroughly in the below sections.

9.3.2 PLA/Cellulose Nanocrystals (CNC) Composites

PLA/CNC nanocomposites have been extensively investigated with various types of CNCs. The final properties of nanocomposite depend on factors like dispersion of CNC, their proportion, percolation path, and processing method. Oguz et al. (2021) prepared PLA/CNC bio-nanocomposites by high shear thermo-kinetic mixing. They reported uniform CNC dispersion even at very high filler content up to 30 wt.%, which results in substantial enhancement in its mechanical and thermal properties such as a nearly 100% increment in the elastic modulus (6.48GPa), 43% enhancement in tensile strength (87.1 MPa), 200% increase in the impact strength (44.2 KJ/m^2), a 113-fold increment in a storage modulus at 90°C (787.8 MPa) after the addition of 30 wt.% of CNC into the PLA. An optimal amount of CNC is required in PLA nanocomposite to improve tensile strength and elastic modulus since CNC beyond optimal level causes aggregation due to incompatible interface between hydrophobic PLA and hydrophilic CNC (Zaaba, Jaafar, and Ismail 2021). Still, Oguz et al. showed enhancement in overall mechanical properties of the composite after the addition of 30 wt.% unmodified CNC. According to Oguz et al., the homogeneous distribution of CNC and robust interactions between the filler and the matrix phases are promoted by high shear mixing. The uniform distribution of CNC in the PLA matrix can be seen in Figure 9.4(a). The strong interactions allow efficient stress transfer, which leads to enhancement in the final mechanical properties of PLA/ CNC composite, as shown in Figure 9.4(b).

In another study by Zhang and co-workers (Zhang et al. 2021), composites of PLA and CNC were prepared by the solution casting method. Here the CNC was obtained by sulfuric acid hydrolysis of office waste paper (OWP). The CNCs thus obtained were directly solution mixed with PLA followed by melt mixing to get nanocomposite. Tensile strength and elongation at break enhanced by 9.7% and 5.8%, respectively, with the addition of 4 wt.% of CNC.

Mariano et al. (2017) prepared CNC reinforced PLA nanocomposites by first preparing a masterbatch via solution casting. This was then diluted with PLA by melt compounding and further injection-molded into tensile specimens. The CNC's used were isolated from ramie fibers through sulfuric acid hydrolysis. Two surfactants containing different polar heads, namely, PEG-b-PLLA and ionic liquid (Imidazolium)-b-PLLA, were synthesized. A masterbatch consisting of CNCs and the desired PLLA-based surfactant was prepared by solution mixing with the

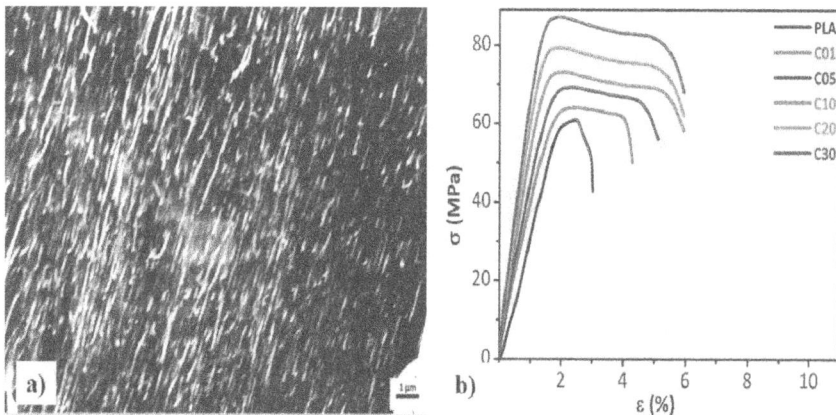

FIGURE 9.4 (a) SEM micrograph of 30 wt.% of CNC in PLA showing uniform dispersion of CNC particles. (b) Stress-strain curve of all PLA/CNC composites with varying CNC content (C01 = 1 wt.% CNC in PLA matrix, the same analogy applies for C05, C10, C20, and C30). (Reproduced from Oguz et al. (2021). With permission from John Wiley and Sons.)

hypothesis that the amphipathic surfactants will enhance PLA/CNC compatibility. This masterbatch was diluted with PLA to enhance the interfacial compatibility.

Despite increased interfacial interactions, a significant improvement in mechanical properties was not observed. It could probably be due to the reduction in average molecular weight of PLA chains during melt compounding, resulting in a plasticization effect and impacting the crystallization of the matrix. In one of the studies conducted by Dhar et al. (2016), PLA was grafted to CNC using Di-cumyl peroxide (DCP) as a reactive crosslinker in a single step reactive extrusion process, as shown in Figure 9.5. They showed that this nanocomposite could be recycled without significant change in its molecular structure. Furthermore, the grafted PLA chains obscure the sulfate and hydroxyl groups of CNCs, increasing the compatibility of the PLA-g-CNC nanocomposite. This resulted in a 40% and 490% increment in tensile strength and young`s modulus, respectively.

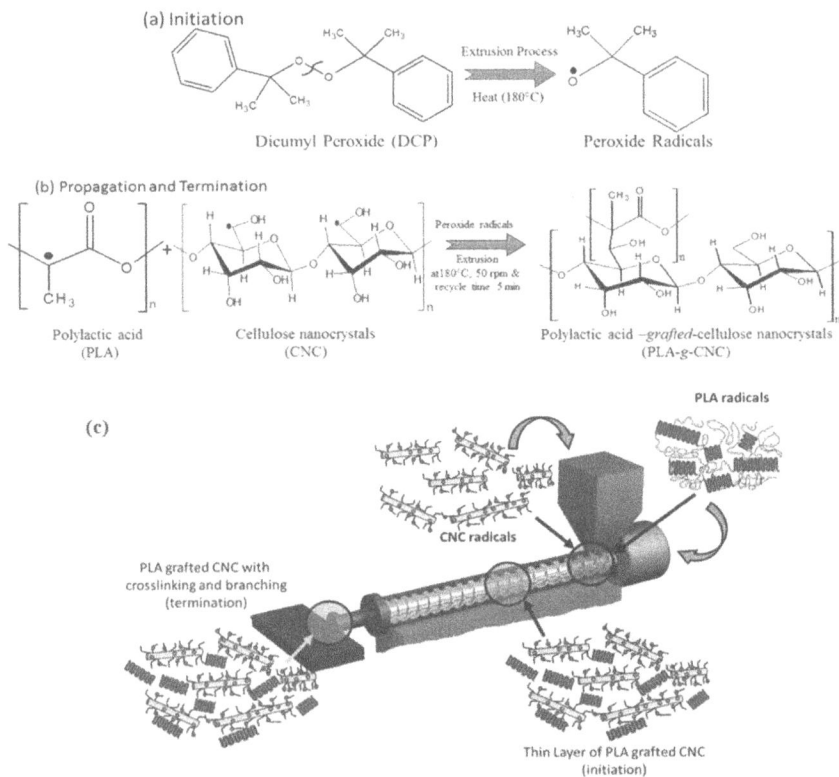

FIGURE 9.5   (a) Thermal decomposition mechanism of the DCP into peroxide radicals during extrusion at T = 180°C (initiation step). (b) Generation of CNC and PLA radicals followed by reactive extrusion at screw speed = 50 rpm and recycle time = 2 min (propagation step) leading to the formation of PLA grafted CNC structures (termination step). (c) Pictorial representation of the grafting mechanism of initiation, propagation, and termination of the reactive extrusion process for PLA-g-CNC along the different zones of the extruder. (Reprinted from Dhar et al. (2016). With permission from Elsevier.)

However, a 70% reduction in the elongation at break was observed for the nano-composite. This was ascribed to the chemical crosslinking between CNC's and polymer chains through covalent bonds. Jin and co-workers (Jin et al. 2020) have prepared PLA/CNC bio-nanocomposites, wherein the CNC was surface modified with silane (KH-550). The nanocomposite films were prepared via a solution casting process. 53.87% enhancement in tensile strength of PLA was reported with 0.5 wt.% CNC. However, further addition of silanized CNC in PLA led to a reduction in tensile strength. The decline in tensile strength can be mainly ascribed to the agglomeration of silanized CNC and hence the weak interaction between silanized CNC and PLA.

The summarized mechanical properties of PLA/CNC nanocomposites from various literature are presented in Table 9.2.

9.3.3   PLA/Cellulose Nanofibers (CNF) Composites

CNFs, like CNCs, are a bio-reinforcement alternative for developing a biobased and biocompatible PLA/cellulose nanocomposite. CNFs are biobased, non-toxic, and have a high specific strength, high modulus, and a large aspect ratio. Despite this, CNFs exhibit hydrophilicity and polar properties, which could pose problems when blending with hydrophobic polymers. Various reports on PLA/CNF bio-nanocomposites claim that CNF addition to PLA increases the modulus and strength with a concomitant reduction in elongation at break. However, when CNF dispersion is poor, the properties could remain unchanged or be adversely affected (Eyholzer et al. 2012). Jonoobi and co-workers (Jonoobi et al. 2010) prepared PLA/CNF composites by twin-screw extrusion. They reported 24% and 21% enhancement in tensile modulus and tensile strength, respectively, after adding 5 wt.% of CNF to PLA. However, elongation at break of 5 wt.% PLA/CNF composite dropped significantly. Iwatake, Nogi, and Yano (2008) claimed that the mechanical percolation network helped to increase young's modulus and tensile strength by 40% and 25%, respectively, when 5 wt.% CNF was used. For better compatibility and interfacial adhesion between PLA and CNF, Frone et al. (2016) prepared PLA/CNF nanocomposites using silane-modified CNF. They showed that silane treated CNF led to a substantial increment in Young's modulus (from 3.0 to 4.5 GPa) and tensile strength (from 31 to 41.4 MPa) of PLA. In another work, Yuan Lu et al. (2015) used amine-functionalized CNF and prepared green composites of PLA. Figure 9.6 shows the schematic representation of steps involved in amine functionalizing CNF and the amide linkage between PLLA matrix and amine-functionalized CNF.

The amine content can be tuned. mCNF-G1 denotes the 1st generation amine functionalization, and mCNF-G2 (higher amine content than mCNF-G1) denotes the 2nd generation amine functionalization. The aminolysis of PLA by amine-functionalized CNF allows for the grafting of PLA chains on the CNF. Stress-strain curves for neat PLLA and amine-functionalized 10 wt.% PLA/CNF nanocomposites are shown in Figure 9.6(2). They reported that for 10wt.% amine-functionalized (mCNF-G1) fiber content within PLA, tensile strength increased more than 100% compared to neat PLA. In the case of 10 wt.% mCNF-G2, there is not much enhancement in the mechanical properties of the nanocomposite, which could be due to the

TABLE 9.2

Summary of Mechanical Properties of PLA/CNC Nanocomposites

Researchers	Processing Method	CNC Content (wt.%)	Modification	Other Components (Surfactants, Compatibilizers, Plasticizers, etc.)	Mean Tensile Modulus (GPa)		Mean Tensile Strength (MPa)		Mean Elongation at Break (%)	
					PLA	PLA/CNC	PLA	PLA/CNC	PLA	PLAPLA/CNC
(Oksman et al. 2006)	Liquid feeding with TSE	5	-	PLA-MA	2.9	3.9	40.9	77.9	1.9	2.7
(Oksman et al. 2006)	Liquid feeding with TSE	5	-	PLA-MA PEG	2.9	2.6	40.9	48.4	1.9	17.8
(Bondeson and Oksman 2012)	TSE	5	-	-	2.65	2.7	62.8	55.5	19.5	9.7
(Pei, Zhou, and Berglund 2010)	Solution casting	2	Silylation	-	1.1	1.4	48.3	58.6	31.1	8.3
(Fortunati et al. 2012)	TSE	5	-	Beycostat A B09	2.1	4.4	43	46.1	90	18
(Espino-Pérez et al. 2013)	Solution casting	2.5	CNN-ICN	-	-	-	40	52	5.6	5.6
(Pracella, Haque, and Puglia 2014)	Hybrid	3	-	PVAc	2.18	2.61	48	55	4	2.3
(Robles et al. 2015)	TSE	1	Esterification	-	2.7	2.8	45	42	11	10
(Arias et al. 2015)	IM	3	-	PEO	3.4	3.5	61	60	2.7	2.3
(Herrera et al. 2016)	Liquid feeding	1	-	Triethyl citrate (TEC)	0.6	0.9	15.8	19.9	19.6	21.4
(Xu et al. 2016)	Solution casting	3	Acetylation	-	1.8	2.1	57	62	3.3	4.2
(Dhar et al. 2016)	Reactive TSE	1	PLA-g-CNC	DCP	0.4	2.1	35	51	8.4	1.9
(Bagheriasl et al. 2018)	Solution	3	-	-	2.9	3.3	65	67	2.9	2.3
(Gwon et al. 2016)	Solution casting	5	Urethanization	-	3.1	3.5	65	74	-	-

(Continued)

TABLE 9.2 (Continued)
Summary of Mechanical Properties of PLA/CNC Nanocomposites

Researchers	Processing Method	CNC Content (wt.%)	Modification	Other Components (Surfactants, Compatibilizers, Plasticizers, etc.)	Mean Tensile Modulus (GPa) PLA	PLA/CNC	Mean Tensile Strength (MPa) PLA	PLA/CNC	Mean Elongation at Break (%) PLA	PLA/CNC
(Mariano, Pilate, Oliveira, et al. 2017)	SC + TSE	20	-	PEG-b-PLLA	2	2.2	40	33	3.5	1.9
(Pal et al. 2017)	Solution casting	1	-	-	2.6	3.5	28.6	34.2	2.2	1.3
(Qian and Sheng 2017)	Solution casting	4	Silylation	-	0.42	0.11	19.8	14	13	250
(Yin et al. 2017)	Hot press	1	CNN-TCT-HA	-	–	–	25	48	1.9	3.1
(Yin et al. 2018)	Hot press	1	CNN-TCT-HA	-	–	–	25	43.6	1.9	2.9
(Shojaeiarani, Bajwa, and Stark 2018b)	SC + TSE	1	Esterification	-	3.3	4.7	57.4	59.9	3.5	3.6
(Shojaeiarani, Bajwa, and Stark 2018a)	SC + TSE	3	-	-	3.3	3.6	59	63.8	3.5	4.7
(Shojaeiarani, Bajwa, and Stark 2018a)	Spin coating + TSE	3	-	-	3.3	3.8	59	62.3	3.5	6

Abbreviation: SC = solution casting, TSE = twin screw extruder, IM = internal mixer, PLA-MA = maleated PLA, PVAc = poly(vinyl acetate), PEO = polyethylene oxide, PEG = polyethylene glycol, CNC-ICN = CNC modified with n-octadecyl-isocyanate, CNC-GMA = CNC grafted with glycidyl methacrylate, TCT = 2,4,6-trichloro-1,3,5-triazine, HA hexylamine, DA = dodecylamine, PEG-b-PLLA, Im = imidazolium, DCP = dicumyl peroxide.

Source: Reprinted from Vatansever, Arslan, and Nofar (2019), Copyright (2019), with permission from Elsevier.

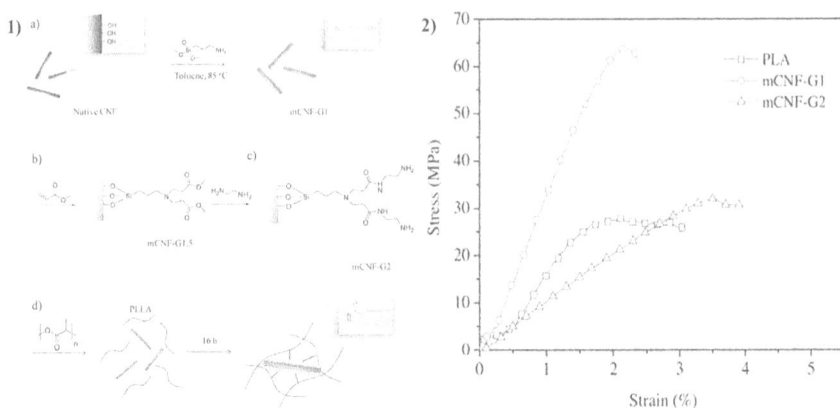

FIGURE 9.6 (1) Schematic illustration of the synthesis of amine-functionalized CNFs (a) mCNF-G1, (b) mCNF-G1.5, (c) mCNF-G2), and (d) the formation of amide linkage between PLLA matrix and the amine-modified CNFs. (2) Stress–strain curves for PLLA, PLLA/10%mCNF-G1, and PLLA/10%mCNF-G2 nanocomposite films. (Reprinted from Lu et al. (2015). Copyright (2015) with permission from Elsevier.)

degradation of PLLA by aminolysis. According to Lu et al., 10 wt.% of mCNF-G1 showed the strongest reinforcing effect, good dispersion, and better interfacial adhesion. Moreover, amine functionalization is simple and can be scaled up easily to get fully biodegradable PLA/CNF nanocomposites with enhanced mechanical properties.

The summarized mechanical properties of PLA/CNF nanocomposites from various literature are presented in Table 9.3.

9.3.4 SOLUTION CASTING VS MELT COMPOUNDING

9.3.4.1 Processing Method for PLA/Cellulose Nanocomposite Preparation and Its Effect on Mechanical Properties

The final mechanical properties are very much dependent on the processing method used to prepare the nanocomposite of PLA/cellulose. Solution casting, direct melt compounding, and dilution of solution mixed masterbatch through melt mixing (mTSE) are the most commonly used processing methods to prepare PLA/cellulose nanocomposites. Solution mixing is a facile processing technique to avoid CNC aggregation and does not need expensive equipment or high processing temperature. This minimizes polymer degradation. Nonetheless, using a large amount of solvent and recycling/disposal issues makes the process economically unviable. Alternatively, melt compounding using a twin-screw extruder is an industrially viable process to prepare PLA/cellulose nanocomposites. However, in the case of direct melt compounding, severe CNC agglomeration is a prime concern. The low thermal stability of CNC restricts processing temperature, and the orientation of CNC precludes the formation of a percolated network, which ultimately influences

TABLE 9.3
Summary of Mechanical Properties of PLA/CNF Nanocomposites

Researchers	Modification	CNC Content (wt.%)	Mean Elastic Modulus (GPa)		Mean Tensile Strength (MPa)		Mean Elongation at Break (%)	
			PLA	PLA/CNF	PLA	PLA/CNF	PLA	PLA/CNF
Iwatake, Nogi, and Yano (2008)	–	10	–	–	50	70	–	–
Okubo, Fujii, and Thostenson (2009)	–	2	3.5	4.75	45	53	–	–
Jonoobi et al. (2010)	–	5	2.9	3.6	59	72	3.4	2.7
Kowalczyk et al. (2011)	–	2	–	–	48	51	21	12
Eyholzer et al. (2012)	Carboxymethylation esterification (surface modification)	5	3.6	3.75	66	63	–	–
Kiziltas et al. (2016)	PHB (carrier system)	10	2.98	3.6	60	57	–	–
Feng et al. (2017)	In situ chemical grafting (surface modification)	5	2.5	2.55	58	58.5	8	7

Source: Reprinted from Vatansever, Arslan, and Nofar (2019). Copyright (2019), with permission from Elsevier.

the CNC reinforcement. Dilution of solution-mixed masterbatch through melt mixing (mTSE) is an excellent approach to prepare PLA/cellulose nanocomposites. According to Arslan et al., compared to direct melt compounding, the dispersion of CNC in PLA is good in the approach of mTSE (Arslan et al. 2020). However, when compared with solution mixed films, the dispersion and final mechanical properties are still inferior for the approach of mTSE. According to the author, the reason could be the re-agglomeration of already well-dispersed CNC when subjected to melt mixing. In Figure 9.7, it can be seen that dispersion of CNC in PLA is better when solution casting alone is used. Even though dispersion of CNC is good in the solution casting approach, residual solvent leads to a significant reduction in tensile strength when composites are prepared by solution casting, as depicted in Figure 9.7.

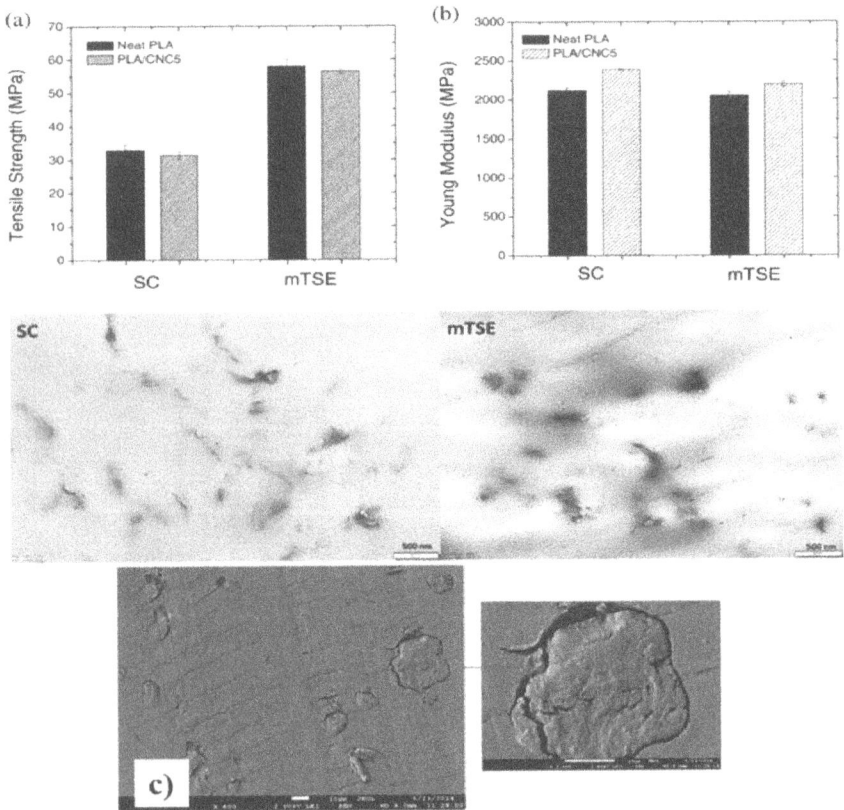

FIGURE 9.7 (a) Tensile strength and (b) Young's modulus of PLA and PLA/5 wt.% CNC nanocomposites prepared through solution cast (SC) or masterbatch-diluted (mTSE) polylactide (PLA) nanocomposites with 5 wt.% cellulose nanocrystal (CNC). (c) TEM micrographs of SC and mTSE (top row) and SEM images of the directly melt mixed PLA with 5 wt.% CNC through a twin-screw extruder. (Reproduced from Arslan et al. (2020). With permission from John Wiley and Sons.)

In another study, Bagheriasl et al. (2019) compared nanocomposites prepared by direct melt mixing (DMM) with that of a combined solution mixing and melt compounding process (mTSE). PLA/CNC prepared by mTSE showed good dispersion of CNC into PLA matrix and enhanced crystalline content, storage modulus, thermal and mechanical properties.

9.4 SUMMARY AND FUTURE OUTLOOK

PLA/cellulose nanocomposites are being investigated extensively due to their eco-friendliness and appealing properties of both the matrix and reinforcement. However, the primary challenge in PLA/cellulose nanocomposites is realizing good dispersion of hydrophilic nanocellulose in the hydrophobic PLA matrix. The presence of strong hydrogen bonding in CNC and CNF inhibits the uniform dispersion of nanocellulose in the PLA matrix. The effects of processing strategies on the final mechanical characteristics of PLA/cellulose nanocomposites were reviewed in this chapter, which covered solution mixing, direct melt compounding, and the strategy of combining solution mixing and melt compounding. This chapter also discussed how surface modification of nanocellulose could be a strategy for increasing nanocellulose dispersibility and improving mechanical properties. However, PLA/cellulose nanocomposites are still a long way off from being used in real-world applications. Though solution mixing can provide good dispersion of CNC/CNF into PLA, the process is not environmentally friendly or industrially viable. Direct melt compounding, an industrially viable process, leads to poor dispersion of CNC/CNF resulting in poor mechanical properties of PLA/cellulose nanocomposites. Surface modification of CNC/CNF can help overcome dispersion issues, but it hinders forming a rigid percolation network of CNC/CNF and adds processing cost. The most difficult challenge is to modify the surface of nanocellulose using a sustainable process while retaining its properties. There's still much more to discover about PLA/cellulose nanocomposites, and efficient processing methods are required to prepare the same. As previously stated, the addition of nanocellulose to PLA increases tensile strength and modulus while reducing elongation at break. PLA/cellulose nanocomposite with balanced properties, such as good tensile strength, good modulus and good elongation at break, is a target that is being widely explored.

ACKNOWLEDGMENT

The authors thank the Council of Scientific and Industrial Research, National Chemical Laboratory (CSIR-NCL), Pune for providing facilities, and CSIR-GATE for a senior research fellowship.

REFERENCES

Aaliya, Basheer, Kappat Valiyapeediyekkal Sunooj, and Maximilian Lackner. 2021. "Biopolymer Composites: A Review." *International Journal of Biobased Plastics* 3 (1). Taylor & Francis: 40–84. doi:10.1080/24759651.2021.1881214

Afrin, Sadaf, and Zoheb Karim. 2017. "Isolation and Surface Modification of Nanocellulose: Necessity of Enzymes over Chemicals." *ChemBioEng Reviews* 4 (5). John Wiley & Sons, Ltd: 289–303. doi:10.1002/CBEN.201600001

Alam, Fahad, Vishnu Raj Shukla, K. M. Varadarajan, and S. Kumar. 2020. "Microarchitected 3D Printed Polylactic Acid (PLA) Nanocomposite Scaffolds for Biomedical Applications." *Journal of the Mechanical Behavior of Biomedical Materials* 103: 103576. doi:10.1016/j.jmbbm.2019.103576

Arias, Andrea, Marie-Claude Claude Heuzey, Michel A. Huneault, Gilles Ausias, and Abdelkader Bendahou. 2015. "Enhanced Dispersion of Cellulose Nanocrystals in Melt-Processed Polylactide-Based Nanocomposites." *Cellulose* 22 (1). Springer: 483–498. doi:10.1007/s10570-014-0476-z

Arslan, Dogan, Emre Vatansever, Deniz Sema Sarul, Yusuf Kahraman, Gurbuz Gunes, Ali Durmus, and Mohammadreza Nofar. 2020. "Effect of Preparation Method on the Properties of Polylactide/Cellulose Nanocrystal Nanocomposites." *Polymer Composites* 41 (10): 4170–4180. doi:10.1002/pc.25701

Bagheriasl, Davood, Pierre J. Carreau, Bernard Riedl, and Charles Dubois. 2018. "Enhanced Properties of Polylactide by Incorporating Cellulose Nanocrystals." *Polymer Composites* 39 (8). John Wiley & Sons, Ltd: 2685–2694. doi:10.1002/PC.24259

Bagheriasl, Davood, Fatemeh Safdari, Pierre J. Carreau, Charles Dubois, and Bernard Riedl. 2019. "Development of Cellulose Nanocrystal-Reinforced Polylactide: A Comparative Study on Different Preparation Methods." *Polymer Composites* 40 (January). John Wiley and Sons Inc.: E342–E349. doi:10.1002/PC.24676

Bondeson, Daniel, and Kristiina Oksman. 2012. "Dispersion and Characteristics of Surfactant Modified Cellulose Whiskers Nanocomposites." *Composite Interfaces* 14 (7–9). Taylor & Francis Group: 617–630. doi:10.1163/156855407782106519

Carvalho, Tiago, Gabriela Guedes, Filipa L. Sousa, Carmen S. R. Freire, and Hélder A. Santos. 2019. "Latest Advances on Bacterial Cellulose-Based Materials for Wound Healing, Delivery Systems, and Tissue Engineering." *Biotechnology Journal* 14 (12). John Wiley & Sons, Ltd: 1900059. doi:10.1002/BIOT.201900059

Chamas, Ali, Hyunjin Moon, Jiajia Zheng, Yang Qiu, Tarnuma Tabassum, Jun Hee Jang, Mahdi Abu-Omar, Susannah L Scott, and Sangwon Suh. 2020. "Degradation Rates of Plastics in the Environment." *ACS Sustainable Chemistry & Engineering* 8: 3511. doi:10.1021/acssuschemeng.9b06635

Cheng, Yanling, Shaobo Deng, Paul Chen, and Roger Ruan. 2009. "Polylactic Acid (PLA) Synthesis and Modifications: A Review." *Frontiers of Chemistry in China 2009* 4 (3). Springer: 259–264. doi:10.1007/S11458-009-0092-X

Chirayil, Cintil Jose, Jithin Joy, Lovely Mathew, Miran Mozetic, Joachim Koetz, and Sabu Thomas. 2014. "Isolation and Characterization of Cellulose Nanofibrils from *Helicteres isora* Plant." *Industrial Crops and Products* 59 (August). Elsevier: 27–34. doi:10.1016/J.INDCROP.2014.04.020

Daud, Jannah B., and Koon-Yang Lee. 2017. "Surface Modification of Nanocellulose." *Handbook of Nanocellulose and Cellulose Nanocomposites*, March. John Wiley & Sons, Ltd: 101–122. doi:10.1002/9783527689972.CH3

Deepa, B., Eldho Abraham, Bibin Mathew Cherian, Alexander Bismarck, Jonny J. Blaker, Laly A. Pothan, Alcides Lopes Leao, Sivoney Ferreira de Souza, and M. Kottaisamy. 2011. "Structure, Morphology and Thermal Characteristics of Banana Nano Fibers Obtained by Steam Explosion." *Bioresource Technology* 102 (2). Elsevier: 1988–1997. doi:10.1016/J.BIORTECH.2010.09.030

Dhar, Prodyut, Debashis Tarafder, Amit Kumar, and Vimal Katiyar. 2016. "Thermally Recyclable Polylactic Acid/Cellulose Nanocrystal Films through Reactive Extrusion Process." *Polymer* 87 (March). Elsevier Ltd: 268–282. doi:10.1016/j.polymer.2016.02.004

El-Hadi, Ahmed M. 2017. "Increase the Elongation at Break of Poly (Lactic Acid) Composites for Use in Food Packaging Films." *Scientific Reports* 7 (March). Nature Publishing Group: 1–14. doi:10.1038/srep46767

Espino-Pérez, Etzael, Julien Bras, Violette Ducruet, Alain Guinault, Alain Dufresne, and Sandra Domenek. 2013. "Influence of Chemical Surface Modification of Cellulose Nanowhiskers on Thermal, Mechanical, and Barrier Properties of Poly(Lactide) Based Bionanocomposites." *European Polymer Journal* 49 (10). Pergamon: 3144–3154. doi:10.1016/J.EURPOLYMJ.2013.07.017

Eyholzer, C., P. Tingaut, T. Zimmermann, and K. Oksman. 2012. "Dispersion and Reinforcing Potential of Carboxymethylated Nanofibrillated Cellulose Powders Modified with 1-Hexanol in Extruded Poly(Lactic Acid) (PLA) Composites." *Journal of Polymers and the Environment* 20 (4). Springer: 1052–1062. doi:10.1007/S10924-012-0508-4

Farah, Shady, Daniel G. Anderson, and Robert Langer. 2016. "Physical and Mechanical Properties of PLA, and Their Functions in Widespread Applications – A Comprehensive Review." *Advanced Drug Delivery Reviews* 107. Elsevier B.V.: 367–392. doi:10.1016/j.addr.2016.06.012

Feng, Jiabin, Yiqi Sun, Pingan Song, Weiwei Lei, Qiang Wu, Lina Liu, Youming Yu, and Hao Wang. 2017. "Fire-Resistant, Strong, and Green Polymer Nanocomposites Based on Poly(Lactic Acid) and Core–Shell Nanofibrous Flame Retardants." *ACS Sustainable Chemistry and Engineering* 5 (9). American Chemical Society: 7894–7904. doi:10.1021/ACSSUSCHEMENG.7B01430

Fortunati, E., I. Armentano, Q. Zhou, A. Iannoni, E. Saino, L. Visai, L. A. Berglund, and J. M. Kenny. 2012. "Multifunctional Bionanocomposite Films of Poly(Lactic Acid), Cellulose Nanocrystals and Silver Nanoparticles." *Carbohydrate Polymers* 87 (2). Elsevier: 1596–1605. doi:10.1016/J.CARBPOL.2011.09.066

Foster, E. Johan, Robert J. Moon, Umesh P. Agarwal, Michael J. Bortner, Julien Bras, Sandra Camarero-Espinosa, Kathleen J. Chan, et al. 2018. "Current Characterization Methods for Cellulose Nanomaterials." *Chemical Society Reviews* 47 (8): 2609–2679. doi:10.1039/C6CS00895J

Frone, Adriana Nicoleta, Denis Mihaela Panaitescu, Ioana Chiulan, Cristian Andi Nicolae, Zina Vuluga, Catalin Vitelaru, and Celina Maria Damian. 2016. "The Effect of Cellulose Nanofibers on the Crystallinity and Nanostructure of Poly(Lactic Acid) Composites." *Journal of Materials Science* 51 (21). Springer US: 9771–9791. doi:10.1007/s10853-016-0212-1

George, Johnsy, and S. N Sabapathi. 2015. "Cellulose Nanocrystals: Synthesis, Functional Properties, and Applications." *Nanotechnology, Science and Applications* 8 (November). Dove Press: 45–54. doi:10.2147/NSA.S64386

Gopi, S., P. Balakrishnan, D. Chandradhara, D. Poovathankandy, and S. Thomas. 2019. "General Scenarios of Cellulose and Its Use in the Biomedical Field." *Materials Today Chemistry* 13 (September). Elsevier: 59–78. doi:10.1016/J.MTCHEM.2019.04.012

Gwon, Jae-Gyoung, Hye-Jung Cho, Sang-Jin Chun, Sun-Young Soo Lee, Qinglin Wu, Mei-Chun Li, and Sun-Young Soo Lee. 2016. "Mechanical and Thermal Properties of Toluene Diisocyanate-Modified Cellulose Nanocrystal Nanocomposites Using Semi-Crystalline Poly(Lactic Acid) as a Base Matrix." *RSC Advances* 6 (77). The Royal Society of Chemistry: 73879–73886. doi:10.1039/C6RA10993D

Hazwan Hussin, M., Djalal Trache, Caryn Tan Hui Chuin, M. R. Nurul Fazita, M. K. Mohamad Haafiz, and Md. Sohrab Hossain. 2019. "Extraction of Cellulose Nanofibers and Their Eco-Friendly Polymer Composites." *Sustainable Polymer Composites and Nanocomposites*, January. Springer, Cham: 653–691. doi:10.1007/978-3-030-05399-4_23

He, Wen, Xiaochuan Jiang, Fengwen Sun, and Xinwu Xu. 2014. "Extraction and Characterization of Cellulose Nanofibers from *Phyllostachys nidularia* Munro via a Combination of Acid Treatment and Ultrasonication." *BioResources* 9 (4): 6876–6887.

Herrera, Natalia, Asier M. Salaberria, Aji P. Mathew, and Kristiina Oksman. 2016. "Plasticized Polylactic Acid Nanocomposite Films with Cellulose and Chitin Nanocrystals Prepared Using Extrusion and Compression Molding with Two Cooling Rates: Effects on Mechanical, Thermal and Optical Properties." *Composites Part A: Applied Science and Manufacturing* 83 (April). Elsevier: 89–97. doi:10.1016/J. COMPOSITESA.2015.05.024

Huang, Jin, Xiaozhou Ma, Guang Yang, and Dufresne Alain. 2019. "Introduction to Nanocellulose." In Jin Huang, Alain Dufresne, Ning Lin (Eds). *Nanocellulose: From Fundamentals to Advanced Materials*. Wiley Online Library, 1–20.

Islam, Mohammad Tajul, Mohammad Mahbubul Alam, and Marina Zoccola. 2013. "Review on Modification of Nanocellulose for Application in Composites." *International Journal of Innovative Research in Science, Engineering and Technology* 2 (10): 5444–5451.

Iwatake, Atsuhiro, Masaya Nogi, and Hiroyuki Yano. 2008. "Cellulose Nanofiber-Reinforced Polylactic Acid." *Composites Science and Technology* 68 (9). Elsevier: 2103–2106. doi:10.1016/j.compscitech.2008.03.006

Jin, Kaiyan, Yanjun Tang, Xianmei Zhu, and Yiming Zhou. 2020. "Polylactic Acid Based Biocomposite Films Reinforced with Silanized Nanocrystalline Cellulose." *International Journal of Biological Macromolecules* 162 (November). Elsevier: 1109–1117. doi:10.1016/J.IJBIOMAC.2020.06.201

Jonoobi, Mehdi, Jalaluddin Harun, Aji P. Mathew, and Kristiina Oksman. 2010. "Mechanical Properties of Cellulose Nanofiber (CNF) Reinforced Polylactic Acid (PLA) Prepared by Twin Screw Extrusion." *Composites Science and Technology* 70 (12). Elsevier Ltd: 1742–1747. doi:10.1016/j.compscitech.2010.07.005

Kaboorani, Alireza, and Bernard Riedl. 2015. "Surface Modification of Cellulose Nanocrystals (CNC) by a Cationic Surfactant." *Industrial Crops and Products* 65 (March). Elsevier: 45–55. doi:10.1016/J.INDCROP.2014.11.027

Kargarzadeh, H., J. Huang, N. Lin, I. Ahmad, M. Mariano, A. Dufresne, S. Thomas, and Andrzej Gałęski. 2018. "Recent Developments in Nanocellulose-Based Biodegradable Polymers, Thermoplastic Polymers, and Porous Nanocomposites." *Progress in Polymer Science* 87: 197–227. doi:10.1016/j.progpolymsci.2018.07.008

Kargarzadeh, Hanieh, Marcos Mariano, Deepu Gopakumar, Ishak Ahmad, Sabu Thomas, Alain Dufresne, Jin Huang, and Ning Lin. 2018. "Advances in Cellulose Nanomaterials." *Cellulose*. Vol. 25. Springer Netherlands. doi:10.1007/s10570-018-1723-5

Kentaro Abe, Shinichiro Iwamoto, and Hiroyuki Yano. 2007. "Obtaining Cellulose Nanofibers with a Uniform Width of 15 nm from Wood." *Biomacromolecules* 8 (10). American Chemical Society: 3276–3278. doi:10.1021/BM700624P

Kiziltas, Alper, Behzad Nazari, Esra Erbas Kiziltas, Douglas J. Gardner, Yousoo Han, and Todd S. Rushing. 2016. "Method to Reinforce Polylactic Acid with Cellulose Nanofibers via a Polyhydroxybutyrate Carrier System." *Carbohydrate Polymers* 140 (April). Elsevier: 393–399. doi:10.1016/J.CARBPOL.2015.12.059

Köse, Kazım, Miran Mavlan, and Jeffrey P. Youngblood. 2020. "Applications and Impact of Nanocellulose Based Adsorbents." *Cellulose* 27 (6). Springer: 2967–2990. doi:10.1007/ S10570-020-03011-1

Kowalczyk, M., E. Piorkowska, P. Kulpinski, and M. Pracella. 2011. "Mechanical and Thermal Properties of PLA Composites with Cellulose Nanofibers and Standard Size Fibers." *Composites Part A: Applied Science and Manufacturing* 42 (10). Elsevier: 1509–1514. doi:10.1016/J.COMPOSITESA.2011.07.003

Li, Meng, Li Jun Wang, Dong Li, Yan Ling Cheng, and Benu Adhikari. 2014. "Preparation and Characterization of Cellulose Nanofibers from De-Pectinated Sugar Beet Pulp." *Carbohydrate Polymers* 102 (1). Elsevier: 136–143. doi:10.1016/J.CARBPOL.2013.11.021

Lin, Ning, and Alain Dufresne. 2014. "Nanocellulose in Biomedicine: Current Status and Future Prospect." *European Polymer Journal* 59 (October). Pergamon: 302–325. doi:10.1016/J.EURPOLYMJ.2014.07.025

Lin, Ning, Jin Huang, and Alain Dufresne. 2012. "Preparation, Properties and Applications of Polysaccharide Nanocrystals in Advanced Functional Nanomaterials: A Review." *Nanoscale* 4 (11). The Royal Society of Chemistry: 3274–3294. doi:10.1039/C2NR30260H

Lu, Yuan, Mario Calderón Cueva, Edgar Lara-Curzio, and Soydan Ozcan. 2015. "Improved Mechanical Properties of Polylactide Nanocomposites-Reinforced with Cellulose Nanofibrils through Interfacial Engineering via Amine-Functionalization." *Carbohydrate Polymers* 131. Elsevier Ltd.: 208–217. doi:10.1016/j.carbpol.2015.05.047

Mariano, Marcos, Nadia El Kissi, Alain Dufresne. 2014. "Cellulose Nanocrystals and Related Nanocomposites: Review of Some Properties and Challenges." *Journal of Polymer Science Part B: Polymer Physics* 52 (12). John Wiley & Sons, Ltd: 791–806.

Mariano, Marcos, Florence Pilate, Franciéli Borges de Oliveira, Farid Khelifa, Philippe Dubois, Jean-Marie Marie Raquez, Alain Dufresne, et al. 2017. "Preparation of Cellulose Nanocrystal-Reinforced Poly(Lactic Acid) Nanocomposites through Noncovalent Modification with PLLA-Based Surfactants." *ACS Omega* 2 (6). American Chemical Society: 2678–2688. doi:10.1021/acsomega.7b00387

Moberg, Tobias, Karin Sahlin, Kun Yao, Shiyu Geng, Gunnar Westman, Qi Zhou, Kristiina Oksman, and Mikael Rigdahl. 2017. "Rheological Properties of Nanocellulose Suspensions: Effects of Fibril/Particle Dimensions and Surface Characteristics." *Cellulose* 24 (6). Springer: 2499–2510. doi:10.1007/S10570-017-1283-0

Mokhena, T. C., and M. J. John. 2019. "Cellulose Nanomaterials: New Generation Materials for Solving Global Issues." *Cellulose* 27 (3). Springer: 1149–1194. doi:10.1007/S10570-019-02889-W

Moohan, John, Sarah A. Stewart, Eduardo Espinosa, Antonio Rosal, Alejandro Rodríguez, Eneko Larrañeta, Ryan F. Donnelly, and Juan Domínguez-Robles. 2019. "Cellulose Nanofibers and Other Biopolymers for Biomedical Applications. A Review." *Applied Sciences 2020* 10 (1). Multidisciplinary Digital Publishing Institute: 65. doi:10.3390/APP10010065

Moon, Robert J., Ashlie Martini, John Nairn, John Simonsen, and Jeff Youngblood. 2011. "Cellulose Nanomaterials Review: Structure, Properties and Nanocomposites." *Chemical Society Reviews* 40 (7). The Royal Society of Chemistry: 3941–3994. doi:10.1039/C0CS00108B

Nam, Joo Young, Suprakas Sinha Ray, and Masami Okamoto. 2003. "Crystallization Behavior and Morphology of Biodegradable Polylactide/Layered Silicate Nanocomposite." *Macromolecules* 36 (19): 7126–7131. doi:10.1021/ma034623j

Naser, Ahmed Z., I. Deiab, and Basil M. Darras. 2021. "Poly(Lactic Acid) (PLA) and Polyhydroxyalkanoates (PHAs), Green Alternatives to Petroleum-Based Plastics: A Review." *RSC Advances* 11 (28). Royal Society of Chemistry: 17151–17196. doi:10.1039/d1ra02390j

Naz, Sania, Joham S. Ali, and Muhammad Zia. 2019. "Nanocellulose Isolation Characterization and Applications: A Journey from Non-Remedial to Biomedical Claims." *Bio-Design and Manufacturing* 2 (3). Springer: 187–212. doi:10.1007/S42242-019-00049-4

Nofar, Mohammadreza, Dilara Sacligil, Pierre J. Carreau, Musa R. Kamal, and Marie Claude Heuzey. 2019. "Poly (Lactic Acid) Blends: Processing, Properties and Applications." *International Journal of Biological Macromolecules* 125 (March). Elsevier: 307–360. doi:10.1016/J.IJBIOMAC.2018.12.002

Nofar, Mohammadreza, Reza Salehiyan, and Suprakas Sinha Ray. 2019. "Rheology of Poly (Lactic Acid)-Based Systems." *Polymer Reviews* 59 (3). Taylor & Francis: 465–509. doi:10.1080/15583724.2019.1572185

Oguz, Oguzhan, Nicolas Candau, Adrien Demongeot, Mehmet Kerem Citak, Fatma Nalan Cetin, Grégory Stoclet, Véronique Michaud, and Yusuf Z. Menceloglu. 2021. "Poly(Lactide)/Cellulose Nanocrystal Nanocomposites by High-Shear Mixing." *Polymer Engineering and Science* 61 (4): 1028–1040. doi:10.1002/pen.25621

Oksman, K., A. P. Mathew, D. Bondeson, and I. Kvien. 2006. "Manufacturing Process of Cellulose Whiskers/Polylactic Acid Nanocomposites." *Composites Science and Technology* 66 (15). Elsevier: 2776–2784. doi:10.1016/J.COMPSCITECH.2006.03.002

Okubo, Kazuya, Toru Fujii, and Erik T. Thostenson. 2009. "Multi-Scale Hybrid Biocomposite: Processing and Mechanical Characterization of Bamboo Fiber Reinforced PLA with Microfibrillated Cellulose." *Composites Part A: Applied Science and Manufacturing* 40 (4). Elsevier: 469–475. doi:10.1016/J.COMPOSITESA.2009.01.012

Oun, Ahmed A., Shiv Shankar, and Jong-Whan Rhim. 2019. "Multifunctional Nanocellulose/ Metal and Metal Oxide Nanoparticle Hybrid Nanomaterials." *Critical Reviews in Food Science and Nutrition* 60 (3). Taylor & Francis: 435–460. doi:10.1080/10408398.2018. 1536966

Pal, Nidhi, Poornima Dubey, P. Gopinath, and Kaushik Pal. 2017. "Combined Effect of Cellulose Nanocrystal and Reduced Graphene Oxide into Poly-Lactic Acid Matrix Nanocomposite as a Scaffold and Its Anti-Bacterial Activity." *International Journal of Biological Macromolecules* 95 (February). Elsevier: 94–105. doi:10.1016/J. IJBIOMAC.2016.11.041

Paolini, Alessandro, Lucà Leoni, Ilaria Giannicchi, Zeinab Abbaszadeh, Valentina D'Oria, Francesco Mura, Antonella Dalla Cort, and Andrea Masotti. 2018. "MicroRNAs Delivery into Human Cells Grown on 3D-Printed PLA Scaffolds Coated with a Novel Fluorescent PAMAM Dendrimer for Biomedical Applications." *Scientific Reports* 8 (1). Nature Publishing Group: 1–11. doi:10.1038/s41598-018-32258-9

Pei, Aihua, Qi Zhou, and Lars A. Berglund. 2010. "Functionalized Cellulose Nanocrystals as Biobased Nucleation Agents in Poly(l-Lactide) (PLLA) – Crystallization and Mechanical Property Effects." *Composites Science and Technology* 70 (5). Elsevier: 815–821. doi:10.1016/J.COMPSCITECH.2010.01.018

Pennells, Jordan, Ian D. Godwin, Nasim Amiralian, and Darren J. Martin. 2019. "Trends in the Production of Cellulose Nanofibers from Non-Wood Sources." *Cellulose* 27 (2). Springer: 575–593. doi:10.1007/S10570-019-02828-9

Phanthong, Patchiya, Prasert Reubroycharoen, Xiaogang Hao, Guangwen Xu, Abuliti Abudula, and Guoqing Guan. 2018. "Nanocellulose: Extraction and Application." *Carbon Resources Conversion* 1 (1). Elsevier: 32–43. doi:10.1016/J.CRCON.2018. 05.004

Pracella, Mariano, Md Minhaz Ul Haque, and Debora Puglia. 2014. "Morphology and Properties Tuning of PLA/Cellulose Nanocrystals Bio-Nanocomposites by Means of Reactive Functionalization and Blending with PVAc." *Polymer* 55 (16). Elsevier: 3720–3728. doi:10.1016/J.POLYMER.2014.06.071

Qian, Shaoping, and Kuichuan Sheng. 2017. "PLA Toughened by Bamboo Cellulose Nanowhiskers: Role of Silane Compatibilization on the PLA Bionanocomposite Properties." *Composites Science and Technology* 148 (August). Elsevier: 59–69. doi:10. 1016/J.COMPSCITECH.2017.05.020

Rajinipriya, Malladi, Malladi Nagalakshmaiah, Mathieu Robert, and Saïd Elkoun. 2018. "Importance of Agricultural and Industrial Waste in the Field of Nanocellulose and Recent Industrial Developments of Wood Based Nanocellulose: A Review." *ACS Sustainable Chemistry & Engineering*. American Chemical Society. doi:10.1021/ACSSUSCHEMENG.7B03437

Rasal, Rahul M., Amol V. Janorkar, and Douglas E. Hirt. 2010. "Poly(Lactic Acid) Modifications." *Progress in Polymer Science (Oxford)* 35 (3). Elsevier Ltd: 338–356. doi:10.1016/j.progpolymsci.2009.12.003

Robles, Eduardo, Iñaki Urruzola, Jalel Labidi, and Luis Serrano. 2015. "Surface-Modified Nano-Cellulose as Reinforcement in Poly(Lactic Acid) to Conform New Composites." *Industrial Crops and Products* 71 (September). Elsevier: 44–53. doi:10.1016/J.INDCROP.2015.03.075

Ruan, Gang, and Si Shen Feng. 2003. "Preparation and Characterization of Poly(Lactic Acid)–Poly(Ethylene Glycol)–Poly(Lactic Acid) (PLA–PEG–PLA) Microspheres for Controlled Release of Paclitaxel." *Biomaterials* 24 (27). Elsevier: 5037–5044. doi:10.1016/S0142-9612(03)00419-8

Saeidlou, Sajjad, Michel A. Huneault, Hongbo Li, and Chul B. Park. 2012. "Poly(Lactic Acid) Crystallization." *Progress in Polymer Science* 37 (12). Pergamon: 1657–1677. doi:10.1016/J.PROGPOLYMSCI.2012.07.005

Salimi, Sina, Rahmat Sotudeh-Gharebagh, Reza Zarghami, Siok Yee Chan, and Kah Hay Yuen. 2019. "Production of Nanocellulose and Its Applications in Drug Delivery: A Critical Review." *ACS Sustainable Chemistry and Engineering* 7 (19): 15800–827. doi:10.1021/acssuschemeng.9b02744

Sharma, Chhavi, and Nishi K. Bhardwaj. 2019. "Bacterial Nanocellulose: Present Status, Biomedical Applications and Future Perspectives." *Materials Science and Engineering* 104 (November). Elsevier: 109963. doi:10.1016/J.MSEC.2019.109963

Shojaeiarani, Jamileh, Dilpreet S. Bajwa, and Nicole M. Stark. 2018a. "Spin-Coating: A New Approach for Improving Dispersion of Cellulose Nanocrystals and Mechanical Properties of Poly (Lactic Acid) Composites." *Carbohydrate Polymers* 190 (November 2017). Elsevier: 139–147. doi:10.1016/j.carbpol.2018.02.069

———. 2018b. "Green Esterification: A New Approach to Improve Thermal and Mechanical Properties of Poly(Lactic Acid) Composites Reinforced by Cellulose Nanocrystals." *Journal of Applied Polymer Science* 135 (27). John Wiley & Sons, Ltd: 46468. doi:10.1002/APP.46468

Su, Shen, Rodion Kopitzky, Sengül Tolga, and Stephan Kabasci. 2019. "Polylactide (PLA) and Its Blends with Poly(Butylene Succinate) (PBS): A Brief Review." *Polymers* 11 (7). Multidisciplinary Digital Publishing Institute: 1193. doi:10.3390/POLYM11071193

Trache, Djalal, M. Hazwan Hussin, M. K. Mohamad Haafiz, and Vijay Kumar Thakur. 2017. "Recent Progress in Cellulose Nanocrystals: Sources and Production." *Nanoscale* 9 (5). The Royal Society of Chemistry: 1763–1786. doi:10.1039/c6nr09494e

Trache, Djalal, Kamel Khimeche, Abderrahmane Mezroua, and Mokhtar Benziane. 2016. "Physicochemical Properties of Microcrystalline Nitrocellulose from Alfa Grass Fibres and Its Thermal Stability." *Journal of Thermal Analysis and Calorimetry* 124 (3). Springer: 1485–1496. doi:10.1007/S10973-016-5293-1

Trache, Djalal, Ahmed Fouzi Tarchoun, Mehdi Derradji, Tuan Sherwyn Hamidon, Nanang Masruchin, Nicolas Brosse, and M. Hazwan Hussin. 2020a. "Nanocellulose: From Fundamentals to Advanced Applications." *Frontiers in Chemistry* 8 (May). Frontiers: 392. doi:10.3389/fchem.2020.00392

Trache, Djalal, Ahmed Fouzi Tarchoun, Mehdi Derradji, Oussama Mehelli, M Hazwan Hussin, and Wissam Bessa. 2020b. "Cellulose Fibers and Nanocrystals: Preparation, Characterization, and Surface Modification." In Vineet Kumar, Praveen Guleria, Nandita Dasgupta, Shivendu Ranjan (Eds). *Functionalized Nanomaterials I.* CRC Press, 171–190.

Trache, Djalal, and Djalal Trache. 2018. "Nanocellulose as a Promising Sustainable Material for Biomedical Applications." *AIMS Materials Science* 5 (2). AIMS Press: 201–205. doi:10.3934/MATERSCI.2018.2.201

Vatansever, Emre, Dogan Arslan, and Mohammadreza Nofar. 2019. "Polylactide Cellulose-Based Nanocomposites." *International Journal of Biological Macromolecules* 137: 912–938. https://doi.org/10.1016/j.ijbiomac.2019.06.205

Vazquez, A., M. Laura Foresti, Juan I. Moran, and Viviana P. Cyras. 2015. "Extraction and Production of Cellulose Nanofibers." In *Handbook of Polymer Nanocomposites. Processing, Performance and Application: Volume C: Polymer Nanocomposites of Cellulose Nanoparticles* January. Springer, 81–118. doi:10.1007/978-3-642-45232-1_57

Vestena, Mauro, Idejan P. Gross, Carmen M. O. Müller, and Alfredo T. N. Pires. 2016. "Nanocomposite of Poly(Lactic Acid)/Cellulose Nanocrystals: Effect of CNC Content on the Polymer Crystallization Kinetics." *Journal of the Brazilian Chemical Society* 27 (5): 905–911. doi:10.5935/0103-5053.20150343

Vineeth, S. K., Ravindra V. Gadhave, Pradeep T. Gadekar, S. K. Vineeth, Ravindra V. Gadhave, and Pradeep T. Gadekar. 2019. "Chemical Modification of Nanocellulose in Wood Adhesive: Review." *Open Journal of Polymer Chemistry* 9 (4). Scientific Research Publishing: 86–99. doi:10.4236/OJPCHEM.2019.94008

Vink, Erwin T. H., Karl R. Rábago, David A. Glassner, and Patrick R. Gruber. 2003. "Applications of Life Cycle Assessment to NatureWorks Polylactide (PLA) Production." *Polymer Degradation and Stability* 80 (3). Elsevier: 403–419. doi:10.1016/S0141-3910(02)00372-5

Wasti, Sanjita, Eldon Triggs, Ramsis Farag, Maria Auad, Sushil Adhikari, Dilpreet Bajwa, Mi Li, and Arthur J. Ragauskas. 2021. "Influence of Plasticizers on Thermal and Mechanical Properties of Biocomposite Filaments Made from Lignin and Polylactic Acid for 3D Printing." *Composites Part B: Engineering* 205 (January). Elsevier: 108483. doi:10.1016/J.COMPOSITESB.2020.108483

Webb, Hayden K, Jaimys Arnott, Russell J Crawford, and Elena P Ivanova. 2013. "Plastic Degradation and Its Environmental Implications with Special Reference to Poly(Ethylene Terephthalate)." *Polymers* 5: 1–18. doi:10.3390/polym5010001

Xi, Wang, Li G, Liu Y, Yu W, and Sun Q. 2017. "Biocompatibility of Biological Material Polylactic Acid with Stem Cells from Human Exfoliated Deciduous Teeth." *Biomedical Reports* 6 (5). Biomed Rep: 519–524. doi:10.3892/BR.2017.881

Xu, Chunjiang, Jianxiang Chen, Defeng Wu, Yang Chen, Qiaolian Lv, and Mengqi Wang. 2016. "Polylactide/Acetylated Nanocrystalline Cellulose Composites Prepared by a Continuous Route: A Phase Interface-Property Relation Study." *Carbohydrate Polymers* 146 (August). Elsevier: 58–66. doi:10.1016/J.CARBPOL.2016.03.058

Yang, Xue, Fuyi Han, Chunxia Xu, Shuai Jiang, Liqian Huang, Lifang Liu, and Zhaopeng Xia. 2017. "Effects of Preparation Methods on the Morphology and Properties of Nanocellulose (NC) Extracted from Corn Husk." *Industrial Crops and Products* 109 (July). Elsevier: 241–247. doi:10.1016/j.indcrop.2017.08.032

Yin, Yuanyuan, Lina Zhao, Xue Jiang, Hongbo Wang, and Weidong Gao. 2017. "Poly(Lactic Acid)-Based Biocomposites Reinforced with Modified Cellulose Nanocrystals." *Cellulose* 24 (11). Springer: 4773–4784. doi:10.1007/S10570-017-1455-Y

———. 2018. "Cellulose Nanocrystals Modified with a Triazine Derivative and Their Reinforcement of Poly(Lactic Acid)-Based Bionanocomposites." *Cellulose* 25 (5). Springer: 2965–2976. doi:10.1007/S10570-018-1741-3

Zaaba, Nor Fasihah, Mariatti Jaafar, and Hanafi Ismail. 2021. "Tensile and Morphological Properties of Nanocrystalline Cellulose and Nanofibrillated Cellulose Reinforced PLA Bionanocomposites: A Review." *Polymer Engineering and Science* 61 (1): 22–38. doi:10.1002/pen.25560

Zhang, Xiaolin, Shaoge Li, Jia Li, Baiqiao Fu, Jingjing Di, Long Xu, and Xiaofeng Zhu. 2021. "Reinforcing Effect of Nanocrystalline Cellulose and Office Waste Paper Fibers on Mechanical and Thermal Properties of Poly (Lactic Acid) Composites." *Journal of Applied Polymer Science* 138 (21): 50462. doi:10.1002/app.50462

Zheng, Ting, and Srikanth Pilla. 2020. "Melt Processing of Cellulose Nanocrystal-Filled Composites: Toward Reinforcement and Foam Nucleation." *Industrial and Engineering Chemistry Research* 59 (18): 8511–8531. doi:10.1021/acs.iecr.0c00170

10 Water Barrier Properties of PLA-based Cellulose Composites

Aswathy Jayakumar, Sabarish Radoor,
and Suchart Siengchin
King Mongkut's University of Technology North Bangkok
Bangkok, Thailand

Jyotishkumar Parameswaranpillai
Alliance University
Bengaluru, India

CONTENTS

10.1 INTRODUCTION

Polylactic acid/polylactide (PLA) is one of the widely used biodegradable polymers having application in diverse fields such as food packaging, medical, textiles, electronics, and others (DeStefano, Khan, & Tabada, 2020). It is derived from renewable sources such as corn starch, tapioca roots and sugar cane (García Ibarra, Sendón, & Rodríguez-Bernaldo de Quirós, 2016). It is developed by the ring-opening polymerization of lactides or condensation of lactic acid monomers such as corn, beet sugar

DOI: 10.1201/9781003160458-10

(Boey et al., 2021). PLA has a poor crystallization rate, low crystallinity and with high cost, and is hydrophobic in nature. They are soluble in chloroform, acetonitrile, dioxane, methylene chloride, dichloroacetic acid, and 1,1,2-trichloroethane while insoluble in water, linear hydrocarbons, and alcohols. It is partially soluble in toluene, ethylbenzene, tetrahydrofuran, and acetone (Casalini et al., 2019). PLA can be generally processed by extrusion, injection molding, thermoforming, fiber spinning, film forming, and blow molding (Byun & Kim, 2014).

The major drawbacks associated with PLA are its low crystallization rate and low melt strength (Vatansever, Arslan, & Nofar, 2019). In order to improve the overall performance of PLA, it is often combined with several biopolymers or synthetic polymers and or nanomaterials of interest. Blending, copolymerizing, physical and chemical treatments are the important methods to improve its properties (Jamshidian et al., 2010). As a biopolymer, several studies suggested the potential of cellulose-based material as a functional filler material in PLA-based composites. Cellulose is a linear biopolymer with β-1,4-linked D-glucose monomers (Bai, Yang, & Ho, 2019). It is generally obtained from plants, rarely by some bacteria and algae. Cellulose is fibrous, tough, and insoluble in water. It plays a major part in preserving the integrity of plant cell wall (Brigham, 2018). Bacterial cellulose is generally obtained and studied in *Acetobacter xylinum* (Saxena & Brown, 2001). The property of bacterial cellulose is that, it is generally pure and does not need to separate from lignin components like in plant cellulose (Havstad, 2020). But it has high porosity, high water-retention potential, and high crystallinity. Hence it can have great promises as a biomaterial scaffold in medical applications (Hickey & Pelling, 2019). The physical or chemical modification of cellulose is extensively used to improve the overall performance of PLA-cellulose composite materials (Fortunati et al., 2012; Rasal, Janorkar, & Hirt, 2010).

The presence of hydroxyl groups in cellulose can be chemically modified to improve its functions. The OH groups in cellulose can be modified with carbonic acid, sulfate, phosphate, aldehyde, thiol, and amino compounds (Tavakolian, Jafari, & van de Ven, 2020). The high thermal and mechanical properties of cellulose are due to the extensive hydrogen bonding between the cellulose chains (Feng et al., 2015). In this chapter, an attempt has been made to review the properties of cellulose nanomaterials, their functionalization, antimicrobial activity, applications, and properties of PLA/cellulose-based composites.

10.2 NANOMATERIALS BASED ON CELLULOSE

Cellulosic nanomaterials are considered as one of the cheapest nanofillers, which are considered sustainable and ecofriendly. They have a light weight, large surface area, and in modified form, they are also able to have good compatibility with polymeric materials (Tayeb et al., 2018). As stated above, nanocellulose can be isolated from both plant and animal origin. They can be classified into cellulose nanocrystals, cellulose nanofibers, bacterial cellulose, algal cellulose, and tunicate cellulose (Mokhena et al., 2018; Sheikhi, 2019). The hydroxyl group present in cellulose nanomaterials undergoes strong hydrogen bonding, which could result in irreversible agglomeration, reorganization, and co-crystallization. Usually, the hydroxyl groups present in cellulose nanomaterials are masked by chemical modification in order to prevent agglomeration. This chemical modification/functionalization may some-times contribute to the antimicrobial activity of cellulose nanomaterials.

10.3 CELLULOSE NANOMATERIALS AND THEIR FUNCTIONALIZATION

The surface morphology of cellulose-based nanomaterials has a great impact in enhancing the properties of polymer composites. This includes the interfacial adhesion between the nanofiller and the matrix. There are mainly three sur-face modification techniques such as "graft to", "graft onto", and substitution of hydroxyl groups. Acetylation, salinization, TEMPO oxidation, cationization, etc., are the commonly employed functionalization methods for cellulose (Mokhena et al., 2018; Tang et al., 2017).

10.4 IMPACT OF NANOPARTICLES ON WATER BARRIER PROPERTIES OF POLYMER COMPOSITES

Nanomaterials have a significant effect in improving barrier properties of polymer-based composites. These nanomaterials can create a tortuous pathway that could result in the altered diffusion of water or other gas molecules (Figure 10.1). The uniform distribution of nanofillers in the polymer matrix was crucial to improving the barrier properties of composites. The immobilization of the polymer matrix by the nanofillers is also important in creating the tortuous pathway (Sarfraz et al., 2020). Both natural and synthetic fillers are used to improve the overall perfor-mance of polymer composites (de Azeredo, Capparelli Mattoso, & Habig, 2011; Honarvar, Hadian, & Mashayekh, 2016; Imai, 2014; Wolf et al., 2018). One of the main challenges associated with the preparation and processing of PLA-cellulose nanocomposite is the dispersion of nanoparticles in the thermoplastic matrix. Due to their hydrophilicity, the cellulose nanofillers are not compatible with hydrophobic polymers (Vatansever et al., 2019). Several studies are ongoing in the field of PLA-cellulose nanocomposites and their analysis on the overall performances in various

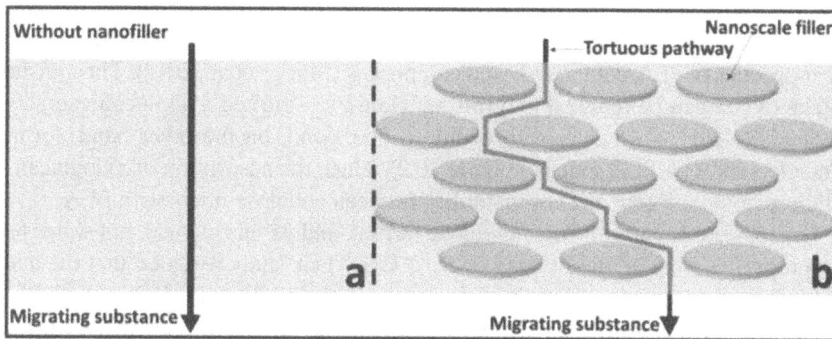

FIGURE 10.1 Represents the impact of with and without nanofiller in polymer composites: (a) polymer alone barrier; (b) polymer nanocomposite barrier. (From Sarfraz et al., 2020.)

applications (Ahmadian, Ghorbani, & Mahmoodzadeh, 2020; Chai et al., 2020; Lu et al., 2020; Patel et al., 2020; Sucinda et al., 2021; Yang et al., 2016; Yu et al., 2021).

10.5 ANTIMICROBIAL PROPERTIES AND APPLICATIONS OF CELLULOSE NANOMATERIALS

As cellulose is not inherently antimicrobial, it is generally modified with other chemical or biological agents like silver nanoparticles, zinc oxide nanoparticles, gold nanoparticles, titanium nanoparticles, proteins, antibiotics, and other materials of interest (Felgueiras et al., 2021; Gao et al., 2021; Limaye et al., 2019; Zhao et al., 2018). The cationic cellulose (modified cellulose by cationization) has antimicrobial properties against both Gram-positive and Gram-negative bacteria (Tavakolian et al., 2020). The cellulose nanofibrils or nanocrystals with cationic nature can have the ability to interact with the negatively charged bacterial cell membrane. This can lead to the leakage of cytoplasmic contents of the bacterial cell membrane and then ultimately lead to the death of the microbial cell.

For use in biomedical applications, the materials should be non-toxic, biocompatible, non-immunogenic and does not cause any problematic reactions. Due to these properties, cellulose nanomaterials can be used for wound healing, drug delivery, and antimicrobial coatings (Amalraj et al., 2018; Mohammed et al., 2018; Moon et al., 2016; Pal et al., 2017; Tayeb et al., 2018). In the food packaging sector, cellulose-based materials are useful to improve the mechanical, thermal, antimicrobial, optical, water, and gas barrier properties. It can also be used for water purification, electronics, and textile applications (Felgueiras et al., 2021; Mohammed et al., 2018; Salah, 2013). In the case of the electronic sector, the self-standing transparent films having improved thermal stability can have great promises as smooth surfaces for printed electronics, conductive material, luminescent materials, optoelectronics, energy harvesting storage devices, and sensors (Dias et al., 2020; Hoeng, Denneulin, & Bras, 2016; Sabo et al., 2016).

10.6 PLA-CELLULOSE COMPOSITES

The reinforcement of plant fibers to PLA-based composites is an emerging area of research and industrial applications. Marais and his coworkers employed the grafting of PLA with xyloglucan (XG) through ring-opening polymerization to improve the properties of PLA/cellulose-based composites (Marais et al., 2012). The moisture uptake of XG can be reduced by grafting. Here PLA-grafted xyloglucan can act as a compatibilizer as it can adsorb cellulose fibers and, on the other hand, formed entanglement with PLA matrix (Figure 10.2). Thus, the adsorption of xyloglucan to cellulose could improve the compatibility between cellulose fibers with PLA.

The reinforcement of formyl cellulose to PLA and its mechanical and water barrier properties were studied (Long et al., 2021). Their study revealed that the addition of formyl groups to cellulose increased the compatibility of cellulose with PLA. It is noted that the water barrier properties such as moisture absorption capacity (40.56%) and water vapor permeability (51.43%) of the composites were decreased than that of neat PLA film. The enhanced interaction between PLA and cellulose

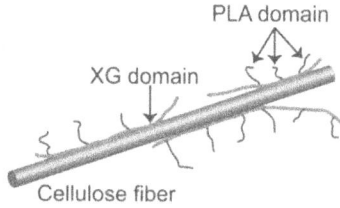

FIGURE 10.2   Represents the adsorption of PLA-grafted xyloglucan to the surface of cellulose fiber. (Reproduced with permission from Elsevier, License Number-5121200753203; Marais et al., 2012.)

by the introduction of hydrophobic formate groups could be the reason behind the improvement in properties.

Espino and his coworkers studied the impact of incorporating cellulose nanocrystals on polylactide-based nanocomposites (Espino-Pérez et al., 2018). The surface grafting of cellulose nanocrystals resulted in a decrease in the water transport rate of the nanocomposite. The bionanocomposites developed through the solvent casting method by the reinforcement of PLA with cellulose nanowhisker have resulted in the improved thermal stability, tensile strength, and tensile modulus (Sucinda et al., 2021). They suggested that the prepared nanocomposites have application in food packaging.

Further, Arteaga and his coworkers utilized the bacterial cellulose obtained from kombucha membranes to develop PLA-based composites (Arteaga-Ballesteros et al., 2020). They prepared a symbiotic colony of bacteria (*Gluconacetobacter xylinus*, *Acetobacter xylinoides*, *Gluconicum* sp., *Acetobacter aceti*, and *Acetobacter pasteurianus*) and yeast (*Candida* sp., *Kloeckera* sp., *Schizosaccharomyce pombe*, *Saccharomycodes ludwigii*, *Saccharomyces cerevisiae*, *Torulospora* sp., *Zygosaccharomyces bailii*, and *Pichia* sp.) for cellulose preparation. They analyzed the effect of different weight percentages of cellulose ranging from 1%, 3%, and 5%. They observed an increase in the crystallinity of composites and mechanical properties while using 3% cellulose. This could be due to the interaction between the bacterial cellulose and PLA matrix. In addition, the composite has shown energy absorption, moderate heat tolerance, and hydrophobicity that suggests its potential in controlling grease and liquid absorption. Some of the studied PLA-cellulose-based composites and their properties are shown in Table 10.1.

10.6.1 PLA-CELLULOSE NANOCRYSTALS-BASED COMPOSITES

It is estimated that the market of nanocellulose is about to grow from 271.26 Million USD (2017) to 1076.43 USD in 2025 (Shojaeiarani, Bajwa, & Chanda, 2021). The nanocrystals based on cellulose are normally less than 100 nm in one dimension and are generally used to prepare polymer-based composites. It is highly crystalline and exhibits surface charge with amphiphilic in nature. Usually, they are hydrophilic in nature, but it does not swell in water. But due to its polar nature and hydrophilic nature, it tends to agglomerate in the polymer matrix. There are several studies that suggest the potential of cellulose nanocrystal-based polymer composites (Mariano et al., 2017).

TABLE 10.1

Represents the PLA-Cellulose-Based Composites and Its Properties

Composites/Nanocomposites	Properties	Antimicrobial Activity	References
PLA/cellulose-based composites	Improved water barrier properties		Marais et al. (2012)
PLA/cellulose nanocrystals	Decreased water transport rate		Espino-Pérez et al. (2018)
PLA/cellulose nanowhisker	Enhanced thermal stability, tensile strength, tensile modulus, and reduced water permeability		Fortunati et al. (2012)
PLA/cellulose nanocrystals	Reduced water vapor permeability		Yu et al. (2016)
PLA/bacterial cellulose	Increased crystallinity, hydrophobicity	*G. xylinus, A. xylinoides, Gluconicum* sp., *A. aceti,* and *A. pasteurianus, Candida* sp., *Kloeckera* sp., *Schizosaccharomyces pombe, S. ludwigii, Saccharomyces cerevisiae, Torulospora* sp., *Zygosaccharomyces bailii,* and *Pichia* sp.	Arteaga-Ballesteros et al. (2020)
PLA/cellulose nano whisker	Improved thermal stability, tensile strength, and tensile modulus		Sucinda et al. (2021)
PLA/cellulose nano whisker	Improved water barrier properties		Sanchez-Garcia and Lagaron (2010)
PLA and cellulose nanocrystals	Decreased water vapor transfer		Shojaeiarani et al. (2020)
PLA/cellulose nanofibril	Increased hydrophobicity and mechanical properties of the composites		Yang et al. (2019)

(Continued)

TABLE 10.1 (*Continued*)
Represents the PLA-Cellulose-Based Composites and Its Properties

Composites/Nanocomposites	Properties	Antimicrobial Activity	References
PLA/acetylated cellulose nanocrystals/zinc oxide nanoparticles	Improved mechanical strength, UV barrier, oxygen barrier, and water barrier properties	*Escherichia coli* and *Staphylococcus aureus*	Yu et al. (2021)
PLA/zinc oxide nanoparticles/cellulose	Higher water vapor permeability	*E. coli, Vibrio parahaemolyticus, S. aureus, Listeria monocytogenes, Salmonella typhimurium* and *Bacillus cereus*	Saedi et al. (2021)
PLA/zinc oxides nanoparticles/cellulose nanocrystals	Reduced water vapor permeability	*E. coli* and *Listeria innocua*	Luzi et al. (2017)
PLA/silver nanoparticles/cellulose nanocrystals	Improved mechanical and crystalline nature	*S. aureus* and *E. coli*	Fortunati et al. (2012)
PLA/silver nanoparticles/cellulose nanofibers	Improved mechanical stability with decreased water vapor permeability		Mohammadalinejhad, Almasi, and Esmaiili (2021)
PLA/silver nanoparticles, cellulose nanofibrils	Decreased water contact angle	*E. coli* and *Bacillus cereus*	Szymańska-Chargot et al. (2020)

For example, PLA-based nanocomposites functionalized with cellulose nano-crystals formates were analyzed for thermal, mechanical, barrier, and antimicro-bial analysis (Yu et al., 2016). The reduction in water vapor permeability could be due to the enhancement in crystallinity and interfacial interaction developed in the composites. Similarly, the incorporation of cellulose nanocrystals to PLA-poly(hydroxybutyrate) has shown application in the packaging (Arrieta et al., 2014). The incorporated cellulose nanocrystals enhanced the interfacial adhesion between PLA and poly(hydroxybutyrate) and this could contribute to the improved water resistance, mechanical stiffness, and stretchability. Sanchez and Lagaron studied the effect of incorporating cellulose nanowhiskers on PLA-based com-posites (Sanchez-Garcia & Lagaron, 2010). The water barrier properties of the developed nanocomposite were found to improve by the addition of cellulose nanowhiskers.

Similarly, Shojaeiarani and his coworkers employed solvent casting and spin coating techniques to develop PLA and cellulose nanocrystals based composites (Shojaeiarani et al., 2020). For this, they prepared composite films with different concentrations (1, 3, and 5 wt.%) of cellulose nanocrystals by both methods. Here, the composites prepared through spin-coated method have shown a lower water vapor transfer rate than solvent-casted nanocomposites. They concluded that the uni-form distribution of cellulose nanocrystals improves the crystallinity and interfacial interaction between PLA and cellulose nanocrystals and this could be the reason for the reduced water vapor transfer rate. The spin-coated films have shown lower hydrophilic character than solvent-cast PLA/cellulose nanocrystals nanocomposites. The contact angle measurement of spin-coated and solvent-cast PLA/cellulose nano-crystals nanocomposites are shown in Figure 10.3.

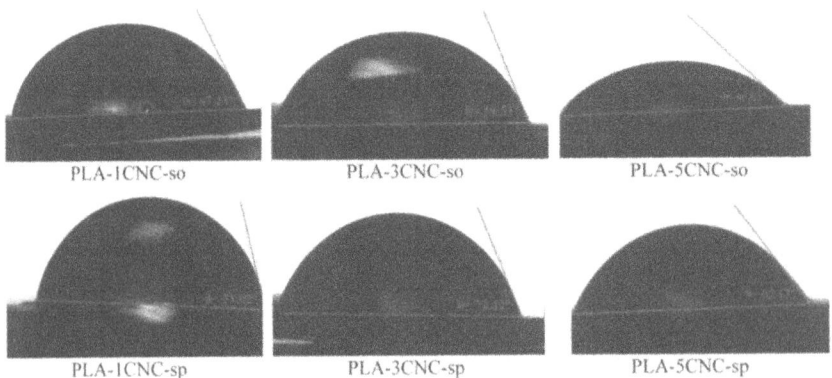

FIGURE 10.3 Contact angle measurement of spin-coated (sp) and solvent-cast (so) PLA/cellulose nanocrystals nanocomposites (PLA-CNC). PLA/1 wt.% cellulose nanocrystals nanocomposites (PLA-1CNC-so), PLA/2 wt.% cellulose nanocrystals (PLA-2CNC-so), PLA/3 wt.% cellulose nanocrystals (PLA-3CNC-so), PLA/1 wt.% cellulose nanocrystals (PLA-1CNC-sp), PLA/2 wt.% cellulose nanocrystals (PLA-2CNC-sp), PLA/3 wt.% cel-lulose nanocrystals (PLA-3CNC-sp). (Reproduced with permission from Elsevier, License Number-5120800940107; Shojaeiarani et al., 2020.)

Yang and his coworkers studied the morphological, mechanical, and thermal barrier properties of PLA-cellulose nanofibril-based composites (Yang et al., 2019). The water contact angle of the developed nanocomposites was found to be decreased by the addition of increasing concentration of cellulose nanofibrils. Here, the hydrophobicity of the membranes was enhanced along with the increase in mechanical properties such as Young's modulus and tensile strength.

10.6.2 PLA-CELLULOSE COMPOSITES BASED ON ZINC OXIDE

Zinc oxide nanoparticles are considered as interesting candidates for reinforcing the PLA-cellulose composites. It is highly stable and low-cost nanoparticles with antimicrobial in nature (Shankar, Wang, & Rhim, 2018; Siddiqi et al., 2018; Tiwari et al., 2018). The reinforcement of PLA with acetylated cellulose nanocrystals and zinc oxide nanoparticles has resulted in improved mechanical strength, UV barrier, oxygen barrier, and water barrier properties (Yu et al., 2021). The developed composites were prepared by solvent casting method and the acetylation of cellulose nanocrystals improved its dispersion in the PLA matrix. The excellent antimicrobial activity against *Escherichia coli* and *Staphylococcus aureus* offers its potential application as active food packaging systems. On the contrary, PLA-based nanocomposites having zinc oxide nanoparticles and cellulose have not shown any improvement in mechanical strength and water vapor permeability of the developed composites (Saedi et al., 2021). But the same has shown improved antimicrobial activity against food-borne pathogens such as *E. coli, Vibrio parahaemolyticus, S. aureus, Listeria monocytogenes, Salmonella typhimurium*, and *Bacillus cereus*. Luzi et al. utilized cellulose nanocrystals obtained from *Posidonia oceanica* for the preparation of zinc oxide nanoparticles based PLA nanocomposites (Luzi et al., 2017). Their studies have shown that the addition of nanoparticles reduced the water vapor permeability of the developed nanocomposites. They concluded that the uniform distribution of nanoparticles and the tortuous pathway created by the nanoparticles might be the reason for the improvement in the water barrier properties of the nanocomposites. In addition to this, the nanocomposites have shown improved mechanical and thermal barriers along with antibacterial against *E. coli* and *Listeria innocua*.

10.6.3 PLA-CELLULOSE COMPOSITES BASED ON SILVER

The slow crystallization behavior of PLA can be improved by the use of nanocrystalline cellulose and silver nanoparticles. Silver nanoparticles are also considered as excellent nanofillers in PLA-based composites. It has excellent antimicrobial properties and low toxic in nature (Rapa et al., 2019; Shameli et al., 2010). Fortunati and his coworkers incorporated PLA with silver nanoparticles and cellulose (Fortunati et al., 2012) nanocrystals. They noticed that the addition of these nanofillers resulted in the improvement in the mechanical and crystallization behavior of PLA-based nanocomposites. Here, they modified the cellulose nanocrystals with acid phosphate ester of ethoxylated nonylphenol (surfactant), which could result in the better dispersion of cellulose nanocrystals in the polymer matrix. And they concluded that the use of surfactant might have led to the improvement in crystalline properties of

PLA-based composites. The developed nanocomposites have shown excellent antimicrobial activity against *S. aureus* and *Escherichia coli* which suggests the use of polymer composite films in food packaging applications. The reinforcement of silver nanoparticles decorated cellulose nanofibers in PLA was studied for food packaging applications (Mohammadalinejhad et al., 2021). Their study revealed its improved mechanical stability with decreased water vapor permeability. The reinforcement of nanofillers has led to the improvement in the mechanical strength of the developed nanocomposites. Szymanska and his coworkers utilized carrot cellulose nanofibrils (CCNF) extracted from carrot pomace to be used for the development of PLA-based nanocomposites (Szymańska-Chargot et al., 2020). For the development of nanocomposites, different weight proportions of PLA and CCNF were used (99:1 or 96:4). They used different concentrations of silver nanoparticles (AgNO$_3$) (0.25 mM and 2 mM) to modify carrot cellulose nanofibrils (CCNF) and studied mechanical, thermal, antibacterial, and hydrophilic properties of the nanocomposites. Their studies suggest that the addition of CCNF or CCNF modified with silver nanoparticles with 99:1 could not have any effect on the water contact angle of nanocomposite when comparing with neat PLA. The addition of pure carrot cellulose nanofibrils to 96:4 proportion caused an increase in water contact angle measurement, whereas CCNF/0.25AgNPs 96:4 and CCNF/2AgNPs 96:4 have shown decreased water contact angle (67.1° and 56.9°) (Figure 10.4). Their study suggested that the addition

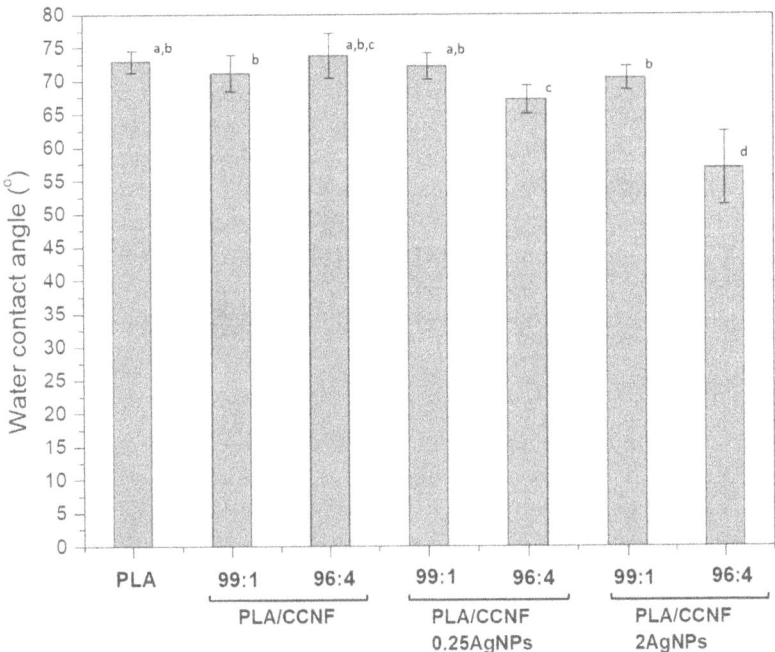

FIGURE 10.4 Water contact angle measurement of bionanocomposites based on PLA (PLA), carrot cellulose nanofibrils (CCNF), and different concentration (0.25 mM and 2 mM) of silver nanoparticles (AgNPs). (From Szymańska-Chargot et al., 2020.)

of silver nanoparticles could be the reason for the increased hydrophilicity of the nanocomposites. The developed nanocomposites also possess antimicrobial activity against *Escherichia coli* and *Bacillus cereus*.

10.7 BIODEGRADATION OF COMPOSITES

The biodegradation of composites/bionanocomposites was considered as a complex reaction that includes biodeterioration, depolymerization, assimilation, and mineralization (Chamas et al., 2020). The microbial degradation initially involves the adhesion of microorganisms to the surface of the polymer (Wagner et al., 1996). This is followed by the colonization of microbes and could result in the decomposition of the polymeric materials. As microbes are not able to carry large polymeric materials, they secrete certain extracellular enzymes and other intracellular depolymerases. These enzymes further degrade the complex polymeric materials into oligomers, dimers, and monomers (Mohanan et al., 2020; Wilkes & Aristilde, 2017). Further, assimilation takes place, which might result in the conversion of these monomers to energy and biomass. When coming to the degradability of PLA and cellulose composites, it meets the American Society for Testing and Materials (ASTM) requirements on biodegradation (Galera Manzano et al., 2021; Muniyasamy et al., 2013). The main factors that contribute to biodegradation include the type of nanofiller, compatibilization, and environmental conditions.

10.8 CONCLUSION

PLA-based cellulose composites are gaining great interest due to their biodegradability, sustainability, and environmental friendliness. It has a wide range of applications in food packaging, biomedical, textiles, and others. The incorporation of various cellulose nanofillers has a great impact on the overall performance of PLA-based composites. The mechanical, water barrier, gas barrier, thermal, crystalline, light barrier, and antimicrobial properties of neat PLA can be improved with the incorporation of nanocellulose. The strong interfacial interaction between PLA and cellulose might be the reason for the improved water barrier properties of PLA/cellulose composites.

ACKNOWLEDGMENT

Authors gratefully thanks for financial support by the King Mongkut's University of Technology North Bangkok (KMUTNB), Thailand through the Post-Doctoral Program (Grant No. KMUTNB-64-Post-01).

REFERENCES

Ahmadian, S., Ghorbani, M., & Mahmoodzadeh, F. (2020). Silver sulfadiazine-loaded electrospun ethyl cellulose/polylactic acid/collagen nanofibrous mats with antibacterial properties for wound healing. *International Journal of Biological Macromolecules, 162*, 1555–1565. doi:10.1016/j.ijbiomac.2020.08.059

Amalraj, A., Gopi, S., Thomas, S., & Haponiuk, J. T. (2018). Cellulose nanomaterials in bio-medical. *Food, and Nutraceutical Applications: A Review. Macromolecular Symposia, 380*(1), 1800115. doi:10.1002/masy.201800115

Arrieta, M. P., Fortunati, E., Dominici, F., Rayón, E., López, J., & Kenny, J. M. (2014). PLA-PHB/cellulose based films: Mechanical, barrier and disintegration properties. *Polymer Degradation and Stability, 107*, 139–149. doi:10.1016/j.polymdegradstab.2014.05.010

Arteaga-Ballesteros, B. E., Guevara-Morales, A., Martín-Martínez, E. S., Figueroa-López, U., & Vieyra, H. (2020). Composite of polylactic acid and microcellulose from kombu-cha membranes. *e-Polymers, 21*(1), 015–026. doi:10.1515/epoly-2021-0001

Bai, F.-W., Yang, S., & Ho, N. W. Y. (2019). Fuel Ethanol Production from Lignocellulosic Biomass, In Murray Moo-Young (Ed). *Comprehensive Biotechnology*, Vol 3, 49–65, Pergamon. doi:10.1016/b978-0-444-64046-8.00150-6

Boey, J. Y., Mohamad, L., Khok, Y. S., Tay, G. S., & Baidurah, S. (2021). A review of the applications and biodegradation of polyhydroxyalkanoates and poly(lactic acid) and its composites. *Polymers, 13*(10), 1544. doi:10.3390/polym13101544

Brigham, C. (2018). Biopolymers: Biodegradable Alternatives to Traditional Plastics. In Béla Török, Timothy Dransfield (Eds). *Green Chemistry*, 753–770. Elsevier. doi:10.1016/ b978-0-12-809270-5.00027-3

Byun, Y., & Kim, Y. T. (2014). Bioplastics for Food Packaging: Chemistry and Physics, In Jung H. Han (Ed). *Innovations in Food Packaging*, 353–368, Academic Press. doi:10.1016/b978-0-12-394601-0.00014-x

Casalini, T., Rossi, F., Castrovinci, A., & Perale, G. (2019). A perspective on polylactic acid-based polymers use for nanoparticles synthesis and applications. *Frontiers in Bioengineering and Biotechnology, 7*. doi:10.3389/fbioe.2019.00259

Chai, H., Chang, Y., Zhang, Y., Chen, Z., Zhong, Y., Zhang, L., Sui, X., Xu, H., & Mao, Z. (2020). The fabrication of polylactide/cellulose nanocomposites with enhanced crystallization and mechanical properties. *International Journal of Biological Macromolecules, 155*, 1578–1588. doi:10.1016/j.ijbiomac.2019.11.135

Chamas, A., Moon, H., Zheng, J., Qiu, Y., Tabassum, T., Jang, J. H., Abu-Omar, M., Scott, S. L., & Suh, S. (2020). Degradation rates of plastics in the environment. *ACS Sustainable Chemistry & Engineering, 8*(9), 3494–3511. doi:10.1021/acssuschemeng.9b06635

de Azeredo, H. M. C., Capparelli Mattoso, L. H., & Habig, T. (2011). Nanocomposites in Food Packaging – A Review. In Boreddy Reddy (Ed). *Advances in Diverse Industrial Applications of Nanocomposites*, IntechOpen. doi:10.5772/14437

DeStefano, V., Khan, S., & Tabada, A. (2020). Applications of PLA in modern medicine. *Engineered Regeneration, 1*, 76–87. doi:10.1016/j.engreg.2020.08.002

Dias, O. A. T., Konar, S., Leão, A. L., Yang, W., Tjong, J., & Sain, M. (2020). Current state of applications of nanocellulose in flexible energy and electronic devices. *Frontiers in Chemistry, 8*. doi:10.3389/fchem.2020.00420

Espino-Pérez, E., Bras, J., Almeida, G., Plessis, C., Belgacem, N., Perré, P., & Domenek, S. (2018). Designed cellulose nanocrystal surface properties for improving barrier properties in polylac-tide nanocomposites. *Carbohydrate Polymers, 183*, 267–277. doi:10.1016/j.carbpol.2017.12.005

Felgueiras, C., Azoia, N. G., Gonçalves, C., Gama, M., & Dourado, F. (2021). Trends on the Cellulose-based textiles: Raw materials and technologies. *Frontiers in Bioengineering and Biotechnology, 9*. doi:10.3389/fbioe.2021.608826

Feng, X., Ullah, N., Wang, X., Sun, X., Li, C., Bai, Y., Chen, L., & Li, Z. (2015). Characterization of bacterial cellulose by *Gluconacetobacter hansenii* CGMCC 3917. *Journal of Food Science, 80*(10), E2217–E2227. doi:10.1111/1750-3841.13010

Fortunati, E., Armentano, I., Zhou, Q., Iannoni, A., Saino, E., Visai, L., Berglund, L. A., & Kenny, J. M. (2012). Multifunctional bionanocomposite films of poly(lactic acid), cellu-lose nanocrystals and silver nanoparticles. *Carbohydrate Polymers, 87*(2), 1596–1605. doi:10.1016/j.carbpol.2011.09.066

Fortunati, E., Peltzer, M., Armentano, I., Torre, L., Jiménez, A., & Kenny, J. M. (2012). Effects of modified cellulose nanocrystals on the barrier and migration properties of PLA nano-biocomposites. *Carbohydrate Polymers*, *90*(2), 948–956. doi:10.1016/j.carbpol.2012.06.025

Galera Manzano, L. M., Ruz Cruz, M. Á., Moo Tun, N. M., Valadez González, A., & Mina Hernandez, J. H. (2021). Effect of cellulose and cellulose nanocrystal contents on the biodegradation, under composting conditions, of hierarchical PLA biocomposites. *Polymers*, *13*(11), 1855. doi:10.3390/polym13111855

Gao, G., Fan, H., Zhang, Y., Cao, Y., Li, T., Qiao, W., Wu, M., Ma, T., & Li, G. (2021). Production of nisin-containing bacterial cellulose nanomaterials with antimicrobial properties through co-culturing *Enterobacter* sp. FY-07 and *Lactococcus lactis* N8. *Carbohydrate Polymers*, *251*, 117131. doi:10.1016/j.carbpol.2020.117131

García Ibarra, V., Sendón, R., & Rodríguez-Bernaldo de Quirós, A. (2016). Antimicrobial Food Packaging Based on Biodegradable Materials. In Jorge Barros-Velázquez (Ed). *Antimicrobial Food Packaging*, 363–384. Academic Press. doi:10.1016/b978-0-12-800723-5.00029-2

Havstad, M. R. (2020). *Biodegradable Plastics*, In Trevor M. Letcher (Ed). *Plastic Waste and Recycling: Environmental Impact, Societal Issues, Prevention, and Solutions*, 97–129. Academic Press. doi:10.1016/b978-0-12-817880-5.00005-0

Hickey, R. J., & Pelling, A. E. (2019). Cellulose biomaterials for tissue engineering. *Frontiers in Bioengineering and Biotechnology*, *7*. doi:10.3389/fbioe.2019.00045

Hoeng, F., Denneulin, A., & Bras, J. (2016). Use of nanocellulose in printed electronics: A review. *Nanoscale*, *8*(27), 13131–13154. doi:10.1039/c6nr03054h

Honarvar, Z., Hadian, Z., & Mashayekh, M. (2016). Nanocomposites in food packaging applications and their risk assessment for health. *Electronic Physician*, *8*(6), 2531–2538. doi:10.19082/2531

Imai, Y. (2014). *Inorganic Nano-Fillers for Polymers*. In S. Kobayashi, K. Müllen (Eds). *Encyclopedia of Polymeric Nanomaterials*, Springer. doi:10.1007/978-3-642-36199-9_353-1

Jamshidian, M., Tehrany, E. A., Imran, M., Jacquot, M., & Desobry, S. (2010). Poly-lactic acid: production, applications, nanocomposites, and release studies. *Comprehensive Reviews in Food Science and Food Safety*, *9*(5), 552–571. doi:10.1111/j.1541-4337.2010.00126.x

Limaye, M. V., Gupta, V., Singh, S. B., Paik, G. R., & Singh, P. (2019). Antimicrobial activity of composite consisting of cellulose nanofibers and silver nanoparticles. *ChemistrySelect*, *4*(41), 12164–12169. doi:10.1002/slct.201901572

Long, S., Zhong, L., Lin, X., Chang, X., Wu, F., Wu, R., & Xie, F. (2021). Preparation of formyl cellulose and its enhancement effect on the mechanical and barrier properties of polylactic acid films. *International Journal of Biological Macromolecules*, *172*, 82–92. doi:10.1016/j.ijbiomac.2021.01.029

Lu, A., Petit, E., Jelonek, K., Orchel, A., Kasperczyk, J., Wang, Y., Su, F., & Li, S. (2020). Self-assembled micelles prepared from bio-based hydroxypropyl methyl cellulose and polylactide amphiphilic block copolymers for anti-tumor drug release. *International Journal of Biological Macromolecules*, *154*, 39–47. doi:10.1016/j.ijbiomac.2020.03.094

Luzi, F., Fortunati, E., Jiménez, A., Puglia, D., Chiralt, A., & Torre, L. (2017). PLA nanocomposites reinforced with cellulose nanocrystals from *Posidonia oceanica* and ZnO nanoparticles for packaging application. *Journal of Renewable Materials*, *5*(2), 103–115. doi:10.7569/jrm.2016.634135

Marais, A., Kochumalayil, J. J., Nilsson, C., Fogelström, L., & Gamstedt, E. K. (2012). Toward an alternative compatibilizer for PLA/cellulose composites: Grafting of xyloglucan with PLA. *Carbohydrate Polymers*, *89*(4), 1038–1043. doi:10.1016/j.carbpol.2012.03.051

Mariano, M., Pilate, F., de Oliveira, F. B., Khelifa, F., Dubois, P., Raquez, J.-M., & Dufresne, A. (2017). Preparation of cellulose nanocrystal-reinforced poly(lactic acid) nanocomposites through noncovalent modification with PLLA-based surfactants. *ACS Omega*, *2*(6), 2678–2688. doi:10.1021/acsomega.7b00387

Mohammadalinejhad, S., Almasi, H., & Esmaiili, M. (2021). Physical and release properties of poly(lactic acid)/nanosilver-decorated cellulose, chitosan and lignocellulose nanofiber composite films. *Materials Chemistry and Physics, 268*, 124719. doi:10.1016/j.matchemphys.2021.124719

Mohammed, N., Grishkewich, N., & Tam, K. C. (2018). Cellulose nanomaterials: promising sustainable nanomaterials for application in water/wastewater treatment processes. *Environmental Science: Nano, 5*(3), 623–658. doi:10.1039/c7en01029j

Mohanan, N., Montazer, Z., Sharma, P. K., & Levin, D. B. (2020). Microbial and enzymatic degradation of synthetic plastics. *Frontiers in Microbiology, 11.* doi:10.3389/fmicb.2020.580709

Mokhena, T., Sefadi, J., Sadiku, E., John, M., Mochane, M., & Mtibe, A. (2018). Thermoplastic processing of PLA/cellulose nanomaterials composites. *Polymers, 10*(12), 1363. doi:10.3390/polym10121363

Moon, R. J., Schueneman, G. T., & Simonsen, J. (2016). Overview of cellulose nanomaterials, their capabilities and applications. *JOM, 68*(9), 2383–2394. doi:10.1007/s11837-016-2018-7

Muniyasamy, S., Anstey, A., Reddy, M. M., Misra, M., & Mohanty, A. (2013). Biodegradability and compostability of lignocellulosic based composite materials. *Journal of Renewable Materials, 1*(4), 253–272. doi:10.7569/jrm.2013.634117

Pal, S., Nisi, R., Stoppa, M., & Licciulli, A. (2017). Silver-functionalized bacterial cellulose as antibacterial membrane for wound-healing applications. *ACS Omega, 2*(7), 3632–3639. doi:10.1021/acsomega.7b00442

Patel, D. K., Dutta, S. D., Hexiu, J., Ganguly, K., & Lim, K.-T. (2020). Bioactive electrospun nanocomposite scaffolds of poly(lactic acid)/cellulose nanocrystals for bone tissue engineering. *International Journal of Biological Macromolecules, 162*, 1429–1441. doi:10.1016/j.ijbiomac.2020.07.246

Rapa, M., Darie-Nita, R. N., Preda, P., Coroiu, V., Tatia, R., Vasile, C., Matei, E., Predescu, A. M., & Maxim, M.-E. (2019). PLA/collagen hydrolysate/silver nanoparticles bionanocomposites for potential antimicrobial urinary drains. *Polymer-Plastics Technology and Materials, 58*(18), 2041–2055. doi:10.1080/25740881.2019.1603999

Rasal, R. M., Janorkar, A. V., & Hirt, D. E. (2010). Poly(lactic acid) modifications. *Progress in Polymer Science, 35*(3), 338–356. doi:10.1016/j.progpolymsci.2009.12.003

Sabo, R., Yermakov, A., Law, C. T., & Elhajjar, R. (2016). Nanocellulose-enabled electronics, energy harvesting devices. *Smart Materials and Sensors: A Review. Journal of Renewable Materials, 4*(5), 297–312. doi:10.7569/jrm.2016.634114

Saedi, S., Shokri, M., Kim, J. T., & Shin, G. H. (2021). Semi-transparent regenerated cellulose/ZnONP nanocomposite film as a potential antimicrobial food packaging material. *Journal of Food Engineering, 307*, 110665. doi:10.1016/j.jfoodeng.2021.110665

Salah, S. M. (2013). Application of nano-cellulose in textile. *Journal of Textile Science & Engineering, 03*(04). doi:10.4172/2165-8064.1000142

Sanchez-Garcia, M. D., & Lagaron, J. M. (2010). On the use of plant cellulose nanowhiskers to enhance the barrier properties of polylactic acid. *Cellulose, 17*(5), 987–1004. doi:10.1007/s10570-010-9430-x

Sarfraz, J., Gulin-Sarfraz, T., Nilsen-Nygaard, J., & Pettersen, M. K. (2020). Nanocomposites for food packaging applications: An overview. *Nanomaterials, 11*(1), 10. doi:10.3390/nano11010010

Saxena, I. M., & Brown, R. M. (2001). Biosynthesis of cellulose. *Progress in Biotechnology 18*, 69–76. doi:10.1016/s0921-0423(01)80057-5

Shameli, K., Mansor Bin, A., Majid, D., Russly Abdul, R., Maryam, J., Wan Md. Zin Wan, Y., & Nor Azowa, I. (2010). Silver/poly (lactic acid) nanocomposites: preparation, characterization, and antibacterial activity. *International Journal of Nanomedicine*, 573. doi:10.2147/ijn.s12007

Shankar, S., Wang, L.-F., & Rhim, J.-W. (2018). Incorporation of zinc oxide nanoparticles improved the mechanical, water vapor barrier, UV-light barrier, and antibacterial properties of PLA-based nanocomposite films. *Materials Science and Engineering, 93*, 289–298. doi:10.1016/j.msec.2018.08.002

Sheikhi, A. (2019). Emerging Cellulose-Based Nanomaterials and Nanocomposites. In Niranjan Karak (Ed). *Nanomaterials and Polymer Nanocomposites Raw Materials to Applications*, 307–351, Elsevier. doi:10.1016/b978-0-12-814615-6.00009-6

Shojaeiarani, J., Bajwa, D. S., & Chanda, S. (2021). Cellulose nanocrystal based composites: A review. *Composites Part C, 5*, 100164. doi:10.1016/j.jcomc.2021.100164

Shojaeiarani, J., Bajwa, D. S., Stark, N. M., Bergholz, T. M., & Kraft, A. L. (2020). Spin coating method improved the performance characteristics of films obtained from poly(lactic acid) and cellulose nanocrystals. *Sustainable Materials and Technologies, 26*, e00212. doi:10.1016/j.susmat.2020.e00212

Siddiqi, K. S., ur Rahman, A., Tajuddin, & Husen, A. (2018). Properties of zinc oxide nanoparticles and their activity against microbes. *Nanoscale Research Letters, 13*(1). doi:10.1186/s11671-018-2532-3

Sucinda, E. F., Majid, M. S. A., Ridzuan, M. J. M., Cheng, E. M., Alshahrani, H. A., & Mamat, N. (2021). Development and characterisation of packaging film from Napier cellulose nanowhisker reinforced polylactic acid (PLA) bionanocomposites. *International Journal of Biological Macromolecules*. doi:10.1016/j.ijbiomac.2021.07.069

Szymańska-Chargot, M., Chylińska, M., Pieczywek, P. M., Walkiewicz, A., Pertile, G., Frąc, M., Cieślak, K. J., & Zdunek, A. (2020). Evaluation of nanocomposite made of polylactic acid and nanocellulose from carrot pomace modified with silver nanoparticles. *Polymers, 12*(4), 812. doi:10.3390/polym12040812

Tang, J., Sisler, J., Grishkewich, N., & Tam, K. C. (2017). Functionalization of cellulose nanocrystals for advanced applications. *Journal of Colloid and Interface Science, 494*, 397–409. doi:10.1016/j.jcis.2017.01.077

Tavakolian, M., Jafari, S. M., & van de Ven, T. G. M. (2020). A review on surface-functionalized cellulosic nanostructures as biocompatible antibacterial materials. *Nano-Micro Letters, 12*(1). doi:10.1007/s40820-020-0408-4

Tayeb, A., Amini, E., Ghasemi, S., & Tajvidi, M. (2018). Cellulose nanomaterials – binding properties and applications: a review. *Molecules, 23*(10), 2684. doi:10.3390/molecules23102684

Tiwari, V., Mishra, N., Gadani, K., Solanki, P. S., Shah, N. A., & Tiwari, M. (2018). Mechanism of anti-bacterial activity of zinc oxide nanoparticle against carbapenem-resistant *Acinetobacter baumannii*. *Frontiers in Microbiology, 9*. doi:10.3389/fmicb.2018.01218

Vatansever, E., Arslan, D., & Nofar, M. (2019). Polylactide cellulose-based nanocomposites. *International Journal of Biological Macromolecules, 137*, 912–938. doi:10.1016/j.ijbiomac.2019.06.205

Wagner, P. A., Little, B. J., Hart, K. R., & Ray, R. I. (1996). Biodegradation of composite materials. *International Biodeterioration & Biodegradation, 38*(2), 125–132. doi:10.1016/s0964-8305(96)00036-4

Wilkes, R. A., & Aristilde, L. (2017). Degradation and metabolism of synthetic plastics and associated products by *Pseudomonas* sp.: capabilities and challenges. *Journal of Applied Microbiology, 123*(3), 582–593. doi:10.1111/jam.13472

Wolf, C., Angellier-Coussy, H., Gontard, N., Doghieri, F., & Guillard, V. (2018). How the shape of fillers affects the barrier properties of polymer/non-porous particles nanocomposites: a review. *Journal of Membrane Science, 556*, 393–418. doi:10.1016/j.memsci.2018.03.085

Yang, W., Fortunati, E., Dominici, F., Giovanale, G., Mazzaglia, A., Balestra, G. M., Kenny, J. M., & Puglia, D. (2016). Effect of cellulose and lignin on disintegration, antimicrobial and antioxidant properties of PLA active films. *International Journal of Biological Macromolecules, 89*, 360–368. doi:10.1016/j.ijbiomac.2016.04.068

Yang, Z., Li, X., Si, J., Cui, Z., & Peng, K. (2019). Morphological, mechanical and thermal properties of poly(lactic acid) (PLA)/cellulose nanofibrils (CNF) composites nanofiber for tissue engineering. *Journal of Wuhan University of Technology-Materials Science Edition, 34*(1), 207–215. doi:10.1007/s11595-019-2037-7

Yu, F., Fei, X., He, Y., & Li, H. (2021). Poly(lactic acid)-based composite film reinforced with acetylated cellulose nanocrystals and ZnO nanoparticles for active food packaging. *International Journal of Biological Macromolecules, 186*, 770–779. doi:10.1016/j.ijbiomac.2021.07.097

Yu, H.-Y., Yang, X.-Y., Lu, F.-F., Chen, G.-Y., & Yao, J.-M. (2016). Fabrication of multifunctional cellulose nanocrystals/poly(lactic acid) nanocomposites with silver nanoparticles by spraying method. *Carbohydrate Polymers, 140*, 209–219. doi:10.1016/j.carbpol.2015.12.030

Zhao, S.-W., Guo, C.-R., Hu, Y.-Z., Guo, Y.-R., & Pan, Q.-J. (2018). The preparation and antibacterial activity of cellulose/ZnO composite: a review. *Open Chemistry, 16*(1), 9–20. doi:10.1515/chem-2018-0006

11 Polylactic Acid and Polylactic Acid-based Composites

Processing Degradation, Physical and Thermal Aging, Water Immersion, and Exposure to Weather Agents

A. Arbelaiz, A. Orue, and U. Txueka
University of the Basque Country UPV/EHU
Donostia-San Sebastian, Spain

CONTENTS

11.1 INTRODUCTION: BACKGROUND AND DRIVING FORCES

In the last decades, research on bio-based polymers and biocomposites has attracted great interest in material science. The replacement of petroleum-based polymers and composites with renewable origin biopolymers and biocomposites is a sustainable option. Among the biopolymers, poly(lactic acid) (PLA), with large-scale industrial production available, shows good mechanical properties similar to many petroleum-based polymers. On the other hand, lignocellulosic fibers are sustainable materials with high mechanical specific properties. In the last decades, research on natural fiber-reinforced composite materials has considerably increased. Biocomposites based on PLA and lignocellulosic fibers are interesting candidates in applications where mechanical specific properties are important. In different applications the material is exposed to different environments that damage/degrade mechanical properties. For example, in outdoor applications the weathering elements such as the rainwater and the sunlight can degrade the material. Besides, the properties of polymers change during material processing because the high temperatures used during processing partially degrade them. In the current work, it was studied the effect of different processes on the mechanical properties of unreinforced poly(lactic acid) and PLA-based biocomposites.

The degradation during processing is undesirable and should be minimized. In processing techniques of thermoplastics, such as extrusion and injection molding, PLA is exposed to high temperatures (above 160°C) under high shear stresses that reduce the molecular weight of the polymer. Concretely in PLA polymer, the combination of high temperatures and the presence of water molecules leads to the hydrolytic degradation of PLA (Figure 11.1) that causes the chain scissions, reducing the molecular weight. During hydrolysis reaction, the number of carboxylic acid groups is increasing and the presence of these acid groups catalyzes the hydrolytic reaction. The hydrolytic reaction of PLA during processing can be reduced by drying the material prior to processing it. However, other degradation mechanisms due to high temperature and shear forces cannot be avoided and some molecular weight reduction happens after the processing step.

Some consequences of polymer degradation during processing could be: (i) the change of rheological properties, (ii) the embrittlement in solid state, and (ii) the color change.

The melt viscosity of polymers is very sensitive to changes in the macromolecular chain structure (1). Therefore, the investigation of rheological properties of polymeric materials at molten state is crucial to understanding the structure–property relationships of polymeric materials (2, 3).

During injection molding, the molten PLA is forced to flow under high pressure through sprues and fill a mold cavity where it cools quickly and solidifies. As a result of the rapid cooling at temperatures below T_g, the molecular mobility was

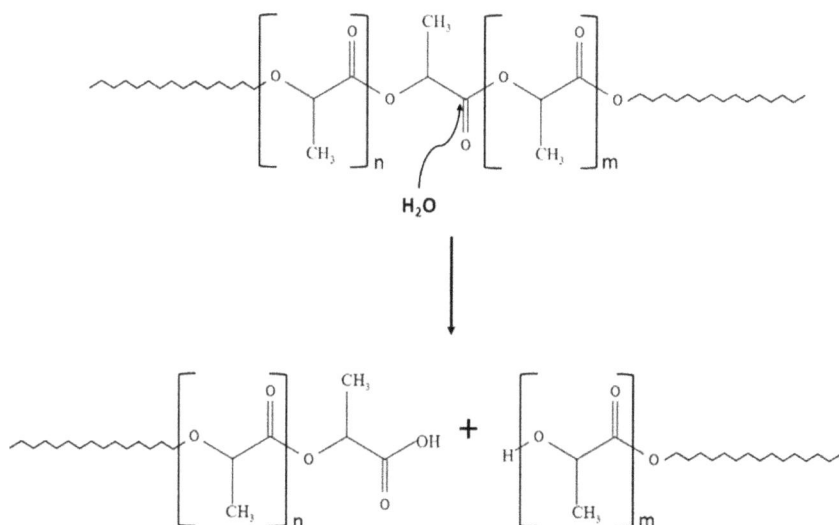

FIGURE 11.1   Schematic representation of the hydrolysis of polylactic acid.

diminished greatly and the crystallinity of PLA is not fully developed, being almost amorphous. The injection-molded material does not achieve the thermodynamic equilibrium because the molecular relaxations were limited. The physical aging phenomenon is related to the glassy state structural changes at temperatures below T_g that involve modifications in material properties. Physical aging could affect the long-term stability of the polymer. Due to the viscoelasticity character of polymer, at temperatures close to polymer T_g and with enough time, PLA amorphous phase could slightly align, increasing the crystallinity degree of the polymer. Physical aging leads to crystallinity degree changes that can influence mechanical properties. The physical aging process can be observed through the time evolution of enthalpy or mechanical properties (4–7).

In outdoor applications, environmental exposure has a significant influence on material properties changes. Polymers exposed outdoor can degrade through the action of agents such as solar ultraviolet (UV) radiation, water, and temperature changes, among others. The polymer degradation mechanism process in polymers aged outdoor is complex since different agents are simultaneously interacting with the polymer. There are different environmental exposures that contribute to the deterioration of PLA-based materials and influence their durability (8–14).

11.2   EXPERIMENT

11.2.1   Materials

PLA 3051D from NatureWorks Co. Ltd. was used in the current work. Its levo:dextro ratio is about 96.7:3.3, and the residual monomer content is less than 0.3 wt.% (15). The flax fibers used in the current work were provided by Arctic Fiber Company (Finland). The flax fibers were obtained by a retting process (Figure 11.2).

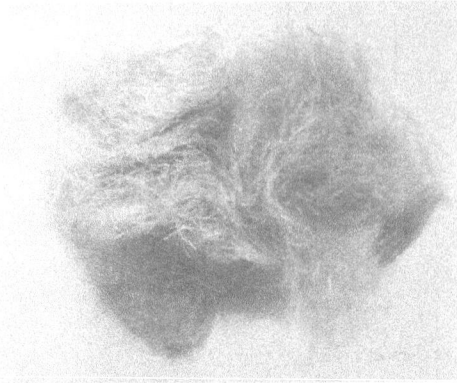

FIGURE 11.2   Flax fibers used for composite preparation.

11.2.2   PLA/Flax Fiber Composites: Compounding and Processing

Compounding of the materials was carried out as using a Haake Rheomix 600 with two Banbury rotors. The mixture was pelletized and dried; afterward, by using the injection molding process (Battenfeld Plus 250), tensile tests according to ASTM D638 standard were obtained. Composite with 20 wt.% fiber loading was fabricated, and for comparison purposes, also raw PLA pellets were injection molded and these specimens were used to study the degradation of PLA at the molten state and the physical and thermal aging of PLA-based systems. On the other hand, the specimens used for the one-year aging were molded in a HAAKE Minijet II machine where the selected injection and mold temperatures were 185 and 80 °C, respectively. Tensile specimens according to the ASTM-D638-10 standard were obtained. Composites with different fiber loadings, from 20 to 40 wt.%, were prepared.

11.2.3   Ageing of PLA and PLA-Based Composites

The specimens were aged for one year and two different aging processes were carried out: (i) aging in water at ~20°C and (ii) aging outdoor (Figure 11.3). For outdoor weathering, the tensile specimens were placed on the roof of a shack located in Pasai Donibane Village (Basque Country) that is geolocated at a latitude of 43.32457 and a longitude of −1.91461. Pasai Donibane Village is located in the north of the Basque Country, where the climate is mesothermic, i.e., moderate in terms of temperatures, and very rainy. Despite the fact that the summers are mild, during summer, short episodes with temperatures up to 40°C are possible (16).

11.2.4   Characterization Techniques

11.2.4.1   Rheological Characterization

Rheological measurements of PLA were conducted using an Advanced Rheometric Expansion System, Ares, using 25 mm parallel plates. Tests were performed at

FIGURE 11.3 The shack where the specimens were outdoor aged.

different temperatures (160, 170, 180, and 190°C) under air atmosphere. All samples were melted at 170°C and the gap between parallel plates was set 1 mm.

Using steady shear mode, steady shear rate sweep tests over a shear rate range of 0.01–100 s^{-1} were carried out. In steady shear tests, the strain was applied using continuous rotation. After determining the region of the linear viscoelastic response of samples, a time sweep of 1 s^{-1} rate was carried out and at different time intervals (3, 5, 10, 15, 20, and 30 min) and the shear viscosity values were recorded in order to evaluate the stability of the sample at different temperatures.

11.2.4.2 Thermogravimetric Analysis (TGA)

Thermogravimetric analysis was carried out in a TGA/SDTA851^e (Mettler Toledo) with a heating rate of 5°C/min from 25 to 700°C in air atmosphere.

11.2.4.3 Gel Permeation Chromatography

Gel permeation chromatography measurements were carried out in a Perkin-Elmer LC-295 chromatograph equipped with a refractive index detector. The molecular weight and molecular weight distribution of PLA before and after processing were determined. The mobile phase was tetrahydrofuran at a flow rate of 1 mL/min at room temperature.

11.2.4.4 Tensile Test Characterization

Tensile properties of PLA and its composites were determined using Instron Model 4206 Universal Testing Instrument. All tensile tests were carried out at 5 mm/min deformation rate. The average values of five specimens were reported.

11.2.4.5 Thermal Analysis

Differential scanning calorimetry (DSC) studies of samples were performed using Mettler Toledo DSC 822 equipment. Samples were heated at the rate of 10°C/min from 25°C to 180°C and from thermograms, characteristic temperatures were determined. For aged specimens, instead of 10°C/min heating rate, 3°C/min was selected. The crystallinity degree (χ) of PLA was determined using the following equation:

$$\chi = \left(\frac{\Delta H_m - \Delta H_{cc}}{\Delta H_{100\%} \times \omega_{PLA}} \right) \times 100 \qquad (11.1)$$

ΔH_m, ΔH_{cc}, and $\Delta H_{100\%}$ are the melting, cold crystallization, and theoretically 100% crystalline PLA melting enthalpies, respectively. In the current work, the value of 93 J/g was used for $\Delta H_{100\%}$ (17). On the other hand, for composite samples, the weight fraction of PLA, ω_{PLA}, was 0.8.

11.2.4.6 Attenuated Total Reflection-Fourier Transform Infrared (ATR-FTIR)

ATR-FTIR spectroscopy was used to analyze the changes in water immersed and outdoor aged specimens. Spectra of the samples were obtained in the range from 600 to 4000 with a resolution of 4 cm^{-1}.

11.3 DEGRADATION OF PLA DUE TO PROCESSING AT HIGH TEMPERATURES

11.3.1 RHEOLOGICAL PROPERTIES

Viscoelastic properties of polymer melts can be characterized by shear viscosity (η) and normal stress difference data (N_1 and N_2). N_1 values are positive, whereas N_2 values are accepted to be negative but significantly smaller than N_1 in magnitude; therefore N_2 values can be neglected (18). The N_1 is the difference between the stress in the flow direction and the perpendicular direction. Figures 11.4 and 11.5 show shear viscosity and first normal stress difference data as a function of shear rate at different temperatures.

PLA shows typically pseudoplastic behavior of shear viscosity, with a Newtonian zone that extends up to a certain critical value of strain rate from which the viscosity is reduced. The deviation of such linearity implies the destruction of internal structures (18) such as entanglements, which are physical links between long polymer chains.

Regarding the elastic character, the non-zero values of N_1 indicate that the polymeric melt possesses elasticity (19) and the elastic character of polymer decreases as increasing the temperature of measurement (18). The elastic character of molten polymer is responsible for different phenomena not found in Newtonian liquids such as die swelling and the Weissenberg effect, among others. The first normal stress difference increases as the shear rate value is increased. Besides, at very low shear rates, polymer melts are predominantly viscous since the viscous values are higher than N_1 value ones. However, the situation is totally reversed at higher shear rates, which is in agreement with results reported for polymer melts (19).

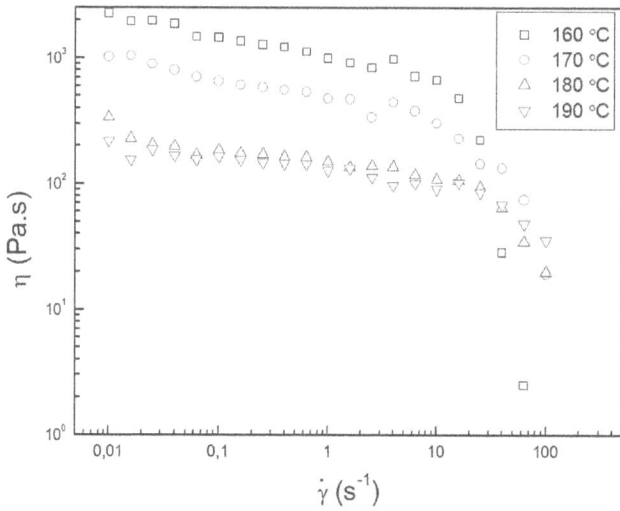

FIGURE 11.4 Dependence of shear viscosity upon shear rate for PLA at different temperatures.

The Carreau–Yasuda model (equation 11.2) was used to fit the non-Newtonian viscosity curves, as shown in Figure 11.6.

$$\eta = C_1 \left[1 + \left(C_2 \dot{\gamma} \right)^{C_3} \right]^{\frac{C_4 - 1}{C_3}} \tag{11.2}$$

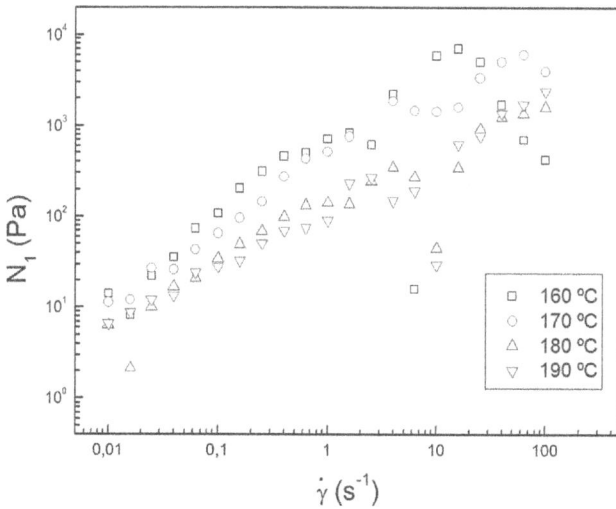

FIGURE 11.5 Dependence of the first normal stress difference with shear rate for PLA at different temperatures.

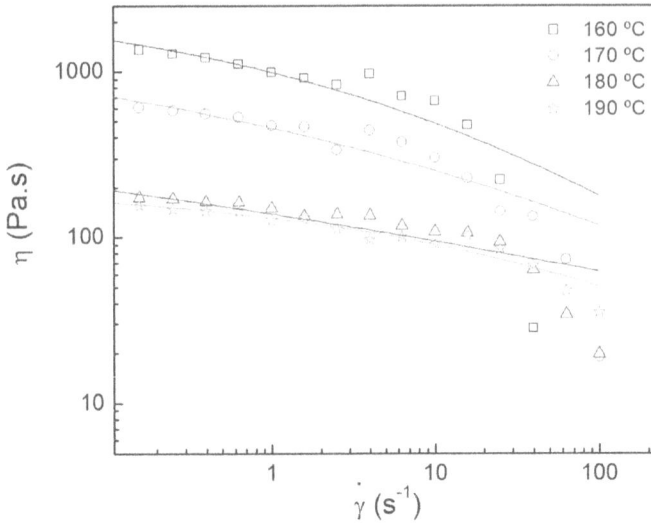

FIGURE 11.6 Dependence of shear viscosity with shear rate for PLA at different temperatures, experimental values (open symbols), and Carreau–Yasuda model values (lines).

where C_1, C_2, C_3, and C_4 are material-dependent parameters. C_1 is the zero-shear-rate viscosity, C_2 and C_3 define the transition from Newtonian to the power-law region, and C_4 is the power-law index. The constants for the model are summarized in Table 11.1. The viscosity experimental data fit well with the Carreau–Yasuda model.

At temperatures above melting temperature, the viscosity of the Newtonian region, C_1, of PLA polymer decreased with increasing temperature. The relationship between $\ln \eta_o$ and $1/T$ is linear according to Arrhenius formula (equation 11.3):

$$\text{Ln} \, \eta_o = \text{Ln} \, A + \frac{E_a}{RT} \qquad (11.3)$$

where A is a constant that is related to material properties; E_a, the activation energy for the flow process; R is the universal gas constant; and T is the absolute temperature (18). Figure 11.7 shows the relationship between logarithmic apparent melt viscosity and reciprocal temperature in PLA. E_a was calculated by using C_1 data of the Carreau–Yasuda model at different temperatures. The regression coefficient

TABLE 11.1

Carreau–Yasuda Model Constants for PLA at Different Temperatures

Temperature (°C)	C_1 (Pa.s)	C_2 (s)	C_3	C_4
160	3148	116×10^{-4}	0.24767	1.04×10^{-15}
170	2304	3.76×10^{-4}	0.15721	2.43×10^{-14}
180	2000	4.72×10^{-4}	0.2556	7.5494×10^{-10}
190	221	3.99×10^{-4}	0.2500	1.49697×10^{-16}

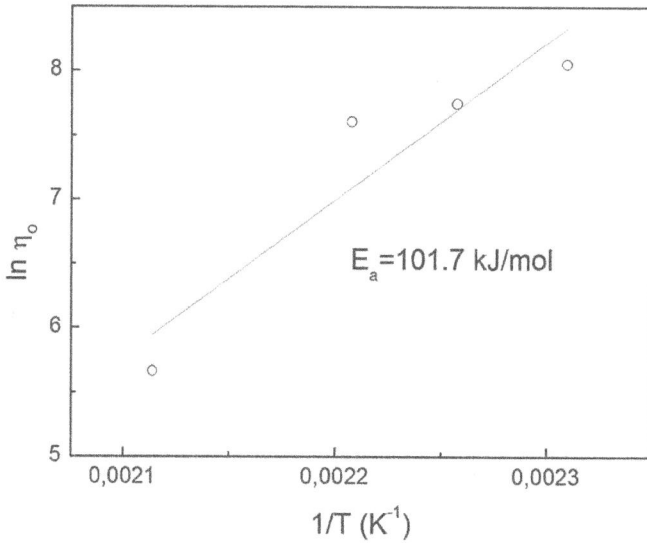

FIGURE 11.7   Relationship of logarithmic apparent melt viscosity and the reciprocal of temperature in PLA polymer.

value of 0.82 indicates that the dependence of η_o on temperature obeys the Arrhenius equation. In the literature (3, 20) for PLA activation energy, different values were reported ranging from 91 to 167 kJ/mol. Thus the value calculated in the current work, 101.7 kJ/mol, is in the same value range.

In order to determine the effect of degradation on molten PLA rheological properties, the Newtonian viscosity was measured along time at different temperatures (Figure 11.8). The viscosity–time curves of molten PLA display a maximum at time

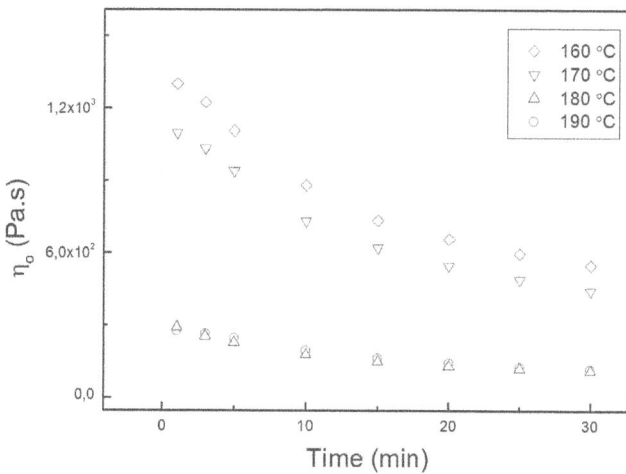

FIGURE 11.8   Change of Newtonian viscosity along the time of PLA samples at different temperatures.

zero and decrease until a horizontal plateau is reached at longer times. This behavior can be explained by the effect of a thermo-oxidative degradation process, implying chain scission leading to a molar mass decrease at residence times in the rheometer.

Dorgan et al. (21) found for PLA the following relationship between the zero shear viscosities and the average molecular weight, equation 11.4:

$$H_o = 2.3 \times 10^{-14} M_w^{3.7} \tag{11.4}$$

They found a scaling exponent of 3.7 rather than the usual 3.4 power normally associated with linear polymers. The loss in zero shear viscosity value at a selected temperature can be related to a loss in molecular weight. Therefore, rheology is a practical and fundamental characterization technique for the study of the degradation process of polymers (1).

Thermogravimetry measurements provide complementary data for the analysis of polymer thermal degradation (22). Thermogravimetric curves (TG and DTG) in air atmosphere for PLA are shown in Figure 11.9. PLA shows one weight-loss stage at the temperature of around 230°C, which is in agreement with other reported studies (23, 24).

Even though the rheological measurements indicated that PLA degrades at temperatures lower than 230°C, by TGA the weight loss due to polymer degradation is noticeable at around 230°C. This fact indicates that the rheological technique is more sensitive than TGA because it can detect low degradation levels due to chains scission (22).

Finally, in order to evaluate the degradation degree that PLA can suffer during processing, dried PLA was injection molded at 195°C and the molecular weight of the polymer was determined by the GPC technique before and after processing. Figure 11.10 shows GPC chromatograms for PLA before and after processing by injection molding. The pristine PLA shows one peak at a retention time of 24.9 min, whereas the processed PLA peak appears at slightly higher retention times. This fact suggests that during the processing step of dried PLA, a few chains scission

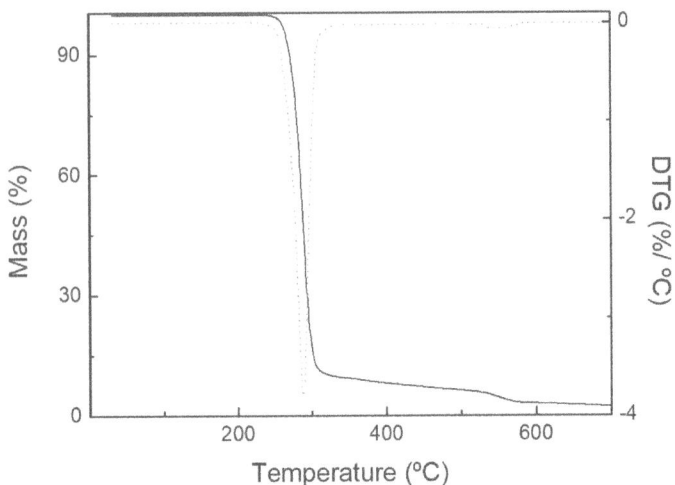

FIGURE 11.9   TG and DTG thermograms of PLA at a heating rate of 5°C/min.

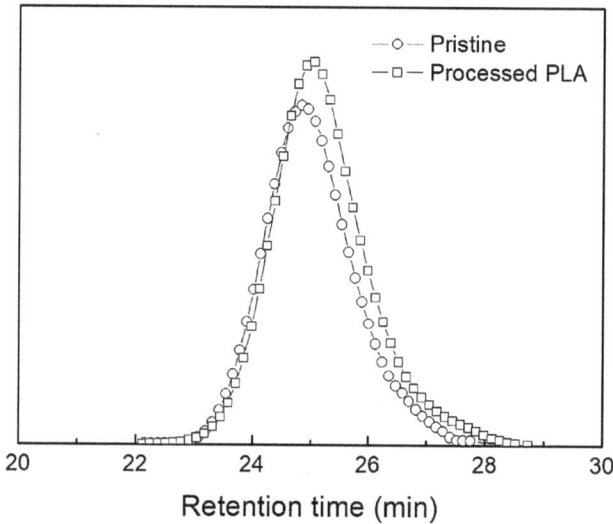

FIGURE 11.10   GPC chromatograms for neat PLA before and after processing.

occurred. Table 11.2 shows the molecular weights and polydispersity index of pristine PLA and processed polymer. The molecular weights were calculated using a universal calibration method with polystyrene standards. Even though the time held by the polymer at 195°C during injection molding was short, the molecular weight reduced by around 10%. According to Lim et al. (25), injection-molded pieces made from properly dried good-quality PLA resins and optimal processes should exhibit molecular weight loss of 10% or less.

11.4   THE EFFECT OF PHYSICAL AND THERMAL AGING ON THERMAL AND MECHANICAL PROPERTIES

11.4.1   Physical Aging at Room Temperature along the Time

After specimens had been molded, specimens were tested mechanically at different times, and the average values of tensile properties are reported in Table 11.3. In Figure 11.11, the stress–strain curves of injection-molded PLA specimens aged

TABLE 11.2

Retention Time and Number-Average and Weight-Average Molecular Weights as Well as Polydispersity Index of Original PLA and Processed Samples Using a Universal Calibration Method with Polystyrene Standards

Sample	Time (min)	M_n (g/mol)	M_w (g/mol)	M_v (g/mol)	I_{PD}
Original PLA	24.9	121,410	172,310	164,180	1.4
After processing	25.1	106,510	154,770	147,070	1.4

TABLE 11.3

Strength, Modulus, and Deformation at Break Values at Different Times

Time	σ_t (MPa)	E_t (MPa)	ε_{break} (%)
1 h	63.5 ± 2.0	1150 ± 55.5	>50
24 h	62.8 ± 0.6	1225 ± 13.5	6.5 ± 0.83
48 h	66.6 ± 1.8	1209 ± 53.5	6.2 ± 0.81
1 week	62.3 ± 1.0	1211 ± 32.5	5.7 ± 0.34
1 month	64.0 ± 0.6	1336 ± 22.9	6.5 ± 0.72

for different times at 20°C are shown. The specimen tested one hour later, immediately following injection molding, shows a ductile performance. However, the specimen tested after 24 hours or later shows a brittle performance. This ductile–brittle transition was observed in amorphous polylactide tensile bars aged at room temperature by Witzke et al. (26). They performed tensile tests every hour immediately following injection molding and they observed the ductile–brittle transition in amorphous polylactide occurred within eight hours at room temperature aging. Similarly, Cui et al. (27) observed that deformability of poly(lactic acid) decreases drastically from 250%, after injection molding, to a few percent after less than a day of aging.

In Figure 11.12, DSC thermograms of injection-molded PLA specimens aged at different times are shown. The specimen tested one hour later immediately following injection molding does not show relaxation enthalpy at T_g, indicating that the physical aging hardly occurred. Increasing the ageing time, no significant difference

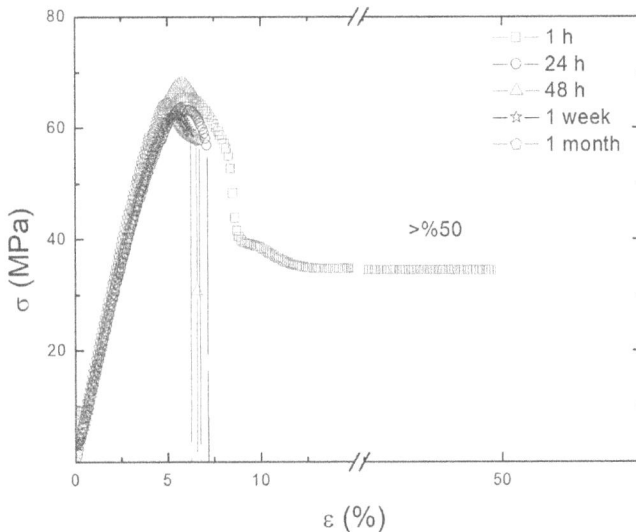

FIGURE 11.11 Stress–strain curves of injection-molded PLA specimens mechanically tested at different times.

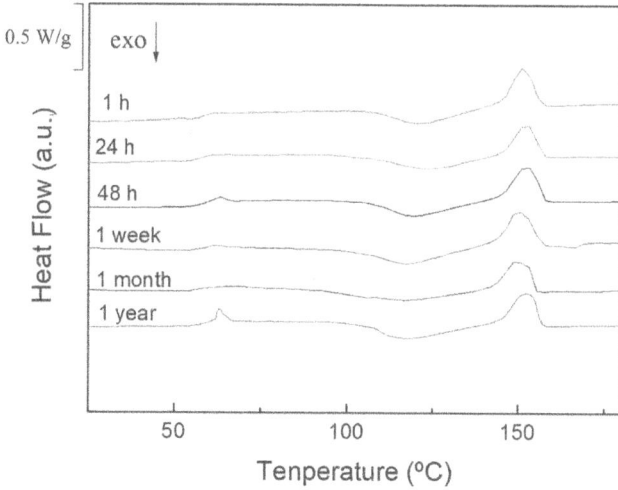

FIGURE 11.12 DSC thermograms of injection-molded PLA specimens aged at different times.

in the thermograms is observed until 1 year aged specimen. The aging temperature affects the time necessary to approach the thermodynamical equilibrium. Moreover, it is often not possible to reach equilibrium when the aging is more than 15°C below the T_g of a polymer (28, 29).

In the literature, it was observed that the relaxation enthalpy increases with decreasing the difference between the glass transition temperature (T_g) and the aging temperature (T_a), i.e., the degree of subcooling ($T_g - T_a$) (30, 31). In the current study, an aging temperature of 20°C was used and the degree of subcooling was about 34°C. Consequently, very slow relaxation kinetics was observed and only the relaxation enthalpy was appreciated in the specimen aged for one year. After one year the enthalpy relaxation at T_g is clearly observed, indicating that the physical aging occurred. In Table 11.4, the thermal characteristics temperatures and the crystallinity degree values are reported.

TABLE 11.4

Thermal Characteristics Temperatures and the Crystallinity Degree Values of Injection-Molded Specimens Aged at Different Times

Time	$T_{g\ onset}$ (°C)	T_c (°C)	T_m (°C)	X (%)
1 h	54.2	121.7	151.0	0.4
24 h	54.8	124.0	153.1	1.9
48 h	55.9	119.5	153.1	1.6
1 week	55.8	117.2	150.9	2.4
1 month	54.6	117.2	148.6	3.4
1 year	56.5	117.6	152.3	4.4

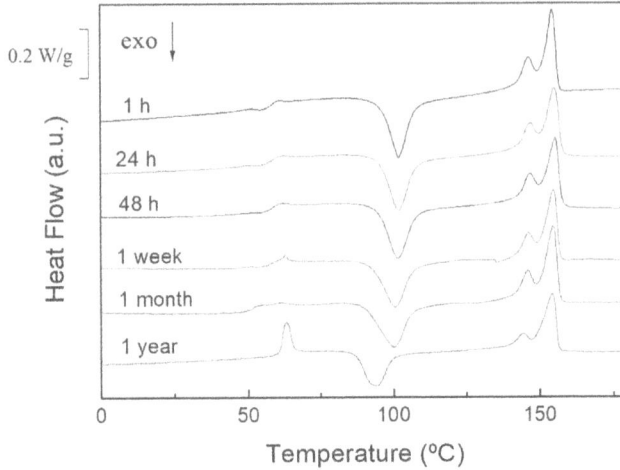

FIGURE 11.13   DSC thermograms of injection-molded composite specimens aged at different times.

The glass transition temperature increases slightly after aging the specimen for one year. In the literature, a similar trend in T_g was observed for aged PLA (32). The increase of crystallinity led to a restriction of the mobility of the amorphous phase and consequently increased the glass transition temperature of the specimen aged for one year. The results also showed that the aging process affects the final degree of crystallinity of PLA. As the aging time was increased, the crystallinity degree was increased slightly.

To study the effect of lignocellulosic fiber addition to PLA on the physical ageing, composites with 20 wt.% of flax fiber were prepared. In Figure 11.13, DSC thermograms of injection-molded composite specimens aged at different times are shown. Similar to the unreinforced system, only the relaxation enthalpy was appreciated clearly in the specimen aged for one year. In Table 11.5 the thermal characteristics temperatures and the crystallinity degree values of composites aged at different times are reported.

TABLE 11.5
Thermal Characteristics Temperatures and the Crystallinity Degree Values of Injection-Molded Composite Specimens Aged at Different Times

Time	$T_{g\ onset}$ (°C)	T_c (°C)	T_m (°C)	X (%)
1 h	56.0	101.8	154.3	4.1
24 h	55.1	101.8	155.1	4.3
48 h	55.6	102.0	155.4	4.2
1 week	55.9	100.7	154.8	3.8
1 month	49.8	100.2	154.6	3.7
1 year	60.9	93.0	154.3	8.5

After the incorporation of fibers, the crystallization temperature decreased considerably with respect to the unreinforced one, indicating that fibers act as nucleating agents. However, the crystallinity degree of the PLA matrix hardly increased because the mobility of polymer chains was very limited at temperatures below T_g. Compared with unreinforced systems, in general, the glass transition temperature of the reinforced system is higher. The incorporation of fibers seems to restrict the mobility of PLA chains. McKenna (33) mentioned that the aging of polymeric matrix composites would be the same as that of the neat resin unless the fibers/matrix interphase plays a significant role. In agreement with this statement, it seemed that the incorporation of flax fibers altered the physical aging process of the PLA matrix.

11.4.2 THERMAL AGING AT DIFFERENT TEMPERATURES

In Figure 11.14, tensile stress–strain curves of injection-molded PLA specimens aged at different temperatures (36°C, 50°C, 85°C, and 110°C) are shown. For comparison purposes, the stress–strain curve of the specimen aged at room temperature, 20°C, was included. The injection-molded PLA specimens were held for 24 hours in an oven at the selected aging temperature, and afterward, the specimens were tested mechanically. Observing all curves, they can be differentiated into two groups: (i) curves of specimens aged at temperatures below T_g and (ii) curves of specimens aged at temperatures above T_g.

The specimens aged at 85°C and 110°C showed higher strength and modulus values than specimens aged at temperatures below T_g. On the other hand, the deformation at break reduced in specimens aged at 85°C and 110°C with respect to specimens aged at temperatures below T_g.

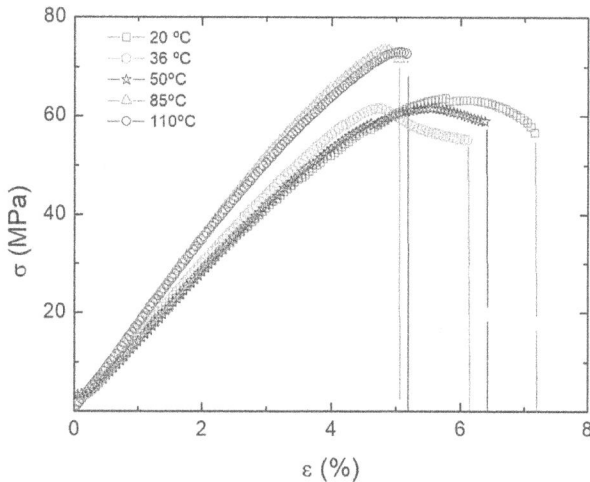

FIGURE 11.14 Stress–strain curves of injection-molded PLA specimens aged at different temperatures.

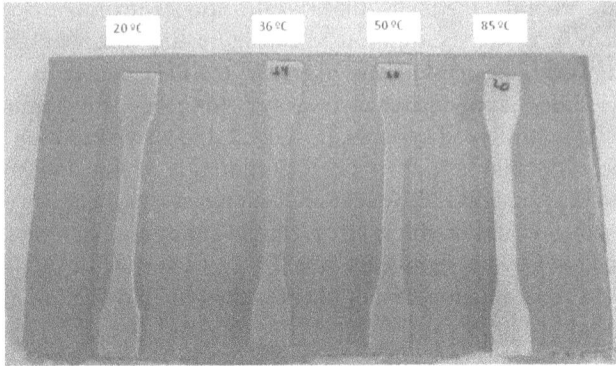

FIGURE 11.15 The photo of tensile specimens aged at different temperatures.

Figure 11.15 shows the photo of tensile specimens aged at different temperatures. The specimens aged at temperatures below T_g are translucent, indicating a low degree of crystallinity. Regarding the specimen aged at 85°C, the sample is opaque, indicating a higher crystallinity degree than specimens aged at lower temperatures. DSC thermograms were carried out to study the effect of aging temperature on thermal properties as well as the degree of crystallinity of aged specimens (Figure 11.16). As in mechanical performance, the same two different groups can be observed. Specimens aged at temperatures below T_g were almost amorphous with a low crystallinity degree (Table 11.6), which is in agreement with the translucent appearance observed in Figure 11.15. When the temperature was raised at temperatures above the glass transition, the molecular mobility of the amorphous phase increased and resulted in crystal growth and formation by the cold crystallization process.

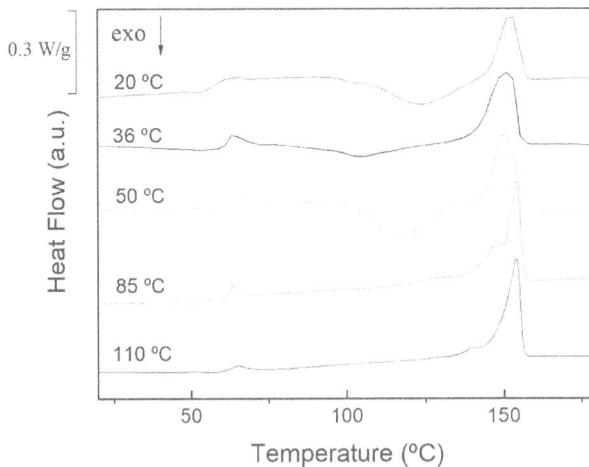

FIGURE 11.16 DSC thermograms of injection-molded composite specimens aged at different temperatures.

TABLE 11.6

Thermal Characteristics Temperatures and the Crystallinity Degree Values of Injection-Molded PLA Specimens Aged at Different Temperatures

Aging Temperature (°C)	$T_{g\,onset}$ (°C)	T_c (°C)	T_m (°C)	X (%)
20	54.2	124.0	153.1	1.9
36	59.9	103.7	150.9	5.2
50	58.2	117.3	150.9	2.4
85	60.5	–	154.0	34.8
110	59.8	–	153.9	36.9

On the other hand, specimens aged at temperatures above T_g did not show the cold crystallization process because, at the aging temperatures selected, the amorphous chains have arranged and crystallized prior to the DSC heating scan. Regarding the glass transition temperature, in general, specimens aged at temperatures above T_g showed slightly higher values than specimens aged at temperatures below glass transition temperatures (Table 11.6). This fact could be ascribed to the crystallinity degree increment, since the crystalline phase restricts the mobility of the amorphous phase.

In Table 11.7, the mechanical properties and the degree of crystallinity of aged specimens are summarized. The increments of strength and modulus values of specimens aged at temperatures above T_g directly correlate with the increment of crystallinity degree. The strength and modulus increment of specimens aged above T_g were around 10% and 13%, respectively, with respect to specimens aged at temperatures below T_g. These increments are ascribed to the crystallinity degree increment from 2% to 35%. Regarding deformation at break, even though specimens aged at temperatures of 85°C and 110°C showed slightly lower values than specimens aged at temperatures below T_g, all systems showed low deformation capability since the tensile test was carried out at temperatures below T_g.

TABLE 11.7

Tensile Properties and the Degree of Crystallinity Degree of Aged Specimens

Aging Temperature (°C)	σ_t (MPa)	E_t (MPa)	ε_{break} (%)	Crystallinity (%)
20	62.8 ± 0.6	1225 ± 13.5	6.5 ± 0.8	1.9
36	62.5 ± 0.9	1275 ± 24.1	6.1 ± 0.2	5.2
50	61.6 ± 0.3	1225 ± 78.9	6.2 ± 0.4	2.4
85	67.9 ± 5.8	1401 ± 100.7	4.5 ± 0.6	34.8
110	69.8 ± 4.2	1395 ± 93.1	5.3 ± 0.4	36.9

11.5 THE EFFECT OF WATER IMMERSION AND OUTDOOR WEATHERING ON COMPOSITE PROPERTIES

In composites immersed in water (Figure 11.17), the color of water changed to brownish, in fact, the color of water gets darker as the fiber loading of the composite was increased. This fact indicates that some water-soluble compounds of fibers were dissolved in water. The FTIR spectra of the control PLA and aged PLA are shown in Figure 11.18. Regarding the control PLA spectrum, the small band at 3508 cm^{-1} is related to hydroxyl groups in adsorbed H$_2$O and ending OH group in the PLA chain. The band at 1755 cm^{-1} is related to carbonyl (C=O) in ester functional groups and the bands related to C–O are 1096 and 1190 cm^{-1}. The absorption bands in the range of 1353 to 1457 cm^{-1} are attributed to aliphatic CH bending vibrations. The stretching vibrations of aliphatic C–H are located at 2943 and 2990 cm^{-1} (34). After the aging

FIGURE 11.17 Injection-molded specimens immersed in water at room temperature.

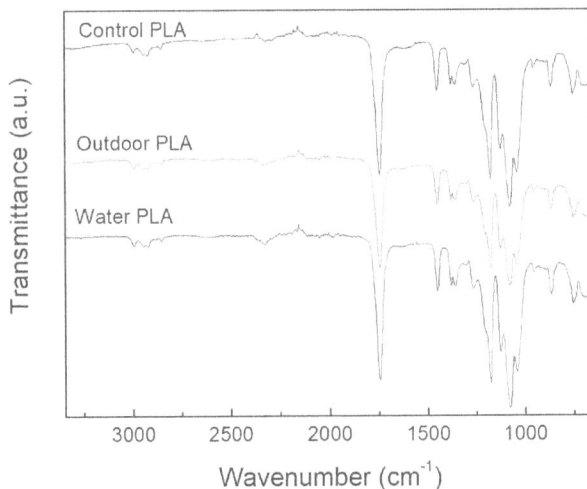

FIGURE 11.18 FTIR spectra of control PLA and aged samples.

processes, immersed in water or outdoor exposure, no significant differences can be observed in FTIR spectra, indicating that the aging processes did not change the composition of PLA specimens.

In Figure 11.19, the DSC thermograms of control and aged PLA samples are shown. The main difference between the control PLA thermogram and aged samples is that the cold crystallization process is not observed in aged samples. During the aging processes, the specimens crystallized totally.

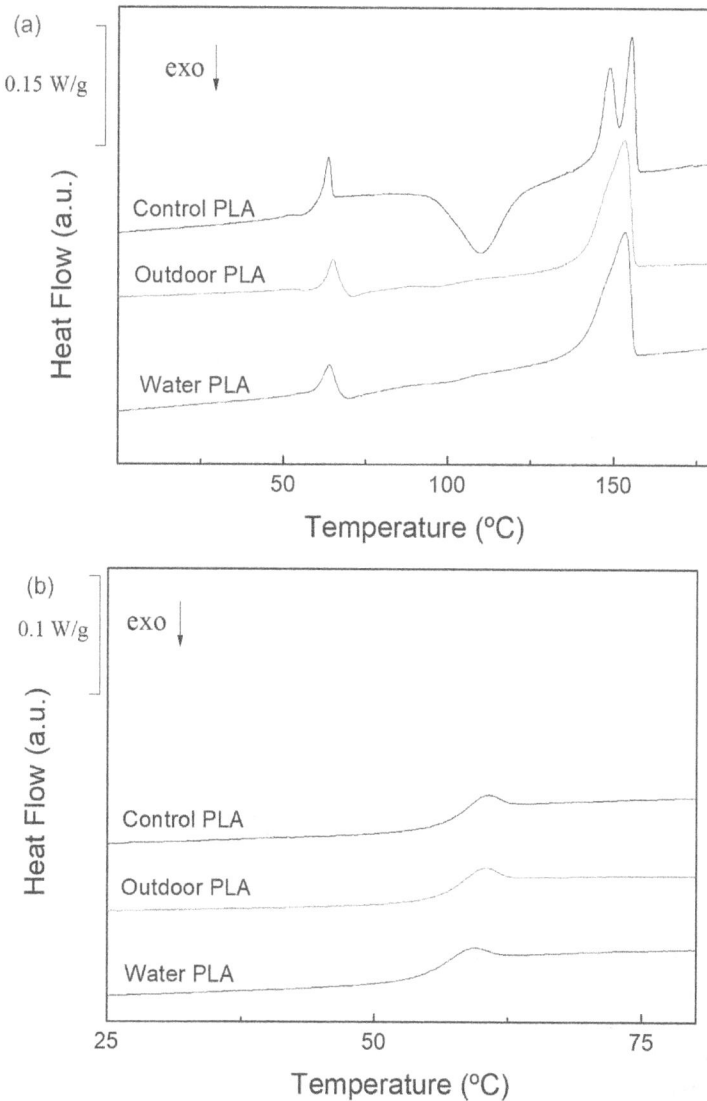

FIGURE 11.19 DSC thermograms of control and aged PLA samples: (a) first heating run and (b) the magnified view of the glass transition region in the second heating run.

TABLE 11.8

Thermal Characteristics Temperatures and the Crystallinity Degree Values of Control PLA and Aged Samples

	Second Heating	First Heating		
System	T_g Onset (°C)	T_c (°C)	T_m (°C)	X (%)
Control PLA	55.9	110.0	155.0	0.9
Outdoor PLA	55.3	–	153.1	34.7
Water PLA	53.7	–	153.2	31.8

Table 11.8 summarizes the thermal properties of control PLA and aged samples. The control PLA is amorphous and after aging the crystallinity degree increased up to 32–35%.

A possible explanation for the outdoor-aged PLA specimens having crystallized during the aging process could be that the temperatures in summer could be up to 40°C, and consequently, the increase of amorphous chains mobility leads to crystallization. Regarding water-immersed PLA, one possible explanation could be the plasticizing effect of water (14). The increase in the local mobility of amorphous chains could favor crystallization.

In the magnified view of the glass transition region in the second heating run (Figure 11.19b), it is observed that PLA aged in water showed a T_g value lower, around 2°C, than control PLA. It must be highlighted that the water-immersed-aged PLA sample was dried prior to the DSC measurements, and consequently, in the absence of water, PLA chains cannot be plasticized. The local mobility increase can be ascribed to the molecular mass reduction due to the hydrolysis reaction of PLA with water molecules. This reduction in T_g can be related to the molar mass reduction according to the Flory–Fox equation (11.5):

$$T_g = T_{g,\infty} - \frac{K}{M_n} \qquad (11.5)$$

where M_n and $T_{g,\infty}$ are the number-average molecular weight and the maximum glass transition temperature, respectively. K is an empirical parameter related to the free volume present in the polymer.

Figure 11.20 compares the transparency of control and aged PLA specimens.

The PLA specimens aged outdoor and in water were characterized by a loss of transparency. This fact can be related partially to the crystallinity degree increment. In the literature, it was suggested that the opacification and whitening phenomenon taking place when PLA samples are immersed in water is ascribed to crystallization and to crazes created in PLA due to water sorption. Pantani et al. confirmed the presence of these voids by density measurements, small angle x-ray scattering, and direct scanning electron microscopy observations (14).

FIGURE 11.20 The transparency comparison of the PLA specimens, from left to right: control, outdoor aged, and water aged.

In Figure 11.21, the visual appearance of control composite and aged composites with 30 wt.% of flax fiber are shown. The change in color of aged composites indicated the adverse impact of aging conditions in composites. Changes in the color of the composites are more obvious in the outdoor-aged composite than in the water-immersed one. The drastic color change of outdoor-aged composite could be related to photooxidative degradation. Lignin of flax fiber contains chromophoric structures, aromatic compounds, responsible for the brown color of the control composite. Lignin absorbs light at a short wavelength from 295 to 400 nm. After aging outdoor, due to the degradation of chromophoric structures, probably, other chromophoric structures are formed, consequently leading to the brown color changing to silver patina. Photodegradation is a surface phenomenon and even though the lignin compound is the most exposed, cellulose and hemicelluloses are also degraded (35–37).

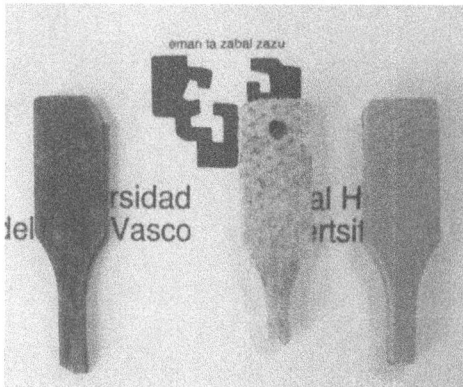

FIGURE 11.21 The visual appearance of control composite and aged composites with 30 wt.% of flax fiber, from left to right: control, outdoor aged, and water aged.

In the literature, the degradation of wood that contains more lignin than flax fiber was studied. FTIR-ATR analysis of the surface allows the study of chemical modifications due to photodegradation. It is a technique used by several authors to study the mechanism and the extent of wood degradation (37–39). Cogulet et al. (37) concluded that the appearance of the silver patina in wood could be correlated to the generation of carbonyl compounds conjugated with double bonds at 1615 cm^{-1} and the birefringence of cellulose.

The FTIR spectra of control and aged composites are shown in Figure 11.22. All spectra are identical and no differences are observed with respect to the unreinforced PLA spectra. The attenuated total reflection technique acquires a spectrum from the surface of the sample with a maximum depth penetration of 5 μm. The spectra of composites suggested that the fibers are embedded with PLA matrix and by FTIR-ATR analysis, only bands related with the matrix can be observed, i.e., the band modifications due to photodegradation of flax fiber lignin cannot be observed.

Regarding water-immersed composites, the color changed to light brown (Figure 11.21). The color change of composites after water immersion could be related to the leaching out of water-soluble fraction of fibers since the water-soluble compounds gave water a brownish color (Figure 11.17).

In Figure 11.23, the DSC thermograms of control and aged PLA composites are shown. In all thermograms, an endotherm ascribed to the total excess enthalpy of the polymer can be observed at around 60°C. In contrast to unreinforced aged PLA systems, aged composites did not crystallize totally and showed a cold crystallization peak. Table 11.9 summarized the thermal properties of control and aged composites with 30 wt.% of flax fiber.

Figure 11.24 shows the mechanical properties of control and aged specimens. Tensile strength and modulus values of unreinforced aged PLA increased slightly or maintained similar respect with control PLA. The slight improvement could be

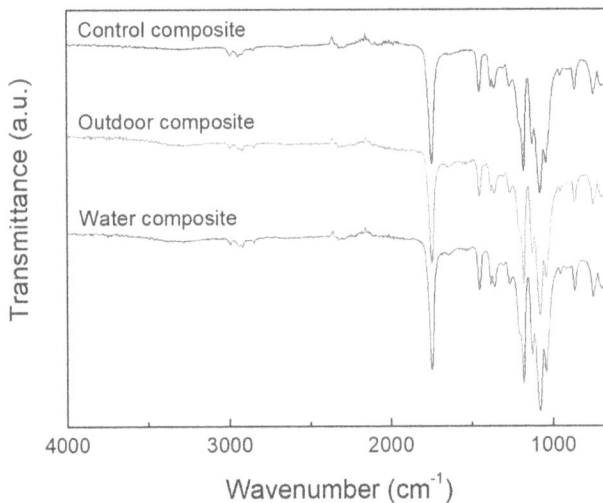

FIGURE 11.22 The FTIR spectra of control and aged composites.

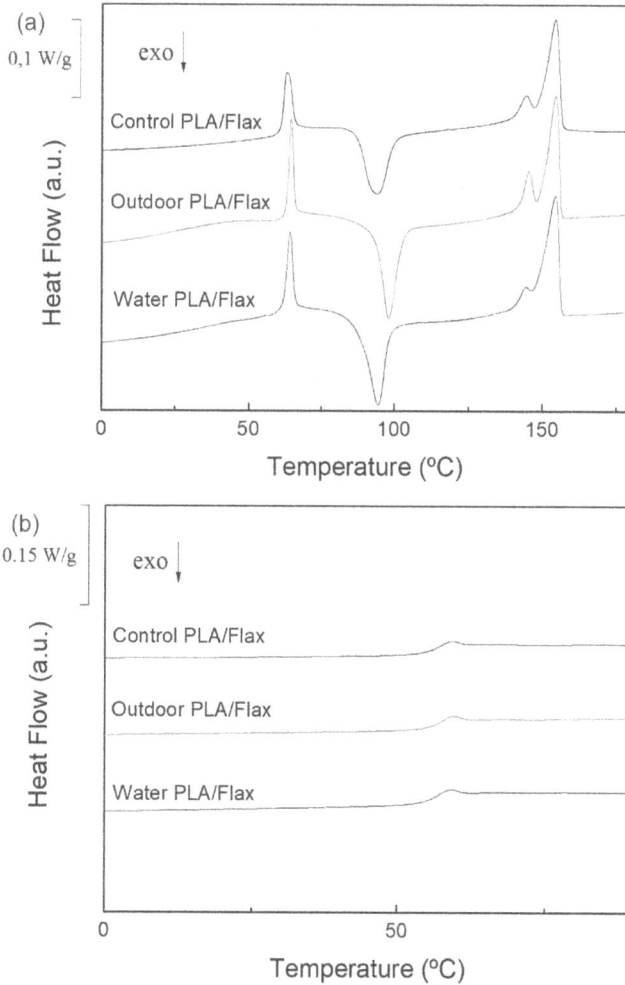

FIGURE 11.23 DSC thermograms of control and aged composites: (a) First heating run and (b) the magnified view of the glass transition region in the second heating run.

TABLE 11.9

Thermal Properties and the Crystallinity Degree of Control and Aged Composites with 30 wt.% of Flax Fiber

	Second Heating	First Heating		
System	T_g Onset (°C)	T_c (°C)	T_m (°C)	X (%)
Control composite	54.4	93.7	154.3	11.2
Outdoor composite	54.7	98.1	154.3	5.0
Water composite	54.0	94.5	154.2	8.4

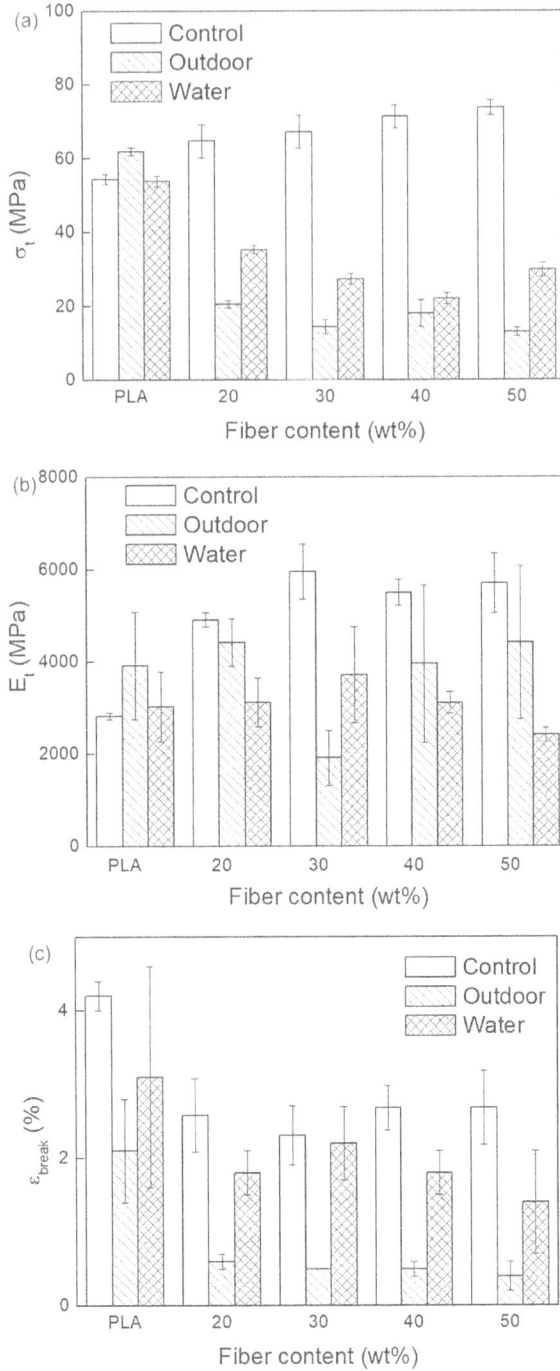

FIGURE 11.24   The tensile properties of control and aged composites: (a) strength, (b) modulus, and (c) deformation at break.

related to the crystallinity degree increment of unreinforced aged PLA as observed in DSC thermograms. Contrarily, aged composites showed lower values, especially strength, than control composites. Probably, in aged composites, the fiber/matrix adhesion was damaged and this could be the main reason to reduce the mechanical properties of aged composites. Zandvliet et al. (40) suggested that the degradation of the lignocellulosic fiber/matrix adhesion is provoked by water absorption of the fiber, resulting in a diameter expansion and then reduction after drying. It contributes to the loss of compatibilization between fibers and matrix, which results in debonding and weakening of the interfacial adhesion.

11.6 CONCLUSIONS

Properties testing of PLA-based systems after aging in different environments is crucial to understanding the possibilities and limitations of these materials. In the current study, it can be concluded that the outdoor and water immersion exposures, in addition to changing the visual aspect of PLA-based systems, also reduced the mechanical properties of flax fiber/PLA composite drastically. Results suggested that flax fiber/matrix adhesion strength was poor and to prevent a drastic mechanical deterioration of composites in environments in contact with water, improved fiber/matrix adhesion is necessary.

REFERENCES

1. Li, Hongbo; Huneault, Michel A. Effect of nucleation and plasticization on the crystallization of poly(lactic acid). *Polymer* (2007), 48(23), 6855–6866.
2. Sinha Ray, Suprakas; Yamada, Kazunobu; Okamoto, Masami; Ueda, Kazue. New polylactide-layered silicate nanocomposites. Concurrent improvements of material properties, biodegradability and melt rheology. *Polymer* (2002), 44(3), 857–866.
3. Gu, Shu-Ying; Zou, Cun-Yang; Zhou, Kai; Ren, Jie. Structure-rheology responses of polylactide/calcium carbonate composites. *Journal of Applied Polymer Science* (2009), 114(3), 1648–1655.
4. Acioli-Moura, Ricardo; Sun, Xiuzhi Susan. Thermal degradation and physical aging of poly(lactic acid) and its blends with starch. *Polymer Engineering and Science* (2008), 48(4), 829–836.
5. Pan, Pengju; Zhu, Bo; Inoue, Yoshio. Enthalpy relaxation and embrittlement of poly(L-lactide) during physical aging. *Macromolecules* (2007), 40(26), 9664–9671.
6. Castaldo, R.; Ambrogi, V.; Avella, M.; Avolio, R.; Carfagna, C.; Cocca, M.; Errico, M. E.; Gentile, G. Effect of physical ageing on properties of PLA plasticized with oligomeric esters of lactic acid. *AIP Conference Proceedings* (2014), 1599 (1, Times of Polymers (TOP) and Composites 2014), 138–141.
7. Mark J. Parker. Test methods for physical properties, In Anthony Kelly, Carl Zweben (Eds). *Comprehensive Composite Materials*, Pergamon (2000), 183–226.
8. Regazzi, Arnaud; Corn, Stephane; Ienny, Patrick; Benezet, Jean-Charles; Bergeret, Anne. Reversible and irreversible changes in physical and mechanical properties of biocomposites during hydrothermal aging. *Industrial Crops and Products* (2016), 84, 358–365.
9. Porfyris, A; Vasilakos, S; Zotiadis, Chr; Papaspyrides, C; Moser, K; Van der Schueren, L; Buyle, G; Pavlidou, S; Vouyiouka, S. Accelerated aging and hydrolytic stabilization of poly(lactic acid) (PLA) under humidity and temperature conditioning. *Polymer Testing* (2018), 68, 315–332.

10. Kamau-Devers, Kanotha; Kortum, Zachary; Miller, Sabbie A. Hydrothermal aging of bio-based poly(lactic acid) (PLA) wood polymer composites: Studies on sorption behavior, morphology, and heat conductance. *Construction and Building Materials* (2019), 214, 290–302.

11. Ndazi, B. S.; Karlsson, S. Characterization of hydrolytic degradation of polylactic acid/ rice hulls composites in water at different temperatures. *eXPRESS Polymer Letters* (2011), 5(2), 119–131.

12. Rocca-Smith, J. R.; Chau, N.; Champion, D.; Brachais, C.-H.; Marcuzzo, E.; Sensidoni, A.; Piasente, F.; Karbowiak, T.; Debeaufort, F. Effect of the state of water and relative humidity on ageing of PLA films. *Food Chemistry* (2017), 236, 109–119.

13. Gil-Castell, O.; Badia, J. D.; Kittikorn, T.; Stromberg, E.; Martinez-Felipe, A.; Ek, M.; Karlsson, S.; Ribes-Greus, A. Hydrothermal ageing of polylactide/sisal biocomposites. Studies of water absorption behaviour and physico-chemical performance. *Polymer Degradation and Stability* (2014), 108, 212–222.

14. Pantani, Roberto; De Santis, Felice; Auriemma, Finizia; De Rosa, Claudio; Di Girolamo, Rocco. Effects of water sorption on poly(lactic acid). *Polymer* (2016), 99, 130–139.

15. Wu, Defeng; Wu, Liang; Zhou, Weidong; Sun, Yurong; Zhang, Ming. Relations between the aspect ratio of carbon nanotubes and the formation of percolation networks in bio-degradable polylactide/carbon nanotube composites. *Journal of Polymer Science, Part B: Polymer Physics* (2010), 48(4), 479–489. (https://www.euskadi.eus/gobierno-vasco/ contenidos/informacion/cla_clasificacion/es_7264/es_cliclasificacion.html)

16. Fischer, E.W., Sterzel, H.J., Wegner, G., Investigation of the structure of solution grown crystals of lactide copolymers by means of chemical reactions. *Kolloid-Zeitschrift und Zeitschrift für Polymere* 251 (1973) 980–990.

17. Gupta, R.K. *Polymer and Composite Rheology*, 2nd Edition, Marcel Dekker, Inc., New York (2000).

18. Choi, Joon Soon; Lim, Sung Taek; Choi, Hyoung Jin; Hong, Soon Man; Mohanty, Amar K.; Drzal, Lawrence T.; Misra, Manjusri; Wibowo, Arief C. Rheological, thermal, and morphological characteristics of plasticized cellulose acetate composite with natural fibers. *Macromolecular Symposia* (2005), 224 (Bio-Based Polymers), 297–307.

19. Schramm, G. *A Practical Approach to Rheology and Rheometry*, 1st Edition, Gebrueder HAAKE GmbH, Karlsruhe (1994).

20. Ramkumar, D.H.S; Bhattacharya, M. Steady shear and dynamic properties of biode-gradable polyesters. *Polymer Engineering and Science* (1998), 38(9), 1426–1435.

21. Dorgan, John R; Lehermeier, Hans; Mang, Michael. Thermal and rheological proper-ties of commercial-grade poly(lactic acid)s. *Journal of Polymers and the Environment* (2000), 8(1), 1–9.

22. Arraiza, A.L.; Sarasua, J.R.; Verdu, J.; Colin, X. Rheological behavior and model-ing of thermal degradation of poly(ε-caprolactone) and poly(L-lactide). *International Polymer Processing* (2007), 22(5), 389–394.

23. Wang, Kuo-Hsiung; Wu, Tzong-Ming; Shih, Yeng-Fong; Huang, Chien-Ming. Water bamboo husk reinforced poly(lactic acid) green composites. *Polymer Engineering and Science* (2008), 48(9), 1833–1839.

24. Pluta, Miroslaw. Melt compounding of polylactide/organoclay: structure and proper-ties of nanocomposites. *Journal of Polymer Science, Part B: Polymer Physics* (2006), 44(23), 3392–3405.

25. Lim, L.-T.; Auras, R.; Rubino, M. Processing technologies for poly(lactic acid). *Progress in Polymer Science* (2008), 33(8), 820–852.

26. Witzke, David Roy. Introduction to properties, engineering, and prospects of polylac-tide polymers. PhD Thesis, Michigan State University (1997).

27. Cui, L.; Imre, B.; Tatraaljai, D.; Pukanszky, B. Physical ageing of poly(lactic acid): Factors and consequences for practice. *Polymer* (2020), 186, 122014.

28. Hutchinson, J.M. (1995) Physical ageing of polymers. *Progress in Polymer Science* 20, 703–760.
29. Toft, Michael. The effect of crystalline morphology on the glass transition and enthalpic relaxation in poly (ether-ether-ketone). Thesis, University of Birmingham (2012).
30. Montserrat, S. Physical aging studies in epoxy resins. I. Kinetics of the enthalpy relaxation process in a fully cured epoxy resin. *Journal of Polymer Science: Part B: Polymer Physics* 32 (1994), pp. 509–522.
31. Chung, Hyun-Jung; Woo, Kyung-Soo; Lim, Seung-Taik. Glass transition and enthalpy relaxation of cross-linked corn starches. *Carbohydrate Polymers* (2004), 55(1), 9–15.
32 Quan, Daping; Liao, Kairong; Zhao, Jianhao. Effects of physical aging on glass transition behavior of poly(lactic acid)s. *Gaofenzi Xuebao* (2004), 5, 726–730.
33. McKenna, G.B. Physical aging in glasses and composites. In: Pochiraju, K.V., Tandon, G., Schoeppner, G.A. (eds.) *Long-Term Durability of Polymeric Matrix Composites*, 1st Edition, pp. 237–309, Springer, Berlin (2011).
34. Hajibeygi, Mohsen; Shafiei-Navid, Saeid. Design and preparation of poly(lactic acid) hydroxyapatite nanocomposites reinforced with phosphorus-based organic additive: Thermal, combustion, and mechanical properties studies. *Polymers for Advanced Technologies* (2019), 30(9), 2233–2249.
35. Hon, David N.S.; Chang, Shang Tzen. Surface degradation of wood by ultraviolet light. *Journal of Polymer Science, Polymer Chemistry Edition* (1984), 22(9), 2227–2241.
36. Evans, P.D.; Thay, P.D.; Schmalzl, K.J. Degradation of wood surfaces during natural weathering. Effects on lignin and cellulose and on the adhesion of acrylic latex primers. *Wood Science and Technology* (1996), 30(6), 411–422.
37. Cogulet, Antoine; Blanchet, Pierre; Landry, Veronic. Wood degradation under UV irradiation: A lignin characterization. *Journal of Photochemistry and Photobiology, B* (2016), 158, 184–191.
38. Colom, X.; Carrillo, F.; Nogues, F.; Garriga, P. Structural analysis of photodegraded wood by means of FTIR spectroscopy. *Polymer Degradation and Stability* (2003), 80(3), 543–549.
39. Pandey, K.K.; Pitman, A.J. FTIR studies of the changes in wood chemistry following decay by brown-rot and white-rot fungi. *International Biodeterioration & Biodegradation* (2003), 52(3), 151–160.
40. Zandvliet, C; Bandyopadhyay, N.R; Ray, D. Water absorption of jute/polylactic acid composite intended for an interior application and comparison with wood-based panels. *Journal of the Institution of Engineers (India): Series D* (2014), 95(1), 49–55.

12 Biocompatibility, Biodegradability, and Environmental Safety of PLA/Cellulose Composites

M. N. F. Norrrahim and N. M. Nurazzi
Universiti Pertahanan Nasional Malaysia (UPNM)
Kuala Lumpur, Malaysia

S. S. Shazleen and S. U. F. S. Najmuddin
Universiti Putra Malaysia (UPM)
Shah Alam, Malaysia

T. A. T. Yasim-Anuar
Nextgreen Pulp & Paper Sdn. Bhd.
Pahang, Malaysia

J. Naveen
Vellore institute of Technology
Vellore, India

R. A. Ilyas
Universiti Teknologi Malaysia
Skudai, Malaysia

CONTENTS

DOI: 10.1201/9781003160458-12

12.1   INTRODUCTION

PLA is a type of compostable biopolymer thermoplastic originated from natural fibers such as sugarcane, corn, potato starch, and beet sugar. Moreover, it offers an advantage over the non-renewable petroleum-based materials [1]. PLA is known to have good mechanical properties and easy processability compared to other biopolymers. PLA performs much like traditional plastics (i.e., its tensile strength and elastic modulus are comparable with those of polyethylene terephthalate [PET] and polystyrene [PS]) but with the crucial benefit of being 100% compostable in commercial compost facilities hence, diverting waste that would otherwise be landfilled [2]. However, PLA also has some drawbacks, especially on brittleness, which limits its applications [3, 4]. Previous studies have taken multiple steps of improvements in increasing the properties of this PLA to suit their needs in several other applications. One of them is by reinforcing other polymers or fibers to the PLA to produce a composite.

Cellulose, including nanocellulose from renewable biomass, has attracted much interest as a reinforcement in composite materials. Similar to PLA, cellulose materials are renewable in nature (i.e., could be extracted from plant cell walls such as oil palm, coconut fibers, cotton fibers, and bamboo), biodegradable, and eco-friendly [5, 6]. Cellulose and nanocellulose are not just sought out for their low cost, desirable fiber aspect ratio, biodegradability, and flexibility during processing but also for their good mechanical properties, low density as well as high specific strength and stiffness, which are useful as a reinforcing agent in composites [7, 8]. Meanwhile, nanocellulose is a referring to nano-structured cellulose. By referring to Figure 12.1, there are three types of nanocellulose which are cellulose nanocrystal (CNC), cellulose nanofibers (CNF), or bacterial nanocellulose (BNC) [9–12]. Nanocellulose can be synthesized from cellulose by using several pretreatments, which are chemical, enzymatical, biological, physical, or combination thereof [13–17]. Nanacelluose is known to have a larger specific surface area and is more versatile than cellulose.

Blending other biodegradable polymers like cellulose as functional filler to PLA improves its properties such as its brittleness, poor crystallization rate, and low thermal stability. Therefore, this could expand their application in the industry [18]. Besides that, the amounts of PLA used can be reduced [19]. The PLA/cellulose composites have received much interest in recent years and have already been widely utilized in food packaging, paper coating, tableware, films, and fiber preparations due to environmental concerns [20]. A number of studies have shown that the reinforcement of cellulose in PLA successfully improved the performance of composite. For example, Ariffin et al. [39] discovered that PLA reinforced with CNF (3 and 5 wt.%) had better tensile strength and Young's modulus properties as compared to the neat PLA. Moreover, the crystallinity of the polymer was also increased by 43% with the addition of 5 wt.% CNF. Meanwhile, in terms of the thermal stability behavior, Hossain et al. [21], reported that the PLA reinforced with CNC was found to be more stable in the temperature region between 20°C and 210°C as compared to the neat PLA.

Biomass residues   Microbes

Cellulose fibers

Synthesis method   Synthesis methods
a. Chemical   a. Mechanical
                       b. Physical
                       c. Combination thereof

Cellulose nanocrystals   Cellulose nanofiber   Bacterial cellulose

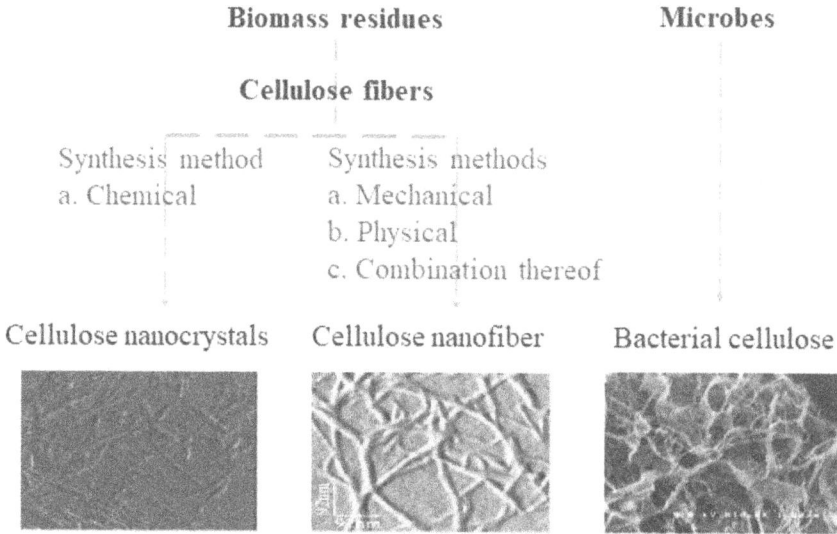

FIGURE 12.1 The preparation and classification of nanocellulose. (Reproduced with permission from ref. [10], with permission from The Royal Society of Chemistry.)

Instead of the improvement on the mechanical, crystallinity, and thermal properties of PLA/cellulose composites, factors such as biocompatibility, biodegradability, and environmental safety concern are important to be tackled. These three factors are scarcely being discussed in most literature. In the next section, a detailed discussion on these three factors of PLA/cellulose composite is presented. This will help to increase the applicability of PLA/cellulose composites into several related fields and overcome their environmental concerns that may arise in the future.

12.2 BIOCOMPATIBILITY

One of the most interesting features of PLA/cellulose composites is their biocompatibility. This unique characteristic of biomaterials has become an essential part of daily human life. However, the use of these biomaterials should carefully be monitored from time to time as interactions between materials and biological environments could result in a variety of local and systemic responses, which can be classified as curative, neutral, or toxic depending on the situation [22]. According to Cohn et al. [23], the ability of an implant material to work *in vivo* without eliciting harmful local or systemic responses in the body is known as biocompatibility. Biomaterials are subjected to tissue and animal testing before being used in the treatment of human fractures fixation. Biocompatibility is highly dependent on the type of application, with the following basic factors influencing biocompatibility properties [24]:

a. Interaction surroundings such as cytotoxins, toxicological reactions, carcinogenic/mutagenic reactions, inflammatory processes, biodegradation rate, and presence of human blood.

b. Implant operation timeframe.

c. Surface biocompatibility of implant with the host tissue. This can be evaluated based on their chemical, biological, and morphological.

d. Structural biocompatibility where the optimal ability of the implant to adapt its mechanical properties of the host tissue.

e. The optimum friction coefficient and mechanical properties required by the application.

f. Proportion (size and shape).

g. Material selection is important because the synthetic materials are usually against the host tissue and vice versa.

Biodegradable polymers are often used in the biomedical industry due to their excellent degradation behaviors in the human body; they possess mechanical properties similar to those of polyolefins, are environmentally friendly, and have few to no carcinogenic effects have been reported. Among all, PLA is the most commonly used biomedical material as it decomposes primarily into water and carbon dioxide, which can be completely removed from the human body [25]. Liu et al. [26] mentioned that further functional application of PLA remains a challenge due to its low cell adhesion and low degradation rate, which can be related to its hydrophobicity, biological inertness, and inflammation *in vivo* due to acidic degradation products. These flaws limit the applicability of PLA in bone-regeneration treatments where precise interactions between cells and implants are needed. Furthermore, the hydrophobicity of PLA inhibits water and living cells from penetrating the material, causing complications and necrosis. Bacterial adhesion and biofilm formation are also facilitated by hydrophobic PLA surfaces.

Due to its abundant availability, promising biocompatible nature, and biodegradability that can meet the sterilization standards of biomaterials, cellulose has been used as a reinforcement material to overcome the shortcoming of PLA [27]. Table 12.1 summarizes the recent applications of PLA reinforced cellulose composites in the biomedical industry.

12.3 BIODEGRADABILITY

Biodegradability properties have also influenced the use of PLA/cellulose composites commercially. Biodegradability refers to the ability of a polymer to break down naturally, without causing any environmental issues [32]. PLA can be degraded on its own, and several factors can influence its degradation rate and duration. This includes its material properties such as molecular weight, purity, crystallinity, and thermal properties, and environmental factors such as temperature, humidity, pH, and the presence of microorganisms or enzymes [33].

Theoretically, polymers can be degraded in several ways and for PLA, the degradation occurs through ester bonds [34]. Chemically, the ester bonds of PLA will be fragmented into carboxylic acid and alcohol by chemical hydrolysis, and this requires plenty of time and energy. Hence, PLA needs to be treated under certain conditions to expedite the process [35]. Meanwhile, for microbial degradation, an extracellular depolymerase will be extracted by the PLA-degrading microorganisms

TABLE 12.1

Recent Applications of PLA Reinforced Cellulose Composites in the Biomedical Industry

Cellulose Source	Types of Cellulose	Additives/ Modification	Intended Application	Findings	References
Not stated	Cellulose acetate	Deacetylation of CA nanofibers in an alkali solution to obtain regenerated cellulose (RC) nanofibers	Scaffold	• The developed composite had a high water absorption rate, better hierarchical cellular structure, and fast recovery from 80% strain. • The acid degradation produced by PLA was stabilized due to the formation of bonelike apatite. The scaffold-to-bone bonding was also improved during the implantation process.	[27]
Cotton	CNC	PLA grafted with MA (MPLA)	Scaffold	• It was found to be non-toxic to human adult adipose-derived mesenchymal stem cells (hASCs). It is also capable of promoting cell proliferation.	[28]
Cotton	CNC	-	Scaffold	• The composite showed an improved hemocompatibility and protein adsorption capacity than the neat PLA. • The presence of CNC improved the *in vitro* degradation. It also facilitated the deposition of Ca^{2+}, CO_3^{2-}, PO_4^{3-} ions in simulated body fluid. • The composite is capable of improving cell attachment and enhanced cell proliferation than PLA scaffolds. It also has low cytotoxicity and better cytocompatibility.	[29]
Not stated	Cellulose acetate	Incorporated with thymoquinone (TQ)	Wound dressing mat	• The fabricated wound dressings simulated the interactive extracellular matrix (ECM) via the 3D nanofibrous structure. It can promote the cell proliferation due to the hydrophilicity and bioactivity of CA. • Based on the *in vivo* study revealed that TQ-loaded PLA: CA (7:3) scaffolds significantly aided the wound healing process by increasing re-epithelialization and regulating granulation tissue formation.	[30]
Pine	CNC	CNC grafted with poly(ethylene glycol) (PEG), (CNC-g-PEG)	Scaffold	• The developed scaffold has more live cells than neat PLA. This can be attributed to the decreased diameter of nanofibers. Thus, it facilitated cell-matrix interactions by providing more binding sites for cell adhesion. • After 14 days of incubation, the cells spread out well on the scaffolds, especially on the PLA/CNC-g-PEG (5%) composite. This indicates that the cells interacted and integrated well with the developed scaffolds.	[31]

to induce the degradation process. This process needs to be stimulated by inducers such as peptides, amino acids, elastin, and gelatin [33]. Following this, the depolymerase will attack the intramolecular ester links of PLA, resulting in the production of oligomers, dimers, and monomers. This then will allow the low-molecular-weight compounds to be decomposed into water, carbon dioxide, and methane [33, 36]. Nevertheless, as stated earlier, favorable environmental conditions are also required for a successful degradation of PLA.

Looking at the current trend, PLA has been incorporated with various kinds of fillers, including cellulose to enhance its properties [37–39], and this indirectly influences its biodegradability properties. Nevertheless, the degradation mechanism and the association, as well as the role of microorganisms of PLA/cellulose composites, are still poorly understood due to limited reports on the degradation of PLA/cellulose composites. In terms of degradation, previous studies mainly focused on the neat PLA, and further studies have to be conducted to determine the favorable degradation conditions based on the cellulose incorporated into the PLA.

Manzano et al. [40] investigated the effect of nanocellulose on the biodegradability of PLA composites. Based on the results obtained as shown in Figure 12.2, the presence of nanocellulose enhances the PLA biodegradation over the time. They suggested that the addition of the nanocellulose decreases the PLA hydrophobicity, thus enhancing their biodegradability.

Table 12.2 summarizes the degradation studies of several PLA-based composites including cellulose. There are several previous studies had revealed that the cellulose-reinforced PLA composites can accelerate the degradation process [41–43]. Nevertheless, the mechanical, crystallinity and thermal properties of those PLA/cellulose composites should not be neglected, as those properties should remain the top priority.

12.3.1 Biodegradation Routes of PLA Composites for Packaging

As mentioned above, one of the major applications of PLA/cellulose composites are in the packaging industry. The packaging industry is accounted for the highest volumes of plastic wastes as an accumulation of 54% (141 million tons) of the plastic

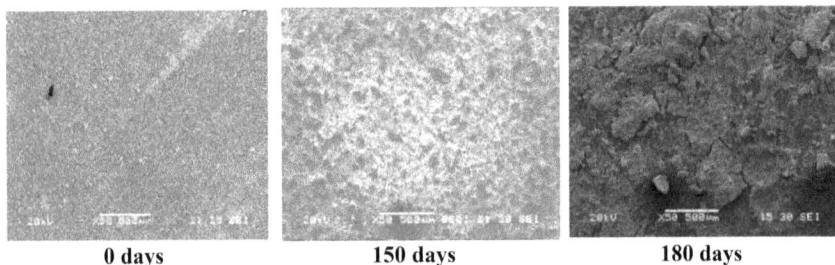

0 days   150 days   180 days

FIGURE 12.2   Micrographs at different composting days (0, 150, 180). (Reproduced from ref. [40], Polymers, MDPI, 2021.)

TABLE 12.2

Summary of Degradation Studies of PLA-Based Composites

Composites	Findings	References
Coir/pineapple leaf fibers/PLA composites	• Soil burial and accelerated weathering tests were conducted to determine the biodegradation properties of coir/pineapple leaf fibers/PLA composites (CF/PALF/PLA). The ratio of CF/PALF/PLA is 9/15/21/30:9/15/21/30:70/100 (w/w). • The results revealed that the degradation rate of neat PLA was slower than the CF/PALF/PLA composites after 250 h of accelerated weathering. In contrast to the neat PLA, the formation of cracks and fractures, changes in the surface roughness and colors can be spotted on the CF/PALF/PLA composites. The CF/PALF/PLA composites also showed weight loss and gradual degradation with burial time. This study shows that CF/PALF/PLA composites degraded after weathering by thermal degradation, hydrolysis. photo-radiation, and oxidation.	[41]
PLA/linseed cake	• This study concluded that the PLA composites filled with 10% (w/w) linseed cake undergo notable changes in their structure and properties which influenced by the UV, humidity, temperature, and humidity. • Unlike the neat PLA, microcracks were observed on the surface of PLA/linseed cake composite, especially after 500 h of weathering duration. It was observed the linseed cake which is not fully miscible with the PLA due to the excessive amounts of linseed oil which leached out of the composite from these cracks. It is believed that this led to the brittleness of the composites, and indirectly cause hydrolytic degradation on the amorphous parts of the PLA. The presence of water, which penetrates the whole PLA/linseed cake composites through cracks and voids also accelerates the entire degradation process.	[42]
PLA/cellulose fiber of *Betula pendula* and *pubescens*	• This study evaluated the weathering properties of PLA reinforced with 10% cellulose fibers from *Betula pendula* and *pubescens*. • Findings revealed that after 600 h of accelerated weathering, the PLA/cellulose composites had obvious voids as a result of the debonding and pull-out of lignocellulosic fibers, as well as from the agglomeration process. • It was also discovered that the crystallinity of PLA/cellulose composites increased. This is due to the combined action of UV irradiation, temperature, and humidity. Therefore, the materials become more fragile, thus, accelerate the degradation process of PLA/cellulose composites.	[43]

wastes were reported in 2015. This worrying large number of plastic wastes generated was mainly due to the negative habits of food preparation and also because of the emerging development of new living areas and markets in the world [1, 44]. Geyer et al. [45] reported that 60% of plastic wastes are sent to landfills, 12% are incinerated, and only 9% are recycled. The disposal of packaging material/container made of fossil-fuel plastics (i.e., as thrash) is difficult to degrade (i.e., normally would take many years). The degraded plastics would turn into microplastics, which can easily enter the food chain leading to the unwanted bio-accumulation event [19]. Incineration of plastic causes air pollution and the toxic fumes that are released from burning plastic pose a threat to human and animal health [46]. Therefore, promoting the application of biodegradable materials represented by PLA/cellulose composites as a suitable alternative to commodity plastics would be an important measure to combat the plastic waste pollution as well as ensure environmental safety and achieve sustainable green materials development.

Biodegradable composites for packaging were developed for single-purpose use only and are bound to follow its 'end of life' route options such as biodegradation in soil, through composting facilities, and aquatic environments [19]. Karamanlioglu et al. [47] stated that the studies on PLA composites degradation coincidentally increase with the growing commercialization of PLA for short-life products.

12.3.1.1 Landfilling/Soil

This conventional waste management approach utilized the presence of a vast number of microorganisms in soils to degrade the disposed materials into natural compounds such as water, carbon dioxide, methane, and several monomers [48]. It is important to note that virgin PLA takes relatively longer to degrade in soil compared to its bio-composites. Meanwhile, Rajesh et al. [49] in their study showed the rate of degradation of PLA and its composites can be regulated by alkali treatment and the amount of fiber loading. In another study, Wilfred et al. [50] revealed that the presence of bio-based fillers improves the biodegradation of PLA when both PLA and starch were buried in soil for 14 and 28 days. The enzymatic hydrolytic and oxidative action (caused by the microorganisms) can disrupt the glycosidic bonds in starch and cellulose. In return, the microorganisms will absorb the resulting metabolites for their energy requirements [48].

12.3.1.2 Compost

Composting involves biodegradation under controlled conditions and can be done at home or in an industrial setup. The latter has a higher rate of biodegradation compared to the former as it involves higher temperatures [51]. Generally, the packaging waste is mixed with mature compost and exposed to optimum oxygen and moisture conditions at a temperature of 58°C for the composting process. Castro-Aguirre et al. [52] reported that bioaugmentation with *Geobacillus* microbes is applied to accelerate the biodegradation rate in compost environments. Composting generates carbon dioxide and water that are released into the atmosphere as well as carbon- and nutrient-rich compost for soil amendment in a variety of agricultural, horticultural, or landscaping applications [53]. Of note, composting resolves the issues of surface water pollution, groundwater pollution, odor, and the increase of harmful insects and animals arising from landfills in the sanitation of urban areas, thus moving toward making a green city [53]. Other than that, the production of compost is beneficial as it could reduce the need for water, fertilizers, and pesticides. Additionally, composting would create valuable humus that improves soil nutrients and quality for growing plants creating a circular economy with positive society-wide benefits [54]. Meanwhile, Rudeekit et al. [55] reported that PLA and PLA/starch composites could not cause harm to the environment. This is because the rate of germination plant growth of monocotyledon and dicotyledon on the resulting compost had no significant difference when compared to blank compost. However, composting sites in Malaysia are still scarce. The comprehensible activity of composting of organic waste accounts for 1% of solid waste from 2000 to 2006 and is expected to increase to 8% by 2020 [56]. By contrast, there are now more than 300 composting sites in the United Kingdom that collectively compost about 2 million tons of waste annually [57].

12.3.1.3 Aquatic Environment

To date, there are not many studies have been done on PLA microbial attack (i.e., biodegradation) in marine environment, and the degradation that occurred as a result of hydrolysis rather than from microbes [19]. Anderson and Shenkar [58] reported that cups and plates made from PLA may have similar effects as fossil-based plastic such as PET in the aquatic environment. Therefore, attention to future research must be given more into this issue.

12.4 ENVIRONMENTAL SAFETY

Ecotoxicity study of nanocellulose and PLA is one of the important criteria to evaluate the effect of these materials on the environment. Several studies have been carried out in the past regarding the fate of nanocellulose material after production and its possible effects on the environment. Vartiainen et al. [59] had done the ecotoxicity tests using aquatic breeds such as *Vibrio fishceri* and *Daphna magna* for evaluation of CNF and CNC. It was revealed the acute toxicity of CNF in *V. fishceri* was only at a very high level (300 mg/L). It also has an impact on flea movement in the *D. magna*, particularly due to mechanical obstacles caused by the cellulose.

In another study, a similar kind of CNF was used for *in vivo* studies on the nematode (*Caernohabditis elegans*), whereby the addition of CNF in soil did not interfere with the reproduction, cause mortality, or affect its movement [60]. Meanwhile, Kovacs et al. [61] reported that CNC produced from bleached kraft pulp with sulfuric acid did not cause serious environmental concerns when toxicity tests were performed on rainbow trout hepatocyte cells and nine aquatic species. CNC affected the reproduction of feathered minnow at 0.29 g/L (i.e., this value is still higher than the concentration would be in receiving water in a worst-case scenario), but no other effects were noticed on endpoints such as survival and growth in the other species at a concentration below 1 g/L. Other than that, eight different CNC and three CNF did not hold any significant difference from control on mortality of the zebrafish at an exposure concentration of 0.2 and 200 mg/L regardless of their aspect ratio, surface chemistry, or charge as revealed by Harper et al. [62] study.

Furthermore, Palsikowski et al. [63] had reported that the PLA is considered not phytotoxic, cytotoxic, genotoxic, or mutagenic for meristematic cells of *Allium cepa*. Moreover, Kong et al. [64] demonstrated that exposure of PLA-graphene with a range of concentrations from 50 µg/mL to 1000 µg/mL to alive adult worms, *Caenorhabdits elegans* did not affect the feeding rate, reproductive rate, and overall lifespan of the nematode. Thus, it indicates that the PLA is safe and not toxic to worms and the environment in general.

12.5   CONCLUSION

The progress of PLA/cellulose composite as a substitution for petroleum-based synthetics has been an area of attention due to its excellent performance for several applications, as listed above. The discussion on biocompatibility, biodegradability, and environmental safety as discussed here will benefit future development in this area. The progress on biocompatibility, biodegradability, and environmental safety of PLA/cellulose composites were summarized as follow:

1. Biocompatibility of PLA/cellulose composites revealed its potential to be used in biomedical industries, especially in the development of scaffolds and wound dressing mats. The presence of cellulose in PLA has overcome the shortcoming of PLA and enhanced the biocompatibility properties.
2. Biodegradability of PLA/cellulose composites is scarcely being reported. However, this composite has an excellent biodegradability property which will benefit several applications especially packaging.
3. Environmental safety of PLA/cellulose composites is one of the important criteria. There is no serious issue related to the ecotoxicity of these materials toward the environment. Therefore, PLA/cellulose composites are considered safe for the environment.

ACKNOWLEDGMENT

We would like to express our gratitude to Universiti Pertahanan Nasional Malaysia (UPNM) for the opportunity and financial support.

REFERENCES

1. L. G. Hong, N. Y. Yuhana, and E. Z. E. Zawawi, "Review of bioplastics as food packaging materials," *AIMS Materials Science*, vol. 8, no. 2, pp. 166–184, 2021.
2. T. C. Mokhena, J. S. Sefadi, E. R. Sadiku, M. J. John, M. J. Mochane, and A. Mtibe, "Thermoplastic processing of PLA/cellulose nanomaterials composites," *Polymers*, vol. 10, no. 12, p. 1363, 2018.
3. R. Ilyas et al., "Properties and characterization of PLA, PHA, and other types of biopolymer composites," in *Advanced Processing, Properties, and Applications of Starch and Other Bio-Based Polymers*, Elsevier, 2020, pp. 111–138.
4. R. A. Ilyas *et al.*, "Polylactic acid (PLA) biocomposite: processing, additive manufacturing and advanced applications," *Polymers*, vol. 13, no. 8, p. 1326, Apr. 2021.

5. N. Lavoine and L. Bergstrom, "Nanocellulose-based foams and aerogels: processing, properties, and applications," *Journal of Materials Chemistry A*, vol. 5, pp. 16105–16117, 2017.

6. M. N. Norizan *et al.*, "Treatments of natural fibre as reinforcement in polymer composites-short review," *Functional Composites and Structures*, vol. 3, p. 024002, 2021.

7. T. A. T. Yasim-Anuar *et al.*, "Well-dispersed cellulose nanofiber in low density polyethylene nanocomposite by liquid-assisted extrusion," *Polymers*, vol. 12, pp. 1–17, 2020.

8. T. A. T. Yasim-Anuar, H. Ariffin, M. N. F. Norrrahim, M. A. Hassan, and T. Tsukegi, "Sustainable one-pot process for the production of cellulose nanofiber and polyethylene/cellulose nanofiber composites," *Journal of Cleaner Production*, vol. 207, pp. 590–599, 2019.

9. M. N. F. Norrrahim *et al.*, "Nanocellulose: a bioadsorbent for chemical contaminant remediation," *RSC Advances*, vol. 11, pp. 7347–7368, 2021.

10. M. N. F. Norrrahim *et al.*, "Nanocellulose: the next super versatile material for the military," *Materials Advances*, vol. 2, pp. 1485–1506, 2021.

11. M. N. F. Norrrahim *et al.*, "Performance evaluation of cellulose nanofiber reinforced polymer composites," *Functional Composites and Structures*, vol. 3, p. 024001, 2021.

12. M. N. F. Norrrahim *et al.*, "Emerging development on nanocellulose as antimicrobial material: an overview," *Materials Advances*, vol. 2, 3538–3551, 2021.

13. M.N.F. Norrrahim, Superheated steam pretreatment of oil palm biomass for improving nanofibrillation of cellulose and performance of polypropylene/cellulose nanofiber composites. Doctoral Thesis, Universiti Putra Malaysia, Selangor, Malaysia, 2018.

14. M. N. F. Norrrahim *et al.*, "Superheated steam pretreatment of cellulose affects its electrospinnability for microfibrillated cellulose production," *Cellulose*, vol. 25, no. 7, pp. 3853–3859, 2018.

15. M. N. F. Norrrahim *et al.*, "Performance evaluation of cellulose nanofiber with residual hemicellulose as a nanofiller in polypropylene-based nanocomposite," *Polymers*, vol. 13, no. 7, p. 1064, 2021.

16. H. Ariffin, T. A. T. Yasim-Anuar, M. N. F. Norrrahim, and M. A. Hassan, "Synthesis of cellulose nanofiber from oil palm biomass by high pressure homogenization and wet disk milling," in *Nanocellulose: Synthesis, Structure, Properties and Applications*, World Scientific, 2021, pp. 51–64.

17. T. A. T. Yasim-anuar, H. Ariffin, M. N. F. Norrrahim, and M. A. Hassan, "Factors affecting spinnability of oil palm mesocarp fiber cellulose solution for the production of microfiber," *BioResources*, vol. 12, no. 1, pp. 715–734, 2017.

18. T. V. Shah and D. V. Vasava, "A glimpse of biodegradable polymers and their biomedical applications," *e-Polymers*, vol. 19, no. 1, pp. 385–410, 2019.

19. E. N. Ogunmuyiwa, R. Zulkifli, I. N. Beas, L. K., Ncube, and A. U., Ude, "Environmental impact of food packaging materials: A review of contemporary development from conventional plastics to polylactic acid based materials," *Materials*, vol. 13, no. 21, p. 4994, 2020.

20. B. E. Arteaga-Ballesteros, A. Guevara-Morales, E. San Martín-Martínez, U. Figueroa-López, and H. Vieyra, "Composite of polylactic acid and microcellulose from kombucha membranes," *e-Polymers*, vol. 21, no. 1, pp. 15–26, 2020.

21. K. M. Z. Hossain *et al.*, "Physico-chemical and mechanical properties of nanocomposites prepared using cellulose nanowhiskers and poly (lactic acid)," *Journal of Materials Science*, vol. 47, no. 6, pp. 2675–2686, 2012.

22. T. Nejatian *et al.*, "Dental biocomposites," in *Biomaterials for Oral and Dental Tissue Engineering*, L. Tayebi and K. Moharamzadeh, Eds., Woodhead Publishing, 2017, pp. 65–84.

23. M. R. Cohn, A. Unnanuntana, T. J. Pannu, S. J. Warner, and J. M. Lane, "Materials in fracture fixation," *Comprehensive Biomaterials II*, vol. 7, pp. 278–297, 2017.

24. L. Cvrček and M. Horáková, "Plasma modified polymeric materials for implant applications," in *Non-Thermal Plasma Technology for Polymeric Materials: Applications in Composites, Nanostructured Materials and Biomedical Fields*, S. Thomas, M. Mozetič, U. Cvelbar, P. Špatenka, and P. K. M., Eds., Elsevier, 2019, pp. 367–407.

25. L. Sha, Z. Chen, Z. Chen, A. Zhang, and Z. Yang, "Polylactic acid based nanocomposites: promising safe and biodegradable materials in biomedical field," *International Journal of Polymer Science*, vol. 2016, pp. 1–11, 2016.

26. S. Liu, S. Qin, M. He, D. Zhou, Q. Qin, and H. Wang, "Current applications of poly(lactic acid) composites in tissue engineering and drug delivery," *Composites Part B*, vol. 199, p. 108238, 2020.

27. J. Chen, T. Zhang, W. Hua, P. Li, and X. Wang, "3D porous poly(lactic acid)/regenerated cellulose composite scaffolds based on electrospun nanofibers for biomineralization," *Colloids and Surfaces A: Physicochemical and Engineering Aspects*, vol. 585, p. 124048, 2019.

28. C. Zhou et al., "Electrospun bio-nanocomposite scaffolds for bone tissue engineering by cellulose nanocrystals reinforcing maleic anhydride grafted PLA," *ACS Applied Materials & Interfaces*, vol. 5, no. 9, pp. 3847–3854, 2013.

29. W. Luo et al., "Preparation, characterization and evaluation of cellulose nanocrystal/poly(lactic acid) in situ nanocomposite scaffolds for tissue engineering," *International Journal of Biological Macromolecules*, vol. 134, pp. 469–479, 2019.

30. S. F. Gomaa, T. M. Madkour, S. Moghannem, and I. M. El-Sherbiny, "New polylactic acid/cellulose acetate-based antimicrobial interactive single dose nanofibrous wound dressing mats," *International Journal of Biological Macromolecules*, vol. 105, pp. 1148–1160, 2017.

31. C. Zhang, M. R. Salick, T. M. Cordie, T. Ellingham, Y. Dan, and L.-S. Turng, "Incorporation of poly(ethylene glycol) grafted cellulose nanocrystals in poly(lactic acid) electrospun nanocomposite fibers as potential scaffolds for bone tissue engineering," *Materials Science and Engineering C*, vol. 49, pp. 463–471, 2015.

32. L. S. Dilkes-Hoffman, P. A. Lant, B. Laycock, and S. Pratt, "The rate of biodegradation of PHA bioplastics in the marine environment: A meta-study," *Marine Pollution Bulletin*, vol. 142, no. March, pp. 15–24, 2019.

33. M. Karamanlioglu, R. Preziosi, and G. D. Robson, "Abiotic and biotic environmental degradation of the bioplastic polymer poly(lactic acid): A review," *Polymer Degradation and Stability*, vol. 137, pp. 122–130, 2017.

34. X. Qi, Y. Ren, and X. Wang, "New advances in the biodegradation of poly(lactic) acid," *International Biodeterioration and Biodegradation*, vol. 117, pp. 215–223, 2017.

35. M. A. Elsawy, K. H. Kim, J. W. Park, and A. Deep, "Hydrolytic degradation of polylactic acid (PLA) and its composites," *Renewable and Sustainable Energy Reviews*, vol. 79, April, pp. 1346–1352, 2017.

36. S. M. Satti, A. A. Shah, R. Auras, and T. L. Marsh, "Isolation and characterization of bacteria capable of degrading poly(lactic acid) at ambient temperature," *Polymer Degradation and Stability*, vol. 144, pp. 392–400, 2017.

37. M. S. Rasidi, L. C. Cheah, and A. M. Nasib, "Mechanical properties and biodegradability of Polylactic Acid/Acrylonitrile Butadiene Styrene with cellulose particle isolated from Nypa fruticans husk," *International Journal of Automotive and Mechanical Engineering*, vol. 17, no. 4, pp. 8351–8359, 2020.

38. X. Wang and D. Wang, "Fire-retardant polylactic acid-based materials: preparation, properties, and mechanism," in *Novel Fire Retardant Polymers and Composite Materials*, Elsevier, 2017, pp. 93–116.

39. H. Ariffin et al., "Oil palm biomass cellulose-fabricated Polylactic acid composites for packaging applications," in *Bionanocomposites for Packaging Applications*, Springer, 2018, pp. 95–105.

40. L. M. G. Manzano, R. M. Á. Cruz, N. M. M. Tun, A. V. González, and J. H. M. Hernandez, "Effect of cellulose and cellulose nanocrystal contents on the biodegradation, under composting conditions, of hierarchical PLA biocomposites," *Polymers*, vol. 13, no. 11, p. 1855, 2021.

41. R. Siakeng, M. Jawaid, M. Asim, and S. Siengchin, "Accelerated weathering and soil burial effect on biodegradability, colour and texture of coir/pineapple leaf fibres/PLA biocomposites," *Polymers*, vol. 12, no. 2, p. 458, 2020.

42. O. Mysiukiewicz, M. Barczewski, K. Skórczewska, J. Szulc, and A. Kloziński, "Accelerated weathering of polylactide-based composites filled with linseed cake: The influence of time and oil content within the filler," *Polymers*, vol. 11, no. 9, pp. 1–21, 2019.

43. I. Spiridon, R. N. Darie, and H. Kangas, "Influence of fiber modifications on PLA/fiber composites. Behavior to accelerated weathering," *Composites Part B: Engineering*, vol. 92, pp. 19–27, 2016.

44. L. Piergiovanni and S. Limbo, "Plastic packaging materials," in *Food Packaging Materials*, Springer, 2016, pp. 33–49.

45. R. Geyer, J. R. Jambeck, and K. L. Law, "Production, use, and fate of all plastics ever made," *Science Advances*, vol. 3, no. 7, p. e1700782, 2017.

46. R. Verma, K. S. Vinoda, M. Papireddy, and A. N. S. Gowda, "Toxic pollutants from plastic waste-a review," *Procedia Environmental Sciences*, vol. 35, pp. 701–708, 2016.

47. M. Karamanlioglu, R. Preziosi, and G. D. Robson, "Abiotic and biotic environmental degradation of the bioplastic polymer poly (lactic acid): A review," *Polymer Degradation and Stability*, vol. 137, pp. 122–130, 2017.

48. S. M. Chisenga, G. N. Tolesa, and T. S. Workneh, "Biodegradable food packaging materials and prospects of the fourth industrial revolution for tomato fruit and product handling," *International Journal of Food Science*, vol. 2020, Article ID 887910, 2020.

49. G. Rajesh, A. R. Prasad, and A. V. S. S. K. S. Gupta, "Soil degradation characteristics of short sisal/PLA composites," *Materials Today: Proceedings*, vol. 18, pp. 1–7, 2019.

50. O. Wilfred et al., "Biodegradation of polylactic acid and starch composites in compost and soil," *International Journal of Nano Research*, vol. 1, no. 2, pp. 1–11, 2018.

51. S. M. Emadian, T. T. Onay, and B. Demirel, "Biodegradation of bioplastics in natural environments," *Waste Management*, vol. 59, pp. 526–536, 2017.

52. E. Castro-Aguirre, R. Auras, S. Selke, M. Rubino, and T. Marsh, "Enhancing the biodegradation rate of poly (lactic acid) films and PLA bio-nanocomposites in simulated composting through bioaugmentation," *Polymer Degradation and Stability*, vol. 154, pp. 46–54, 2018.

53. S. Saheri, M. A. Mir, N. E. A. Basri, R. A. Begum, and N. Z. B. Mahmood, "Solid waste management by considering composting potential in Malaysia toward a green country," *e-Bangi*, vol. 4, no. 1, pp. 48–55, 2009.

54. K. Palaniveloo *et al.*, "Food waste composting and microbial community structure profiling," *Processes*, vol. 8, no. 6, p. 723, 2020.

55. Y. Rudeekit, P. Siriyota, P. Intaraksa, P. Chaiwutthinan, M. Tajan, and T. Leejarkpai, "Compostability and ecotoxicity of poly (lactic acid) and starch blends," *Advanced Materials Research*, vol. 506, pp. 323–326, 2012.

56. A. Johari, H. Alkali, H. Hashim, S. I. Ahmed, and R. Mat, "Municipal solid waste management and potential revenue from recycling in Malaysia," *Modern Applied Science*, vol. 8, no. 4, p. 37, 2014.

57. J. H. Song, R. J. Murphy, R. Narayan, and G. B. H. Davies, "Biodegradable and compostable alternatives to conventional plastics," *Philosophical Transactions of the Royal Society B: Biological Sciences*, vol. 364, no. 1526, pp. 2127–2139, 2009.

58. G. Anderson and N. Senkar, "Potential effects of biodegradable single-use items in the sea: Polylactic acid (PLA) and solitary ascidians," *Environmental Pollution*, vol. 268, no. 115364, 2021.

59. J. Vartiainen *et al.*, "Health and environmental safety aspects of friction grinding and spray drying of microfibrillated cellulose," *Cellulose*, vol. 18, no. 3, pp. 775–786, 2011.

60. V. Väänänen *et al.*, "Evaluation of the suitability of the developed methodology for nanoparticle health and safety studies," in Proceedings of the Scale-up Nanoparticles in Modern Papermaking (SUNPAP 2012), 2012, p. 27.

61. T. Kovacs *et al.*, "An ecotoxicological characterization of nanocrystalline cellulose (NCC)," *Nanotoxicology*, vol. 4, no. 3, pp. 255–270, 2010.

62. B. J. Harper *et al.*, "Impacts of chemical modification on the toxicity of diverse nanocellulose materials to developing zebrafish," *Cellulose*, vol. 23, no. 3, pp. 1763–1775, 2016.

63. P. A. Palsikowski, M. M. Roberto, L. R. Sommaggio, P. M. Souza, A. R. Morales, and M. A. Marin-Morales, "Ecotoxicity evaluation of the biodegradable polymers PLA, PBAT and its blends using *Allium cepa* as test organism," *Journal of Polymers and the Environment*, vol. 26, no. 3, pp. 938–945, 2018.

64. C. Kong, A. I. Aziz, A. B. Kakarla, I. Kong, and W. Kong, "Toxicity evaluation of graphene and poly (lactic-acid) using a nematode model," *Solid State Phenomena*, vol. 290, pp. 101–106, 2019.

13 Recycling and Reuse Issues of PLA and PLA/Cellulose Composites

Kiana Rafiee, Guneet Kaur, and Satinder Kaur Brar
York University
Toronto, Canada

CONTENTS

13.1 INTRODUCTION

A thermoplastic polymer such as polylactic acid (PLA) with a building block of lactic acid is classified as a biodegradable and sustainable polymer (Luckachan and Pillai 2011; Sethi 2017). Moreover, it is called biobased since it is produced by renewable resources such as starch and potato (Agrawal and Bhalla 2003).

Furthermore, cellulose is abundant in the environment and naturally degradable polymers (Masmoudi et al. 2016). Due to the low crystallization, resistance, toughness (Huneault and Li 2007), and thermal degradation in virgin PLA (Taubner & Shishoo, 2001), PLA/cellulose composites with natural cellulosic fibers such as wood (Åkesson et al., 2016), sisal (Chaitanya, Singh, and Song 2019), and flax (Le Duigou et al. 2020) are used as reinforcement to produce biodegradable plastic (Surip and Jaafar 2018).

PLA and PLA/cellulose composites have high mechanical and performance properties besides environmental adaptation (Siracusa et al. 2008). Fast decomposition in PLAs, compared to other plastics such as polystyrene (PS), has increased its applications in bottles, packaging, containers, automotive, and sports (Tsuji 2011; Yun-xuan 2007). On the other hand, PLA can extrude in different shapes e.g., sheets and fibers (Lim, Cink, and Vanyo 2010), which makes it a suitable candidate to reduce solid

DOI: 10.1201/9781003160458-13

265

waste materials by increasing the end-of-life cycle (Ren 2011). Also, recent legislation on producing economically and environmentally friendly products with reusable and recyclable materials (Bismarck, Mishra, and Lampke 2005) have put PLA and PLA composites in the center of attention during the last decade (Van de Velde and Kiekens 2002).

There are a few ways to process the polymer wastes e.g., composting, landfilling, incineration, mechanical (Soroudi and Jakubowicz 2013), and chemical recycling (McKeown and Jones 2020). Composting the PLA in an industrial environment requires high humidity and temperatures of 60°C for about 30 days (Kale et al. 2007). This process is also slow in domestic environments and produces hazardous by-products such as carbon dioxide (Hong and Chen 2017a; Verma et al. 2016). Furthermore, incineration has some negative effects such as producing hazardous exhaust gas. To this end, it is quite important to determine and educate about other two recycling processes (mechanical and chemical) (McKeown and Jones 2020) of PLA and PLA/cellulose composites, which will save the initial resources and carbon content (Le Duigou, Davies, and Baley 2010). Moreover, it will reduce the costs and prevent the depletion of fossil fuels and negative effects on the environment (Fowler, Hughes, and Elias 2007).

In this chapter, different recycling methods, their advantages, and the issues of each method will be discussed.

13.2 RECYCLING METHODS

There are several ways to recycle PLA and PLA/cellulose composites, which include composting and burying in landfills, incineration, mechanical and chemical methods (Figure 13.1) (Hong and Chen 2017b).

13.2.1 COMPOST, LANDFILL, AND INCINERATION

Biocomposites reach the degradation phase at the end of their life cycle. PLA degradation requires specific microorganisms, temperature, moisture, and light. Degradation also depends on the particle size, since smaller particles have a higher surface area, making it easier for microorganisms to attack the biocomposites (Kale et al. 2007). To this end, the addition of natural fibers such as core, kenaf, and bast fibers helps the degradation process, resulting in a 0.3%–1.3% higher weight loss compared to pure PLA composites. A study in a specially designed composting facility determined the degradation of PLA, PLA biocomposites reinforced with soy (70:30) and wheat straw (70:30), and untreated soy straw and wheat straw. Results showed shorter degradation in the following order: soy stalk, wheat straw, PLA/soy straw, PLA/wheat straw, and PLA. Due to the presence of natural biomass, the degradation of the PLA improved, which could have potential in compostable composites with modified components (Pradhan et al. 2010).

Furthermore, landfilling is an extremely slow degradation process of polymer wastes. Methane can be produced during this process emitting hazardous and unsafe gas to the environment. These two methods create environmental pollution issues while there is no or minimum amount of recovery in valued materials (Gross and

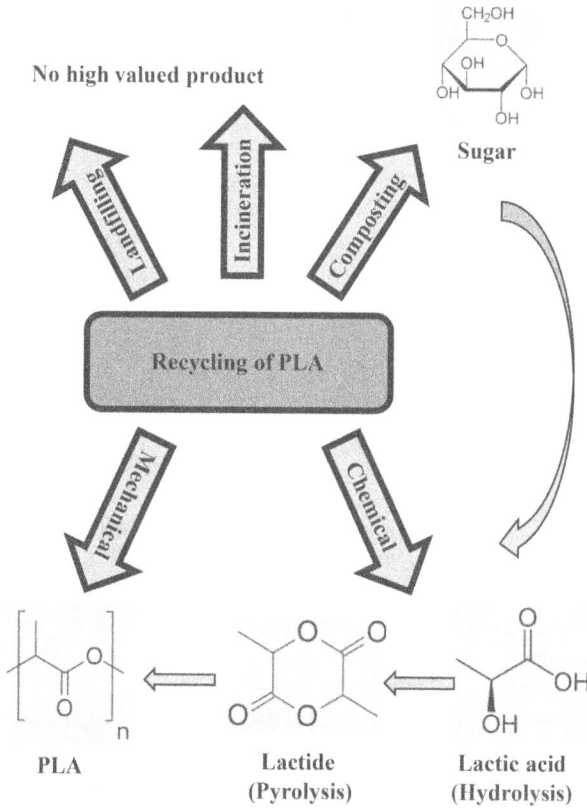

FIGURE 13.1 Different recycling methods of PLA.

Kalra 2002). Moreover, disposing of a large amount of consumed PLA will lessen the material value and increase the amount of plastic wastes in the environment (Dechy-Cabaret, Martin-Vaca, and Bourissou 2004).

Freed (1998) studied the methane production in landfilling of cellulose and hemi-cellulose. The lignin of the cellulose was not compostable in anaerobic conditions. PLA has the same property as lignin and will not degrade in an environment with low humidity and temperature. Same as lignin, PLA produces methane in an anaerobic condition. Studies show that PLA will degrade in water at 25°C after 11 months (Lunt 1998).

Due to the environmental issues, slow degradation, and releasing of toxic gases in compost, landfill, and incineration, mechanical recycling of PLA and PLA/cellulose composites could be an alternate option, whereby they can be transformed into other high-value products.

13.2.2 MECHANICAL RECYCLING (PRIMARY AND SECONDARY RECYCLING)

Since thermoplastic polymers such as PLA can undergo the melting and molding process, they have more recycling value than thermoset polymers. Nevertheless, there is a limit in the number of recycling cycles to keep their performance properties (Bhadra, Al-Thani, and Abdulkareem 2017).

The first step to recycle materials is to identify and collect, transfer to the recycling site, and sort them. However, there are several issues in recycling PLA. Since the PLA can decompose, its presence can contaminate other plastics such as polyethylene terephthalate (PET). Therefore it is important to separate the PLA and recycle it for further usage (Cornell 2007). Also, from an economical point of view, it is important to put recycling materials in the same category to make the recycling process faster and easier. Separating bio-based waste such as PLA bottles from organic wastes which require composting is not a problem (Alaerts, Augustinus, and Van Acker 2018). Nevertheless, there are some issues with separating PLA from other plastics such as PET and polypropylene (PP). Since PLA and PET are both clear and have a similar appearance, they can be visually separated by labeling. To solve this issue, near-infrared spectroscopy (NIR) technology can be used to separate the PET from PLA bottles. Furthermore, due to higher density in PLA compared to PP, they can be separated by sink/float tank separation (Québec Eco Entreprises n.d.). Another separation method is optical sorting technology, which is more costly than other methods and can separate PLA from other plastics (Foster 2008).

Mechanical recycling is a proper and temporary method for thermoplastic polymers such as PLA since they can undergo remelting process (Grigore 2017). This is a physical method that happens by grinding, cutting, and shredding to achieve pellets, granules, and flakes (Bhadra, Al-Thani, and Abdulkareem 2017).

Primary recycling is the simplest recycling method that requires small pieces of wastes by shredding, grinding, or milling followed by extrusion. In this method, the properties of recycled materials are not significantly changed, and they can mix with virgin PLA for further applications (Ignatyev, Thielemans, and Vander Beke 2014). However, there are some disadvantages to this method. Since there is no separation and classification step in primary recycling, contamination can be an issue in mixed plastics (Solis and Silveira 2020). Moreover, continuous and over-recycling will decrease the mechanical properties of the end-product (Thiounn and Smith 2020).

Another mechanical method, which is secondary mechanical recycling, can be classified into the following steps (As seen in Figure 13.2): First, piling up the PLA and PLA/cellulose composites materials which are collected from different places and transferred to the recycling site. Then the materials need to be washed and dried for better properties of the final product. After this step, different polymers need to be separated from each other. Next, the polymers will be shredded into smaller pieces. Finally, the shredded materials go under various techniques such as extrusion, film, injection, and blow molding (Francis 2016; Hopewell, Dvorak, and Kosior 2009). Applying and preparing all the mentioned steps is one of the main disadvantages of this method, which is only cost-effective when no complex separation is needed (Al-Salem, Lettieri, and Baeyens 2009).

There are some logistical and economical barriers in mechanical recycling since there is no net benefit (Craighill and Powell 1996) and the price of recycled polymers is dependent on the virgin one (Hopewell, Dvorak, and Kosior 2009). To have productive mechanical recycling, 200,000 tons/year critical mass of PLA is required (Cornell 2007). This amount has increased in the past few years and

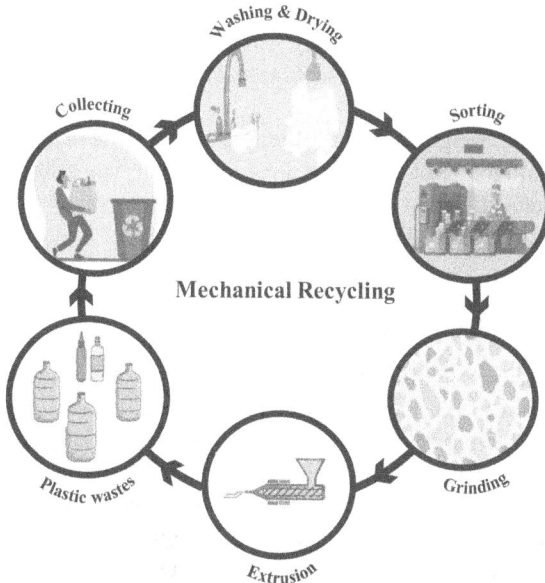

FIGURE 13.2   Mechanical recycling of PLA.

makes mechanical recycling economically efficient. In any case, it is crucial to investigate the performance properties of the PLA after mechanical recycling. If the performance decreases, mechanical recycling is not practical, even though it causes fewer environmental issues (Niaounakis 2013). Besides all this, several factors such as moisture, color, and residual catalysis can reduce the performance and create more challenges in the polymer after the first recycling cycle (Hong and Chen 2017a).

Nascimento et al. (2010) investigated the mechanical and thermal properties of the recycled PLA. Results showed no significant difference in thermal and mechanical properties of the polymer after one recycling cycle. However, studies by Pillin et al. (2008) and Brüster et al. (2016) show that the molecular weight of recycled polymer decreased by 60% and 30%, respectively. After seven and three recycling processes, crystallization increased, with no significant effect on Young's modulus. Moreover, thermal stability and tensile strength were reduced after the ten recycling cycles (Żenkiewicz et al. 2009). To this end, recycled PLA is mixed with neat PLA at industry scale to make the recycling process practical (Rafiee et al. 2021).

There are also several studies on the mechanical recycling of PLA reinforced with cellulose fibers. Chaitanya et al. (2019) investigated the tensile strength of PLA/sisal after eight extrusion cycles. The tensile strength was decreased by 20% after the third extrusion cycle. Furthermore, morphological and thermal properties show the degradation of PLA and cellulosic fiber. Fazita et al. (2015) studied the strength properties of the PLA/bamboo fiber. They reported a decrease in tensile strength and an increase in flexural properties. However, these two factors are still higher in recycled PLA/bamboo compared to virgin PLA.

Even though the thermal and mechanical properties are decreased in the recycled PLA and PLA/cellulose composite, they still have the potential to be used in products with low to medium strength such as packaging materials (Chaitanya et al., 2019; Fazita et al., 2015).

To overcome the weight loss, less thermal and mechanical performance, and increase in crystallization of mechanical recycled materials after mechanical recycling, chemical recycling can be performed, which breaks down the PLA and now can be reset into diverse products.

13.2.3   Chemical Recycling (Tertiary Recycling)

The mechanical recycling of PLA and PLA/cellulose composites reduces the quality of substrate due to the mentioned disadvantages. To overcome the mentioned disadvantages in mechanical recycling, it is important to bring up the research for chemical recycling processes such as depolymerization. This method broke down the polymers into monomers (lactic acid) (Taubner & Shishoo, 2001), which are initial materials for other new applications (As seen in Figure 13.3). It is called chemical recycling since it changes the structure of the waste materials (Al-Salem, Lettieri, and Baeyens 2009).

In chemical recycling, more energy is saved during the lactic acid production compared to the actual fermentation to produce virgin feedstock (Piemonte, Sabatini, and Gironi 2013). Another advantage of chemical recycling compared to composting is recovering the chemicals instead of losing them as carbon dioxide (Hong and Chen 2017a). Furthermore, this method is less costly since the polymer

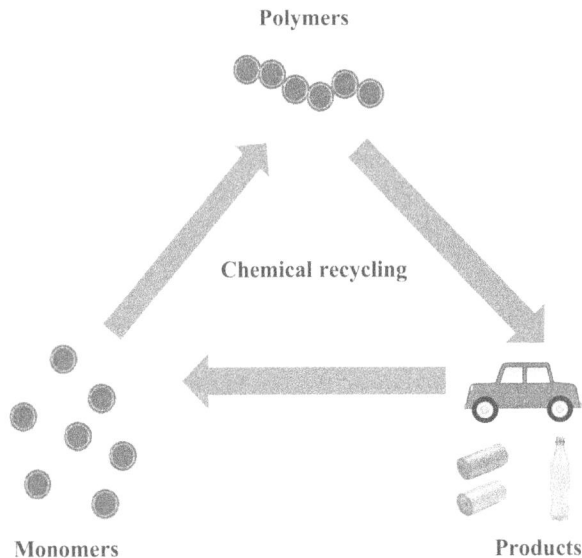

FIGURE 13.3 Chemical recycling of PLA.

is tolerant to contamination by other plastics and does not require to be separated from other plastics.

There are several methods in the chemical recycling of PLA, e.g., hydrolysis and pyrolysis (McKeown and Jones 2020).

13.2.3.1 Hydrolysis

Another chemical recycling method is the hydrolysis of PLA, which breaks down the PLA at high temperatures (250°C–350°C) and high-pressure water to produce lactide (Mohd-Adnan, Nishida, and Shirai 2008; Von Burkersroda, Schedl, and Göpferich 2002) due to the solubility of PLA. The hydrolysis of PLA depends on factors such as temperature, pH, and morphology of the PLA (De Jong et al. 2001). This method is easier and economical than the thermal method since it does not require catalysis and specific environments (Inkinen 2011). Moreover, it is more cost-effective than the fermentation of sugar. Produced lactide has a high level of purity, which could polymerize into the virgin PLA again (Faisal et al. 2006).

Gironi et al. (2016) investigated the solubility of PLA in acetone and ethyl acetate. Acetone provides better solubility while increasing the water concentration decreases the solubility.

Faisal et al. (2006) showed that the efficiency of hydrolysis depends on the reaction time and temperature. The optimum amount of PLA was observed at 250°C for 10–20 min and other produced by-products were harmless. They also concluded that higher temperature and longer reaction time would produce a racemic mixture. Moreover, Tsuji et al. (2003) studied the PLA hydrolysis in an aquatic environment and at high temperatures (120°C–350°C) to find the optimum conditions. Results showed that by increasing the temperature, the reaction time will be decreased. However, a temperature higher than 250°C is not favorable since it will result in the production of a racemic mixture. At 350°C, lactic acid is decomposed into carbon dioxide and methane.

Depending on the recycling environment and final products, additional chemical recycling such as pyrolysis could be applied, which is defined in the next section.

13.2.3.2 Pyrolysis

The purpose of this method is to heat PLA in a specific and oxygen-free environment (200°C–250°C, 4–5 mmHg, and with or without catalysts such as Zn and Sn) to recover lactide using simple and straightforward procedures (Liang et al. 2013; Noda and Okuyama 1999). The temperature at pyrolysis and the type of the reactor are two parameters that affect the final product (Sasse and Emig 1998).

High temperatures can cause side reactions and the production of materials other than lactide. Studies done by McNeill and Leiper (1985a, 1985b) showed additional products such as CO_2, cyclic oligomers, propylene, ethylene, etc., after providing controlled heating in an isothermal environment. Another study by Fan et al. (2004) showed the production of oligomers other than lactide and considerable racemization after PLA thermal degradation. Moreover, Nishida et al. (2003) studied the pyrolysis of PLA with different content of Sn (20–607 ppm). They showed that a higher amount of Sn can reduce the activation energy (E_a) and degradation temperature, resulting in lactide production while less Sn content produced cyclic oligomers.

13.3　FUTURE OUTLOOK

As mentioned earlier, a combination of PLA with other plastic causes contamination and changes the properties of the recycled polymer. To increase the efficiency of PLA and PLA/cellulose recycling, further studies on producing compatibilizers to control the polymers, newly designed strategies to decrease the sorting step, and decontamination techniques in mechanical recycling need to be done. Furthermore, since using various catalysts has different results in chemical recycling procedures, it is crucial to develop a catalyst to optimize the process and choose the low-energy catalysts. Moreover, researches are required to improve the recycling methods for plastics since it decreases waste production, emission of greenhouse gasses, and usage of fossil fuels.

Hence, to have a successful and high rate of recycling, a collaboration of companies, government, and researchers is needed. While new companies try to apply new chemical and sorting methods, government educates people on the advantages of recycling. Meanwhile, researchers are working on data analysis and new programs that can help to optimize the whole recycling process.

13.4　CONCLUSION

Bioplastic polymers such as PLA are sustainable materials that are widely used in various applications such as packaging, automotive, and medical fields. However, PLA production required an intensive amount of initial materials and energy. Therefore, using PLA and its composites such as PLA/cellulose requires sustainable and developed recycling techniques.

In terms of negative environmental impacts, composting is first followed by chemical recycling and then mechanical recycling. On the other hand, the chemical recycling of PLA is an economical way to keep the properties of the recycled materials the same as the initial materials, while mechanical recycling reduces the molecular weight and decreases the thermal and mechanical properties of PLA. To this end, chemical recycling draws extensive attention these days.

ACKNOWLEDGMENTS

This work was supported by the Natural Sciences and Engineering Research Council of Canada (NSERC-Discovery Grant 355254) and James and Joanne Love Chair in Environmental Engineering.

REFERENCES

Agrawal, Ashwini K., and Rahul Bhalla. 2003. "Advances in the Production of Poly (Lactic Acid) Fibers. A Review." *Journal of Macromolecular Science, Part C* 43 (4): 479–503.

Åkesson, Dan, Samaneh Fazelinejad, Ville-Viktor Skrifvars, and Mikael Skrifvars. 2016. "Mechanical Recycling of Polylactic Acid Composites Reinforced with Wood Fibres by Multiple Extrusion and Hydrothermal Ageing." *Journal of Reinforced Plastics and Composites* 35 (16): 1248–1259.

Alaerts, Luc, Michael Augustinus, and Karel Van Acker. 2018. "Impact of Bio-Based Plastics on Current Recycling of Plastics." *Sustainability* 10 (5): 1487.

Al-Salem, S. M., Paola Lettieri, and Jan Baeyens. 2009. "Recycling and Recovery Routes of Plastic Solid Waste (PSW): A Review." *Waste Management* 29 (10): 2625–2643.

Bhadra, J., N. Al-Thani, and A. Abdulkareem. 2017. "Recycling of Polymer-Polymer Composites." In *Micro and Nano Fibrillar Composites (MFCs and NFCs) from Polymer Blends*, 263–277. Elsevier.

Bismarck, Alexander, Supriya Mishra, and Thomas Lampke. 2005. "Plant Fibers as Reinforcement for Green Composites." In Amar K. Mohanty, Manjusri Misra, Lawrence T. Drzal (Eds). *Natural Fibers, Biopolymers, and Biocomposites*, 52–128. CRC Press.

Brüster, B., F. Addiego, F. Hassouna, D. Ruch, J.-M. Raquez, and P. Dubois. 2016. "Thermo-Mechanical Degradation of Plasticized Poly (Lactide) after Multiple Reprocessing to Simulate Recycling: Multi-Scale Analysis and Underlying Mechanisms." *Polymer Degradation and Stability* 131: 132–44.

Chaitanya, Saurabh, Inderdeep Singh, and Jung Il Song. 2019. "Recyclability Analysis of PLA/Sisal Fiber Biocomposites." *Composites Part B: Engineering* 173: 106895.

Cornell, David D. 2007. "Biopolymers in the Existing Postconsumer Plastics Recycling Stream." *Journal of Polymers and the Environment* 15 (4): 295–299.

Craighill, Amelia L., and Jane C. Powell. 1996. "Lifecycle Assessment and Economic Evaluation of Recycling: A Case Study." *Resources, Conservation and Recycling* 17 (2): 75–96.

De Jong, S. J., E. Ruiz Arias, D. T. S. Rijkers, C. F. Van Nostrum, J. J. Kettenes-Van den Bosch, and W. E. Hennink. 2001. "New Insights into the Hydrolytic Degradation of Poly (Lactic Acid): Participation of the Alcohol Terminus." *Polymer* 42 (7): 2795–2802.

Dechy-Cabaret, Odile, Blanca Martin-Vaca, and Didier Bourissou. 2004. "Controlled Ring-Opening Polymerization of Lactide and Glycolide." *Chemical Reviews* 104 (12): 6147–6176.

Faisal, M., T. Saeki, H. Tsuji, H. Daimon, and K. Fujie. 2006. "Recycling of Poly Lactic Acid into Lactic Acid with High Temperature and High Pressure Water." *WIT Transactions on Ecology and the Environment* 92, 225–233.

Fan, Yujiang, Haruo Nishida, Tomokazu Mori, Yoshihito Shirai, and Takeshi Endo. 2004. "Thermal Degradation of Poly (L-Lactide): Effect of Alkali Earth Metal Oxides for Selective L, L-Lactide Formation." *Polymer* 45 (4): 1197–1205.

Fazita, M. R., Krishnan Jayaraman, Debes Bhattacharyya, Md Hossain, M. K. Haafiz, and Abdul Khalil HPS. 2015. "Disposal Options of Bamboo Fabric-Reinforced Poly (Lactic) Acid Composites for Sustainable Packaging: Biodegradability and Recyclability." *Polymers* 7 (8): 1476–1496.

Foster, S. 2008. "Domestic Mixed Plastics Packaging Waste Management Options." Waste and Resources Action Programme (WRAP). Oxford, 1–77.

Fowler, Paul A., J. Mark Hughes, and R. M. Elias. 2007. "Biocomposites from Crop Fibres and Resins." *Iger Innovations*, 66–68.

Francis, Raju. 2016. "Recycling of Polymers: Methods." *Characterization and Applications*. John Wiley & Sons.

Freed, Randy. 1998. "Greenhouse Gas Emissions from Management of Selected Materials in Municipal Solid Waste."

Gironi, F., S. Frattari, and V. Piemonte. 2016. "PLA Chemical Recycling Process Optimization: PLA Solubilization in Organic Solvents." *Journal of Polymers and the Environment* 24 (4): 328–333.

Grigore, Mădălina Elena. 2017. "Methods of Recycling, Properties and Applications of Recycled Thermoplastic Polymers." *Recycling* 2 (4): 24.

Gross, Richard A., and Bhanu Kalra. 2002. "Biodegradable Polymers for the Environment." *Science* 297 (5582): 803–807.

Hong, Miao, and Eugene Y.-X. Chen. 2017a. "Chemically Recyclable Polymers: A Circular Economy Approach to Sustainability." *Green Chemistry* 19 (16): 3692–3706.

———. 2017b. "Chemically Recyclable Polymers: A Circular Economy Approach to Sustainability." *Green Chemistry* 19 (16): 3692–3706.

Hopewell, Jefferson, Robert Dvorak, and Edward Kosior. 2009. "Plastics Recycling: Challenges and Opportunities." *Philosophical Transactions of the Royal Society B: Biological Sciences* 364 (1526): 2115–2126.

Huneault, Michel A., and Hongbo Li. 2007. "Morphology and Properties of Compatibilized Polylactide/Thermoplastic Starch Blends." *Polymer* 48 (1): 270–280.

Ignatyev, Igor A., Wim Thielemans, and Bob Vander Beke. 2014. "Recycling of Polymers: A Review." *ChemSusChem* 7 (6): 1579–1593.

Inkinen, Saara. 2011. *Structural Modification of Poly (Lactic Acid) by Step-Growth Polymerization and Stereocomplexation*. Lab. of polymer technology, Center for funktional materials (FUNMAT), Department of Chemical Engineering, Division for Natural Sciences and Technology, Åbo Akademi University.

Kale, Gaurav, Rafael Auras, Sher Paul Singh, and Ramani Narayan. 2007. "Biodegradability of Polylactide Bottles in Real and Simulated Composting Conditions." *Polymer Testing* 26 (8): 1049–1061.

Le Duigou, A., G. Chabaud, R. Matsuzaki, and M. Castro. 2020. "Tailoring the Mechanical Properties of 3D-Printed Continuous Flax/PLA Biocomposites by Controlling the Slicing Parameters." *Composites Part B* 203: 108474.

Le Duigou, Antoine., P. Davies, and Christophe Baley. 2010. "Life Cycle Analysis of a Flax/PLLA Biocomposite." *Matériaux & Techniques* 98 (2): 143–150.

Liang, Zhenhua., Min Zhang, Xufeng Ni, Xue Li, and Zhiquan Shen. 2013. "Ring-Opening Polymerization of Cyclic Esters Initiated by Lithium Aggregate Containing Bis (Phenolate) and Enolate Mixed Ligands." *Inorganic Chemistry Communications* 29: 145–147.

Lim, Loong-Tak, Kevin Cink, and Tim Vanyo. 2010. "Processing of Poly (Lactic Acid)." *Poly (Lactic Acid) Synthesis, Structures, Properties, Processing, and Applications*, 189–215.

Luckachan, Gisha E., and C. K. S. Pillai. 2011. "Biodegradable Polymers-A Review on Recent Trends and Emerging Perspectives." *Journal of Polymers and the Environment* 19 (3): 637–676.

Lunt, James. 1998. "Large-Scale Production, Properties and Commercial Applications of Polylactic Acid Polymers." *Polymer Degradation and Stability* 59 (1–3): 145–152.

Masmoudi, Fatma, Atef Bessadok, Mohamed Dammak, Mohamed Jaziri, and Emna Ammar. 2016. "Biodegradable Packaging Materials Conception Based on Starch and Polylactic Acid (PLA) Reinforced with Cellulose." *Environmental Science and Pollution Research* 23 (20): 20904–20914.

McKeown, Paul, and Matthew D. Jones. 2020. "The Chemical Recycling of PLA: A Review." *Sustainable Chemistry* 1 (1): 1–22.

McNeill, I. C., and H. A. Leiper. 1985a. "Degradation Studies of Some Polyesters and Polycarbonates—1. Polylactide: General Features of the Degradation under Programmed Heating Conditions." *Polymer Degradation and Stability* 11 (3): 267–285.

———. 1985b. "Degradation Studies of Some Polyesters and Polycarbonates—2. Polylactide: Degradation under Isothermal Conditions, Thermal Degradation Mechanism and Photolysis of the Polymer." *Polymer Degradation and Stability* 11 (4): 309–326.

Mohd-Adnan, Ahmad-Faris, Haruo Nishida, and Yoshihito Shirai. 2008. "Evaluation of Kinetics Parameters for Poly (L-Lactic Acid) Hydrolysis under High-Pressure Steam." *Polymer Degradation and Stability* 93 (6): 1053–1058.

Nascimento, L., J. Gamez-Perez, O. O. Santana, José Ignacio Velasco, M. Ll Maspoch, and E. Franco-Urquiza. 2010. "Effect of the Recycling and Annealing on the Mechanical and Fracture Properties of Poly(Lactic Acid)." *Journal of Polymers and the Environment* 18 (4): 654–660.

Niaounakis, Michael. 2013. "Biopolymers: Reuse." In *Recycling, and Disposal*. William Andrew.

Nishida, Haruo, Tomokazu Mori, Shinya Hoshihara, Yujiang Fan, Yoshihito Shirai, and Takeshi Endo. 2003. "Effect of Tin on Poly (l-Lactic Acid) Pyrolysis." *Polymer Degradation and Stability* 81 (3): 515–523.

Noda, Masaki, and Hisashi Okuyama. 1999. "Thermal Catalytic Depolymerization of Poly (L-Lactic Acid) Oligomer into LL-Lactide: Effects of Al, Ti, Zn and Zr Compounds as Catalysts." *Chemical and Pharmaceutical Bulletin* 47 (4): 467–471.

Piemonte, V., S. Sabatini, and F. Gironi. 2013. "Chemical Recycling of PLA: A Great Opportunity towards the Sustainable Development?" *Journal of Polymers and the Environment* 21 (3): 640–647.

Pillin, Isabelle., Nicolas Montrelay, Alain Bourmaud, and Yves Grohens. 2008. "Effect of Thermo-Mechanical Cycles on the Physico-Chemical Properties of Poly (Lactic Acid)." *Polymer Degradation and Stability* 93 (2): 321–328.

Pradhan, Ranjan., Manjusri Misra, Larry Erickson, and Amar Mohanty. 2010. "Compostability and Biodegradation Study of PLA–Wheat Straw and PLA–Soy Straw Based Green Composites in Simulated Composting Bioreactor." *Bioresource Technology* 101 (21): 8489–8491.

Québec Eco Entreprises. n.d. *Fact Sheet Impact of Packaging on Curbside Recycling Collection and Recycling System: PLA Bottle.* 2012.

Rafiee, Kiana, Helge Schritt, Daniel Pleissner, Guneet Kaur, and Satinder K. Brar. 2021. "Biodegradable Green Composites: It's Never Too Late to Mend." *Current Opinion in Green and Sustainable Chemistry*, 30, 100482.

Ren, Jie. 2011. "Biodegradable Poly (Lactic Acid): Synthesis." *Modification, Processing and Applications.* Springer Science & Business Media.

Sasse, Frank, and Gerhard Emig. 1998. "Chemical Recycling of Polymer Materials." *Chemical Engineering & Technology: Industrial Chemistry-Plant Equipment-Process Engineering-Biotechnology* 21 (10): 777–789.

Sethi, Beena. 2017. "Methods of Recycling." In *Recycling of Polymers: Methods, Characterization and Application*, 55–114.

Siracusa, Valentina, Pietro Rocculi, Santina Romani, and Marco Dalla Rosa. 2008. "Biodegradable Polymers for Food Packaging: A Review." *Trends in Food Science & Technology* 19 (12): 634–643.

Solis, Martyna, and Semida Silveira. 2020. "Technologies for Chemical Recycling of Household Plastics–A Technical Review and TRL Assessment." *Waste Management* 105: 128–138.

Soroudi, Azadeh., and Ignacy Jakubowicz. 2013. "Recycling of Bioplastics, Their Blends and Biocomposites: A Review." *European Polymer Journal* 49 (10): 2839–2858.

Surip, Siti Norasmah, and Wan Nor RaihanWan Jaafar. 2018. "Comparison Study of the Bio-Degradation Property of Polylactic Acid (PLA) Green Composites Reinforced by Kenaffibers." *International Journal of Technology* 6: 1205–1215.

Taubner, V., and R. Shishoo. 2001. "Influence of Processing Parameters on the Degradation of Poly (L-Lactide) during Extrusion." *Journal of Applied Polymer Science* 79 (12): 2128–2135.

Thiounn, Timmy, and Rhett C. Smith. 2020. "Advances and Approaches for Chemical Recycling of Plastic Waste." *Journal of Polymer Science* 58 (10): 1347–1364.

Tsuji, Hideto. 2011. *Poly (Lactic Acid): Synthesis, Structures, Properties, Processing, and Applications.* Wiley.

Tsuji, Hideto, Hiroyuki Daimon, and Koichi Fujie. 2003. "A New Strategy for Recycling and Preparation of Poly (L-Lactic Acid): Hydrolysis in the Melt." *Biomacromolecules* 4 (3): 835–840.

Van de Velde, Kathleen, and Paul Kiekens. 2002. "Biopolymers: Overview of Several Properties and Consequences on Their Applications." *Polymer Testing* 21 (4): 433–442.

Verma, R., K. S. Vinoda, M. Papireddy, and A. N. S. Gowda. 2016. "Toxic Pollutants from Plastic Waste-a Review." *Procedia Environmental Sciences* 35: 701–708.

Von Burkersroda, Friederike, Luise Schedl, and Achim Göpferich. 2002. "Why Degradable Polymers Undergo Surface Erosion or Bulk Erosion." *Biomaterials* 23 (21): 4221–4231.

Yun-xuan, Weng. 2007. "Review of Study of Synthesis, Production, Process and Application of PLA." *China Plastics Industry* 35: 69–73.

Żenkiewicz, Marian, Józef Richert, Piotr Rytlewski, Krzysztof Moraczewski, Magdalena Stepczyńska, and Tomasz Karasiewicz. 2009. "Characterisation of Multi-Extruded Poly (Lactic Acid)." *Polymer Testing* 28 (4): 412–418.

14 Electrospinning and Electrospraying in Polylactic Acid/ Cellulose Composites

*Juliana Botelho Moreira, Suelen Goettems Kuntzler,
Ana Gabrielle Pires Alvarenga,
Jorge Alberto Vieira Costa,*
and

Michele Greque de Morais
Federal University of Rio Grande
Rio Grande, Brazil

Loong-Tak Lim
University of Guelph
Guelph, Canada

CONTENTS

14.1 INTRODUCTION

Global warming and the depletion of natural resources are the main drivers for the need to develop biodegradable and environmentally friendly materials. To this end, the development of new plastics production technologies will contribute to reducing the environmental impact due to improper disposal of polymers of petrochemical origin and limited availability of landfills (Kumar et al., 2019; Stoyanova et al., 2014).

DOI: 10.1201/9781003160458-14

In this context, poly(lactic acid) (PLA) has the potential to replace petrochemical polymers for some packaging applications. PLA is produced from renewable agricultural resources rich in starch (Tiimob et al., 2018; Vorawongsagul et al., 2020). This aliphatic polyester is non-toxic and has biocompatibility, thermoplasticity, and processability. PLA has been used in packaging materials (Shah & Vasava, 2019; Tiimob et al., 2018) and biomedical applications. However, its characteristics, such as brittleness, low biodegradability, and high cost, have restricted the applicability of this polymer (Shah & Vasava, 2019).

Blending with other polymers and the incorporation with nanofillers have been studied to modify and improve PLA properties (Kumar et al., 2020; Vorawongsagul et al., 2020). For example, cellulose is an abundant and low-cost homopolymer, which has advantages related to its mechanical properties and biodegradability (Vorawongsagul et al., 2020). Thus, it has been used as a reinforcing agent for several nanocomposites, including those produced with PLA (Khosravi et al., 2020).

Electrohydrodynamic processes such as electrospinning and electrospraying have received attention in recent years for the production of nanocomposites due to their characteristics of versatility and practicality. The manipulation of the solution parameters (e.g., polarity, viscosity, molecular weight, surface tension, and electrical conductivity), process parameters (e.g., electrical potential, feed rate, spinneret diameter, and spinneret-collector distance), and environmental conditions (e.g., relative humidity and temperature) allow for the development of nanomaterials with desirable morphologies. Different equipment setups as single spinneret, coaxial spinneret, triaxial spinneret, multiple spinnerets, and free-surface have been used to produce nano to micron scaled materials (Jafari, 2017; Lim, 2021).

PLA/cellulose nanocomposites formed by electrohydrodynamic processes have the potential to be used in several areas, especially in biomedical and pharmaceutical applications for tissue regeneration and controlled release of bioactive compounds (Ghafari et al., 2020; Zhang et al., 2020). The chapter discusses the electrospinning and electrospraying processes for producing PLA/cellulose composites, highlighting different technologies, the importance of cellulose as a reinforcing agent, as well as the main applications and challenges in various areas.

14.2 PLA/CELLULOSE BLEND AND NANOCOMPOSITES

PLA is a linear aliphatic thermoplastic polyester produced by polymerization through lactide ring-opening or condensation of lactic acid monomers produced by fermentation of renewable resources such as sugar, corn, potatoes, cane, and beets (Ghafari et al., 2020; Madhavan et al., 2010). PLA has several advantages due to its characteristics such as renewability, compostability, biocompatibility, and market availability. This polyester is resistant to ultraviolet light, has a glossy appearance and high transparency. In addition, the use of this polymer helps to reduce CO_2 emissions (Siakeng et al., 2018). PLA has a lower carbon footprint than conventional fossil-fuel-based plastic materials (Zaaba & Jaafar, 2020). Improvement measures for the production of PLA can further reduce environmental impacts. Furthermore, the use of renewable energy in the production of lactic acid, lactide, and PLA can be increased and the use of chemicals in the production of lactic acid can be reduced (Morão & Bie, 2019). However, this polymer is inherently brittle with a relatively high modulus of elasticity and low elongation at break (Scaffaro et al., 2017), slow crystallization, has a low heat distortion temperature, and high cost (Shah & Vasava, 2019). To overcome these issues, PLA is often being blended with other polymers to improve its properties and increase applicability in fields where its use is still limited.

Cellulose is an abundant and low-cost organic material found in different sources, such as wood pulp, cotton fibers, linen, coconut, and bamboo fibers (Lavoine et al., 2012). Cellulose is a straight-chain polysaccharide consisting of many β(1→4) linked D-glucose units (Gutiérrez & Alvarez, 2017). Cellulose has important characteristics such as versatility, availability, hydrophilicity, biodegradability, and biocompatibility (Xue et al., 2017). Moreover, cellulose has been considered an excellent reinforcing agent due mainly to its high molecular weight, Young's modulus, and specific strength (Nishino & Peijs, 2014). When mixed with other polymers, cellulose contributes to biodegradability, thermal insulator, and thermal resistant properties to the material formed (Herniou-Julien et al., 2019; Khosravi et al., 2020).

Another interesting approach is the preparation of nanocomposites, which are materials composed of additives on the nanometer scale within the polymeric matrix (Armstrong, 2015; Crosby & Lee, 2007). Many PLA-based nanocomposites have been developed (Akbari et al., 2015; Alves et al., 2020; Ortenzi et al., 2015; Zare & Rhee, 2019; Zawawi et al., 2020). Cellulose nanocrystals and cellulose nanofibrils have been reported as promising renewable nano-fillers in polymeric PLA matrices (Ghafari et al., 2020; Kumar et al., 2020). The high area-to-volume ratio provided by the nanometric scale provides an excellent interface between the particle and the polymeric matrix to enhance the material properties of the resulting nanocomposites, provided if good interfacial interaction and adequate dispersion of the filler in the polymer can be achieved (Gazzotti et al., 2019). In addition, the application of nanocellulose in biodegradable polymeric materials, such as PLA, has the potential of improving the sustainable production of environmentally friendly materials (Khosravi et al., 2020).

14.3 ELECTROSPINNING AND ELECTROSPRAYING TECHNOLOGIES

Electrohydrodynamic processes, such as electrospraying and electrospinning, are methods used for the production of particles and fibers, respectively (Garavand et al., 2019; Katouzian & Jafari, 2016; Mendes et al., 2017). These processes allow the production of materials at micron, submicron, and nano scales of various morphology by manipulating the properties of the polymeric solution and the processing parameters. Moreover, these nonthermal processes do not require the application of heat and are hence advantageous for many heat-labile materials (Mendes et al., 2017). The electrohydrodynamic processes are adaptable for different materials, such as natural/synthetic polymers, ceramics, and composites. Typical equipment for electrohydrodynamic processes consists of elements of a high voltage supply, a positive displacement pump for the injection of the polymeric solution, and an electrically grounded collector for the deposition of electrospun fibers/electrosprayed particles (Figure 14.1).

The principle of the method is based on the application of high electrical potential (10–30 kV) between the spinneret and the grounded collector. In this process, the

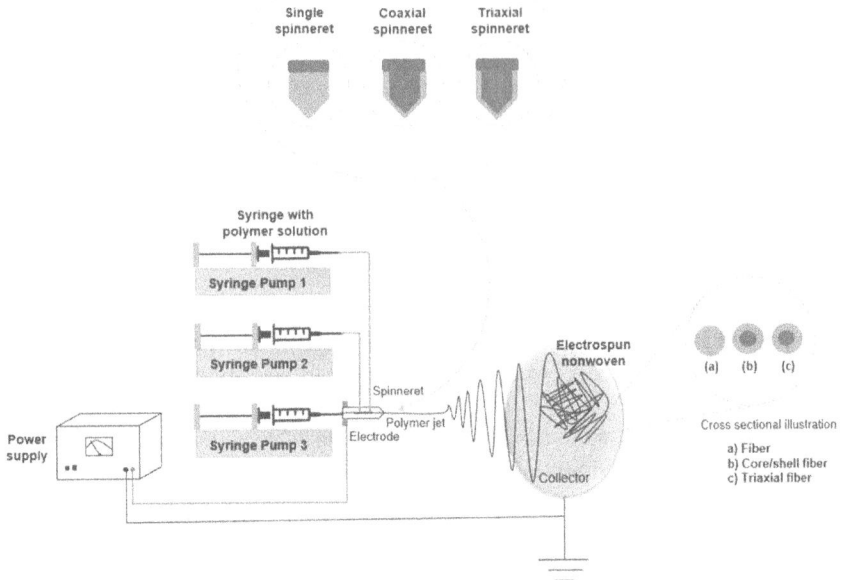

FIGURE 14.1 Schematic drawing of an electrospinning setup with single, coaxial, and triaxial spinnerets.

viscoelastic and surface tension forces tend to counter jet formation. Furthermore, electric field and Coulombic forces aid in jetting. The electric field force drives the jet forward to the target, while Coulombic force works against the surface tension. The electric field generates enough force to overcome the surface tension of the pendant droplet at the tip of the capillary, resulting in the formation of a conical-shaped jet called the Taylor cone. As the polymer jet takes flight in the air toward the target, the solvent evaporates, and thereby forming solidified fibrous nonwoven or particulate materials on the collector that can be static or dynamic (Morais et al., 2014; Moreira et al., 2018a).

In addition to the characteristics inherent to the polymer solution (type of solvent, molecular weight, surface tension, and electrical conductivity), both the environmental conditions (humidity and temperature) and the process parameters (electrical potential, feed rate, capillary diameter, and spinneret-collector distance) must be considered in electrohydrodynamic processes, as they directly influence this system (Jafari, 2017; Moreira et al., 2018a).

The solvent dissolves the polymer and must have a moderate vapor pressure to evaporate during the electrohydrodynamic process (Niu et al., 2020). Highly volatile solvents should be avoided, as premature drying of the spin dope solution can clog the spinneret. On the other hand, low volatile solvents can hinder their evaporation, resulting in wet/fused products. The distance between the capillary and the collector should be optimized for maximal solvent evaporation, allowing the formation of materials with the minimal solvent residual (Mercante et al., 2017).

The molecular weight directly influences the morphological properties of fibers and particles. An increase in molecular weight promotes the entanglement of the polymer chain, thereby stabilizing the polymer jet during electrohydrodynamic processes (Bock et al., 2012). Moreover, higher molecular weight can lead to the formation of electrospun fibers of larger diameters (Haider et al., 2018).

The conductivity of the solution is mainly determined by the type of polymer, solvent, and ionizable salts (Zong et al., 2002). Low conductivity solutions (1 μs cm^{-1}) are recommended for obtaining electrosprayed particles (Niu et al., 2020). High conductivity tends to reduce stretching due to the charge leakage along the fiber jet (Lim et al., 2019). In this case, a higher potential difference can be required (Carroll & Joo, 2006).

The electrical potential influences the morphology and the size of the fibers and particles in the electrohydrodynamic processes. Taylor cone is formed when the applied electrical potential exceeds the surface tension of the polymer solution drop at the tip of the capillary (Songsurang et al., 2011). According to Park and Lee (2009), a critical electrical potential (CEp) is influenced by a solvent-polymer system. In general, to form electrosprayed particles, PLA was used with different solvents including chloroform, chloroform/ethanol, acetone, and chloroform/acetone, and the process parameters could be varied in flow rate (2–40 μL m^{-1}), voltage (10–30 kV), distance from capillary to the collector (10–20 cm) and spinneret diameter (20–22 Gauge) (Ibili & Dasdemir, 2019). Values lower than CEp cause the polymeric solution to drip into the collector. For the formation of electrospun fibers, values far above the CEp provide a greater elongation of the solution and reduction in the diameter, as it increases the repulsive forces in the jet (Park & Lee, 2009). The flow rate together with the solution parameters can control the electrospinning behavior (Niu et al., 2020). Increasing flow rate tends to produce non-spherical particles with a high polydispersity (Ibili & Dasdemir, 2019; Niu et al., 2020). Similarly, at high flow rates, irregular fibers are formed, with larger diameters and the presence of residual solvent (Mercante et al., 2017).

Environmental parameters, such as humidity and temperature, play significant roles in the electrohydrodynamic processes and the morphology of materials. Mit-Uppatham et al. (2004) reported a reduction in fiber diameter with an increase in temperature, due to a decrease in solution viscosity. Relative humidity is another environmental factor that affects the morphology of nanomaterials. In water-based systems, high humidity will reduce the evaporate rate of water, which may result in wet/fused materials on the collector. Besides, excess humidity can dissipate the electrostatic charge, thereby weakening the jetting of the spin dope solution (Lim et al., 2019).

14.4 ELECTROHYDRODYNAMIC SYSTEMS

With the advanced development of technology, electrohydrodynamic processes have innovated to produce nanomaterials with differentiated functionalities. The application of these technologies is based on special collectors that allow the development of products using two or more liquid fluids (Liu et al., 2016). Coaxial/triaxial and side-by-side are examples of electrohydrodynamic systems (Buzgo et al., 2018).

For the two-fluid process, products with side-by-side structure (Janus fibers) or with the internal-external arrangement (core/shell) can be formed, which are generated side-by-side and coaxial electrospinning, respectively (Chen et al., 2015; Walther & Muller, 2013). Coaxial electrospinning consists of using two capillary spinnerets that are placed coaxially. The core pumped by the internal capillary and the shell material distributed by the external capillary receives the application of high electrical potential, forming core/shell fibers (Buzgo et al., 2018).

Janus fibers formed using the side-by-side technique have a potential advantage over the core/shell structure. This structure allows both sides to have direct contact with the surrounding environment. Therefore, it can be explored to design new functional nanomaterials and provide controlled release of several compounds (Chen et al., 2015; Wang et al., 2018). In the side-by-side electrospinning technique for the formation of Janus fibers, the working fluids do not come into contact substantially, as in the coaxial process. Thus, the interfacial interaction between the two working fluids decreases, favoring a successful dual fluid process. In contrast, new dual-compartment structures can be more easily produced by side-by-side electrospinning rather than by coaxial process for certain working system fluids (Yu et al., 2017a).

Triaxial electrospinning is the method for producing multilayer fibers with characteristics for specific applications. Like coaxial, in triaxial electrospinning, three polymeric solutions are loaded by syringe to form three-layer fibers (Khalf & Madihally, 2016). In the standard triaxial electrospinning method, all fluids are spinnable. However, there are modified processes where one or more unspinnable liquids can be used. If two unspinnable solutions are being processed, these must be combined with the spinnable middle fluid for fiber formation. The process does not occur if the outer (or inner) and middle solutions are not spinnable (Yang et al., 2017). In the triaxial electrospinning process, in addition to the high voltage power supply and grounded collector, there are three syringe pumps to drive the three working fluids and three-layer concentric spinneret. The conductivity, surface tension, and viscosity are properties of the fluids that influence the behavior of the high voltage electrostatic field (Yu et al., 2017a).

Free surface electrospinning is a technology based on the production of nanomaterials on a large scale due to the higher yield and simplicity of the technique compared to the capillary approach. In this process, a wire electrode is coated with a spin dope solution using a deposition device. As the solution is being electrified, multiple jets of charged liquid are formed on the surface of the wire and ejected toward a grounded collector (Forward & Rutledge, 2012; Moreira et al., 2018b; Moreira et al., 2019; Xiao & Lim, 2018).

14.5 APPLICATION OF PLA/CELLULOSE COMPOSITES DEVELOPED BY ELECTROHYDRODYNAMIC PROCESSES

The development of PLA/cellulose composites produced by electrohydrodynamic processes has attracted considerable industrial and academic interest. This composite shows improved functional properties such as high thermal and mechanical stability. In addition, the characteristics of biodegradability and biocompatibility with cells and tissues make PLA and cellulose polymers promising materials for applications in various areas (Santoro et al., 2016; Xue et al., 2017) (Figure 14.2). The use of

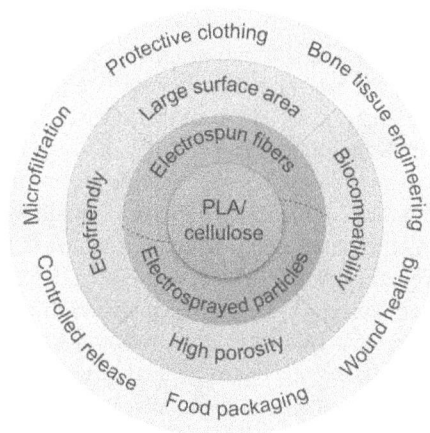

FIGURE 14.2 An overview of PLA/cellulose composite from electrohydrodynamic processes diagram.

PLA/cellulose in the field of biomedicine is the most explored by researchers, including recent studies on tissue engineering in bone and organ regeneration and wound healing (Ghafari et al., 2020; Wang et al., 2021; Yin et al., 2018; Zhang et al., 2020).

Scaffolds for tissue engineering produced by electrospinning are capable of forming nanofibrous networks with structural and mechanical properties similar to the extracellular matrix of the native tissue (Hasan et al., 2014; O'brien, 2011). Patel et al. (2020) fabricated scaffolds of PLA/cellulose nanocrystals (CNC) composites to evaluate the biocompatibility of CNCs and the osteogenic potential of PLA by *in vivo* bone regeneration study. The composite was produced by electrospinning with 10% (w/v) PLA dissolved in chloroform/N, N-dimethylformamide (3:1) solvent mixture under stirring for 3 h. CNC was then added at 1, 2, and 4% (wt.) concentration levels into the PLA solution and stirred for 4 h. The electrospinning parameters were voltage of 16 kV, the distance between spinneret tip and collector was set to 150 mm, and rolling speed 2000 rpm. The results showed that the composite with 2% CNC had higher biocompatibility and high osteoinductivity compared to the other composites. In addition, a greater bone formation was seen in the groups treated with PLA/2% CNC scaffolds compared to the control (without treatment) after 3 weeks of transplantation. Abdelaziz et al. (2021) produced PLA/cellulose acetate (CA)-based scaffolds with hydroxyapatite nanoparticles for application in regenerating damaged periodontal tissue and bone. The scaffolds were prepared with a polymer mixture and nanoparticles were dissolved in dichloromethane/dimethylformamide (7:3). The electrospinning parameters were voltage of 25 kV, a flow rate of 2 mL h^{-1}, and the distance between spinneret tip and collector was set to 150 mm. In their *in vitro* study, the composite showed a 50% improvement in cell viability due to the nanoparticles. However, the highest degradation was observed in the PLA/CA composite resulting in between 40% and 70% of its mass 8 weeks. The authors explained that this high degradation might be due to the absence of the hydroxyapatite nanoparticles, which results in the acceleration of the autocatalysis process.

Wound healing is another segment of biomedicine that searches for new materials that prevent infection and inflammation (Ahmadian et al., 2020; Frykberg & Banks, 2015). Therefore, the use of nanomaterials is alternative for dressings due to the high surface area and porosity that result in better cell adhesion (Ahmadian et al., 2020; Ambekar & Kandasubramanian, 2019). Elsayed et al. (2020) developed PLA/CA nanofiber dressing containing sulfonamide as an antimicrobial and anti-inflammatory agent. Polymer solutions were prepared by dissolving the polymers in the solvent mixture of dichloromethane/dimethylformamide (1:1) for 6 h. The electrospinning parameters were 24–29 kV voltage, 1.1–1.4 mL h^{-1} flow rate, and distance between spinneret tip and collector of 150 mm. *In vitro* study of antibacterial activity showed inhibition of up to 85% of *Staphylococcus aureus* and *Escherichia coli*, and *in vivo* trials confirmed the improvement in the ability to heal skin lesions. Yan et al. (2020) applied PLA/cellulose membranes and silver nanoparticles to ocular epithelial cells in *in vitro* assay. Fibers of 15% (wt.) PLA were prepared by solubilizing the polymer in methylene chloride/N, N-dimethylformamide (7:3) solvent mixture. The electrospinning parameters were flow rate of 3 mL h^{-1}, a distance between spinneret tip and collector of 100 mm, and voltage of 14 kV. The PLA membranes were inserted into vacuum filtration equipment to filter the 0.01% (wt.) cellulose suspension containing 33 ppm silver nanoparticles. The results showed that the scaffolds were non-toxic to the ocular cells, showing cell proliferation after 72 h. The composite was also evaluated for antimicrobial activity, which showed a 95% inhibitory effect on *E. coli* and *S. aureus*.

The controlled release of compounds with bioactivity is an important property of nanoscale materials that can be targeted for various applications. Yu et al. (2017b) studied the incorporation of tetracycline hydrochloride into PLA/CNC/poly(ethylene glycol) composite for drug delivery systems application. Composites loaded with 15–30% (wt.) tetracycline hydrochloride showed a long drug release period, resulting in delivery of 95.7% of its content in 1032 h. Jash and Lim (2018) encapsulated hexanal from an imidazolidine precursor in PLA fibers and ethylcellulose particles by electrospinning and electrospraying processes. The controlled release experiment of the hexanal precursor was carried out at 5, 25, and 45°C. The results showed that the release rate was significantly higher for PLA fibers compared to ethylcellulose particles. Rojas et al. (2020) produced PLA/CNC/polyvinyl alcohol composite with curcumin by coaxial electrospinning. PLA and CNC solutions were prepared under a chloroform/dimethylformamide mixture (7:3), and the polyvinyl alcohol and curcumin solutions were dissolved in 10% ethanol. The flow rate of coaxial electrospinning was 0.6 mL h^{-1} for the core PLA/CNC and 2.25 mL h^{-1} polyvinyl alcohol and curcumin for the shell. The distance between spinneret tip and collector was 140 mm and the applied voltage varied between 10 and 13 kV. The curcumin release kinetics assay was performed using ethanol solutions (10 and 50%) at 40°C as food simulants. Results showed that the lowest values of the curcumin diffusion coefficient were found in the 50% ethanol simulants, due to the affinity of the compound with alcohol. Moreover, the addition of CNC in the composite decreased the diffusion coefficient of curcumin by six times as compared to the sample without CNC. The composites developed from these studies could be promising for controlled release applications of bioactives and preservatives in food packaging systems to extend the shelf-life of food products.

The active system that prevents microorganism growth and spoilage in food packaging is another applicability of nanostructured PLA/cellulose composites. Hajikhani et al. (2021) developed polylactic acid (PLA)/copolymer electrospun nanofiber and evaluated the potential of these fibers for controlled release of lycopene in fatty food simulants. For this, PLA/cellulose acetate (CA) solution was prepared in dichloromethane/N,N-dimethylformamide (7:3) under stirring for 2 h. After, dissolution of the polymers, lycopene was added to the solution and stirred for another 15 min. The electrospinning parameters used for fibers production were 13–20 kV voltage, a flow rate of 0.5 mL h^{-1}, and a distance between the spinneret tip and collector of 100–150 mm. The authors observed that the initial release rate of lycopene decreased with increasing CA concentration. Their results showed that at 48 h the release of lycopene was 48.4, 30.5, and 15.1% for nanofibers developed with 5, 10, and 20% (w/w) of CA, respectively. The material can be applied as active packaging to prevent the oxidation of foods with high fatty acid content. Alvarado et al. (2018) produced the PLA/CNC/polyvinyl alcohol composite with thymol. In this study, nanofibers were formed from CNC/polyvinyl alcohol solution by electrospinning where a voltage of 10 kV, a flow rate of 0.25 mL h^{-1}, and a distance between spinneret tip and collector of 100 mm were used. The nanofibers were mixed with PLA solution in chloroform using the casting process. Thymol was added to the composite by the supercritical impregnation process. The release properties of thymol from the composite were studied in 10 and 95% (w/w) ethanol solutions as aqueous and fatty food simulants, respectively at 40°C. The partition coefficient of thymol in the PLA/CNC/polyvinyl alcohol composite was higher in both food simulants compared to the PLA film. Moreover, the diffusion coefficient of thymol from the composite was lower for the 10% and 95% (w/w) ethanol food simulants compared to the PLA film. The researchers concluded that the composites containing the nanostructures present a tortuous path for thymol release, which may be useful for the controlled release of thymol in active food packaging.

Electrohydrodynamic processes can also form composites with PLA and cellulose polymers for use in microfiltration and clothing protection. Jalvo et al. (2017) developed a PLA/polyacrylonitrile/CNC composite by coaxial electrospinning. The fibers were prepared with a solution of PLA core in chloroform/N,N-dimethylformamide (3:2), and a polyacrylonitrile/CNC shell in N,N-dimethylformamide. The flow rate of both solutions was 0.8 mL h^{-1}, the voltage was 20 kV, and the fibers were deposited on a flat collector plate with a distance between spinneret tip and collector of 200 mm. Microfiltration tests were performed using suspensions of *E. coli* cells and *Aspergillus niger* spores. Their results on microfiltration with the PLA/polyacrylonitrile/CNC composite show that the retention capacity of *A. niger* spores and *E. coli* cells were in the range of 85%–98%. Alam et al. (2020) produced multilayer PLA/CA membranes and evaluated the contact angle and surface energy. Solution containing 15% CA was prepared in acetone and 8% PLA was dissolved in a mixture of chloroform and acetone (1:3). PLA nanofibers and CA fibers were produced separately by the electrospinning process. The flow rate used was 0.05 mL min^{-1}, the applied voltage was 25 kV, and the distance between spinneret tip and collector used was 120 mm. After the nanofibers of each polymer were intercalated and the multilayer membrane was formed. Their results showed that the membranes have a high wetting potential for liquids and solid surfaces, exhibiting an angle of approximately

130° and 75 mN/m of surface free energy. The authors suggest that this material can be used in chemical protective clothing for pesticide applicators.

14.6 CHALLENGES

There are several studies in the literature addressing the techniques of electrospinning/electrospraying for the development of fibers/particles. These materials have adequate functional properties to fill technology gaps and the need for innovative products. However, there are some challenges related to the production of PLA/cellulose composites by electrohydrodynamic processes. Generally, the electrospinning/electrospraying process is associated with another method for the manufacture of composites (Chen et al., 2020; Dicastillo et al. 2017; Somord et al., 2018; Yang et al., 2017; Zhang et al., 2020). Although electrohydrodynamic techniques present different strategies for the production of materials such as multi-jet and coaxial configurations, the development of PLA/cellulose composites is still under-explored.

The application of biodegradable polymers PLA and cellulose has a development perspective and the main remaining challenge is to reach all market segments (Vatansever et al., 2019). The biomedicine area is the most consolidated by PLA/cellulose composites and electrohydrodynamic techniques. The potential impact of nanomaterials on human health and nanotoxicology issues should be the focus of research (Padmanabhan et al., 2018). In the literature, fields that lack studies with PLA/cellulose composites are active and intelligent food packaging, including colorimetric sensors and volatile compound absorbers.

Moreover, studies on the availability of solvents compatible with food, more efficient techniques to quantify the migration of constituent nanomaterials in food packaging, and toxicity should be further investigated (Han et al., 2011; Lim, 2021). In the areas of filtration and bioremediation of contaminants, formulation of biopesticides, and storage of electrochemical energy, electrohydrodynamic processes are well explored, allowing the use of adequate studies that enable the application of PLA/cellulose composites.

14.7 CONCLUSIONS AND FUTURE PERSPECTIVES

The technology of electrospinning/electrospraying has emerged as a promising alternative in the development of polymeric composites. These processes are characterized by high efficiency and versatility to produce nanomaterials with different polymers. Cellulose and PLA are obtained from renewable sources, are biodegradable, biocompatible with cells and tissues, and eco-friendly. These characteristics make these polymers favorable for the production of composites by electrohydrodynamic processes. Many studies are concentrated in the field of biomedicine for tissue and bone regeneration, wound healing, and drug delivery. Although numerous advantages are presented by nanomaterials, there are still areas unexplored by PLA/cellulose composites. However, nanostructured PLA/cellulose composites have to potential to be applied in different areas, such as food packaging, agriculture, the environment, and energy storage devices. Furthermore, the functional properties of the nanostructures and the configurations of the equipment for industrial-scale are fundamental for the development of innovative products.

ACKNOWLEDGMENTS

This study was financed in part by the Coordenação de Aperfeiçoamento de Pessoal de Nível Superior – Brasil (CAPES) – Finance Code 001. This research was developed within the scope of the Capes-PrInt Program (Process # 88887.310848/2018-00).

REFERENCES

Abdelaziz, D., A. Hefnawy, E. Al-Wakeel, A. El-Fallal, and I. M. El-Sherbiny. 2021. New biodegradable nanoparticles-in-nanofibers based membranes for guided periodontal tissue and bone regeneration with enhanced antibacterial activity. *J. Adv. Res.* 28:51–62.

Ahmadian, S., M. Ghorbani, and F. Mahmoodzadeh. 2020. Silver sulfadiazine-loaded electrospun ethyl cellulose/polylactic acid/collagen nanofibrous mats with antibacterial properties for wound healing. *Int. J. Biol. Macromol.* 162:1555–1565.

Akbari, A., M. Majumder, and A. Tehrani. 2015. Polylactic acid (PLA) carbon nanotubes composites. In *Handbook of polymer nanocomposites, processing, performance and application*, eds. K. Kar, J. Pandey, and S. Rana, 283–297. Berlin, Heidelberg: Springer.

Alam, A. K. M., E. Ewaldz, C. Xiang, W. Qu, and X. Bai. 2020. Tunable wettability of biodegradable multilayer sandwich-structured electrospun nanofibrous membranes. *Polymers* 12:2092.

Alvarado, N., J. Romero, and A. Torres et al. 2018. Supercritical impregnation of thymol in poly(lactic acid) filled with electrospun poly(vinyl alcohol)-cellulose nanocrystals nanofibers: Development an active food packaging material. *J. Food Eng.* 217:1–10.

Alves, J. L., P. D. T. V. Rosa, V. Realinho, M. Antunes, J. I. Velasco, and A. R. Morales. 2020. The effect of Brazilian organic-modified montmorillonites on the thermal stability and fire performance of organoclay-filled PLA nanocomposites. *Appl. Clay Sci.* 194:105697.

Ambekar, R. S., and B. Kandasubramanian. 2019. Advancements in nanofibers for wound dressing: a review. *Eur. Polym. J.* 117:304–336.

Armstrong, G. 2015. An introduction to polymer nanocomposites. *Eur. J. Phys.* 36:063001.

Bock, N., T. R. Dargaville, and M. A. Woodruff. 2012. Electrospraying of polymers with therapeutic molecules: State of the art. *Prog. Polym. Sci.* 37:1510–1551.

Buzgo, M., A. Mickova, M. Rampichova, and M. Doupnik. 2018. Blend electrospinning, coaxial electrospinning, and emulsion electrospinning techniques. In *Core-shell nanostructures for drug delivery and theranostics*, eds. M. L. Focarete, and A. Tampiere, 325–347. Duxford: Elsevier.

Carroll, C. P., and Y. L. Joo. 2006. Electrospinning of viscoelastic Boger fluids: Modeling and experiments. *Phys. Fluids* 18:053102.

Chen, G., Y. Xu, D. G. Yu, D. F. Zhang, N. P. Chatterton, and K. N. White. 2015. Structure-tunable Janus fibers fabricated using spinnerets with varying port angles. *Chem. Commun.* 51:4623–4626.

Chen, J., T. Zhang, W. Hua, P. Li, and X. Wang. 2020. 3D Porous poly(lactic acid)/regenerated cellulose composite scaffolds based on electrospun nanofibers for biomineralization. *Colloid Surf. A Physicochem. Eng. Asp.* 585:124048.

Crosby, A. J., and J. Y. Lee. 2007. Polymer nanocomposites: the "nano" effect on mechanical properties. *Polym. Rev.* 47:217–229.

Dicastillo, C. L., L. Garrido, N. Alvarado, J. Romero, J. L. Palma, and M. J. Galotto. 2017. Improvement of polylactide properties through cellulose nanocrystals embedded in poly(vinyl alcohol) electrospun nanofibers. *Nanomaterials* 7:106.

Elsayed, R. E., T. M. Madkour, and R. A. Azzam. 2020. Tailored-design of electrospun nanofiber cellulose acetate/poly(lactic acid) dressing mats loaded with a newly synthesized sulfonamide analog exhibiting superior wound healing. *Int. J. Biol. Macromol.* 164:1984–1999.

Forward, K. M., and G. C. Rutledge. 2012. Free surface electrospinning from a wire electrode. *Chem. Eng. J.* 183:492–503.

Frykberg, R. G., and J. Banks. 2015. Challenges in the treatment of chronic wounds. *Adv. Wound Care* 4:560–582.

Garavand, F., S. Rahaee, N. Vahedikia, and S. M. Jafari. 2019. Different techniques for extraction and micro/nanoencapsulation of saffron bioactive ingredients. *Trends Food Sci. Technol.* 89:26–44.

Gazzotti, S., R. Rampazzo, M. Hakkarainen et al. 2019. Cellulose nanofibrils as reinforcing agents for PLA-based nanocomposites: An *in situ* approach. *Compos. Sci. Technol.* 171:94–102.

Ghafari, R., R. Scaffaro, A. Maio, E. F. Gulino, G. L. Re, and M. Jonoobi. 2020. Processing-structure-property relationships of electrospun PLA-PEO membranes reinforced with enzymatic cellulose nanofibers. *Polym. Test.* 81:106182.

Gutiérrez, T. J., and V. A. Alvarez. 2017. Cellulosic materials as natural fillers in starch-containing matrix-based films: A review. *Polym. Bull.* 74:2401–2430.

Haider, A., S. Haider, and I. K. Kang. 2018. A comprehensive review summarizing the effect of electrospinning parameters and potential applications of nanofibers in biomedical and biotechnology. *Arab. J. Chem.* 11:1165–1188.

Hajikhani, M., Z. E. Djomeh, and G. Askari. 2021. Lycopene loaded polylactic acid (PLA) and PLA/copolymer electrospun nanofibers, synthesis, characterization, and control release. *J. Food Process. Preserv.* 45:15055.

Han, W., Y. J. Yu, N. T. Li, and L. B. Wan. 2011. Application and safety assessment for nanocomposite materials in food packaging. *Chi. Sci. Bull.* 56:1216.

Hasan A., A. Memic, N. Annabi et al. 2014. Electrospun scaffolds for tissue engineering of vascular grafts. *Acta Biomater.* 10:11–25.

Herniou-Julien, C., J. R. Mendieta, and T. J. Gutiérrez. 2019. Characterization of biodegradable/non-compostable films made from cellulose acetate/corn starch blends processed under reactive extrusion conditions. *Food Hydrocoll.* 89:67–79.

Ibili, H., and M. Dasdemir. 2019. Investigation of electrohydrodynamic atomization (electrospraying) parameters' effect on formation of poly(lactic acid) nanoparticles. *J. Mater. Sci.* 54:14609–14623.

Jafari, S. M. 2017. *Nanoencapsulation of food bioactive ingredients: principles and applications.* Amsterdam: Academic Press.

Jalvo, B., A. P. Mathew, and R. Rosal. 2017. Coaxial poly(lactic acid) electrospun composite membranes incorporating cellulose and chitin nanocrystals. *J. Membr. Sci.* 544:261–271.

Jash, A., and L. T. Lim. 2018. Triggered release of hexanal from an imidazolidine precursor encapsulated in poly(lactic acid) and ethylcellulose carriers. *J. Mater. Sci.* 53:2221–2235.

Katouzian, I., and S. M. Jafari. 2016. Nano-encapsulation as a promising approach for targeted delivery and controlled release of vitamins. *Trends Food Sci. Technol.* 53:34–48.

Khalf, A., and S. V. Madihally. 2016. Recent advances in multiaxial electrospinning for drug delivery. *Eur. J. Pharm. Biopharm.* 112:1–17.

Khosravi, A., A. Fereidoon, M. M. Khorasani, et al. 2020. Soft and hard sections from cellulose-reinforced poly(lactic acid)-based food packaging films: A critical review. *Food Packag. Shelf Life.* 23:100429.

Kumar, D., G. Babu, and S. Krishnan. 2019. Study on mechanical & thermal properties of PCL blended graphene biocomposites. *Polímeros.* 29:1–9.

Kumar, S. D., K. Venkadeshwaran, and M. K. Aravindan. 2020. Fused deposition modelling of PLA reinforced with cellulose nanocrystals. *Mater. Today.* 33:868–875.

Lavoine, N., I. Desloges, A. Dufrasne, and J. Bras. 2012. Microfibrillated cellulose – Its barrier properties and applications in cellulosic materials: A review. *Carbohydr. Polym.* 90:735–764.

Lim, L.-T. 2021. Electrospinning and electrospraying technologies for food and packaging applications. In *Electrospun polymers and composites*, eds. Y. Dong, A. Baji, and S. Ramakrishna, 217–259. Sawston: Woodhead Publishing.

Lim, L.-T., A. C. Mendes, and I. S. Chronakis. 2019. Electrospinning and electrospraying technologies for food applications. *Adv. Food Nutr. Res.* 88:167–234.

Liu, Y., Q. Ma, and M. Yang et al. 2016. Flexible hollow nanofibers: Novel one-pot electrospinning construction, structure and tunable luminescence – electricity – magnetism trifunctionality. *Chem. Eng. J.* 284:831–840.

Madhavan, K. N., N. R. Nair, and R. P. John. 2010. An overview of the recent developments in polylactide (PLA) research. *Bioresour. Technol.* 101:8493–8501.

Mendes, A. C., K. Stephansen, and I. S. Chronakis. 2017. Electrospinning of food proteins and polysaccharides. *Food Hydrocoll.* 68:53–68.

Mercante, L. A., V. P. Scagion, F. L. Migliorini, L. H. C. Mattoso, and D. S. Correa. 2017. Electrospinning-based (bio) sensors for food and agricultural applications: A review. *Trends Anal. Chem.* 91:91–103.

Mit-Uppatham, C., M. Nithitanakul, and P. Supaphol. 2004. Ultrafine electrospun polyamide-6 fibers: effect of solution conditions on morphology and average fiber diameter. *Macromol. Chem. Phys.* 205:2327–2338.

Morão, A., and F. Bie. 2019. Life cycle impact assessment of polylactic acid (PLA) produced from sugarcane in Thailand. *J. Environ. Polym. Degrad.* 27:2523–2539.

Morais, M. G. D., B. S. Vaz, E. G. Morais, and J. A. V. Costa. 2014. Biological effects of *Spirulina* (*Arthrospira*) biopolymers and biomass in the development of nanostructured scaffolds. *BioMed Res. Int.* 2014:1–9.

Moreira, J. B., M. G. Morais, E. G. Morais, B. S. Vaz, and J. A. V. Costa. 2018a. Electrospun polymeric nanofibers in food packaging. In *Impact of nanoscience in the food industry*, eds. A. M. Grumezescu and A. M. Holban, 387–417. Amsterdam: Elsevier, Inc.

Moreira, J. B., L. T. Lim, E. R. Zavareze, A. R. G. Dias, J. A. V. Costa, and M. G. Morais. 2018b. Microalgae protein heating in acid/basic solution for nanofibers production by free surface electrospinning. *J. Food Eng.* 230:49–54.

Moreira, J. B., L. T. Lim, E. D. R. Zavareze, E. R. G. Dias, J. A. V. Costa, and M. G. D. Morais. 2019. Antioxidant ultrafine fibers developed with microalga compounds using a free surface electrospinning. *Food Hydrocoll.* 93:131–136.

Nishino, T., and T. Peijs. 2014. All-cellulose composites. In *Handbook of green materials: Bionanocomposites: Processing, characterization and properties*, eds. K. Oksman, A.P. Mathew, A. Bismarck, O. Rojas, and O. Sain, 201–216. Hackensack: World Scientific.

Niu, B., P. Shao, Y. Luo, and P. Sun. 2020. Recent advances of electrosprayed particles as encapsulation systems of bioactives for food application. *Food Hydrocoll.* 99:105376.

O'brien, F. J. 2011. Biomaterials & scaffolds for tissue engineering. *Mater. Today.* 14:88–95.

Ortenzi, M. A., L. Basilissi, H. Farina, G. Di Silvestro, L. Piergiovanni, and E. Mascheroni. 2015. Evaluation of crystallinity and gas barrier properties of films obtained from PLA nanocomposites synthesized via "in situ" polymerization of L-lactide with silane-modified nanosilica and montmorillonite. *Eur. Polym.* 66:478–491.

Padmanabhan, S. C., M. C. Cruz-Romero, J. P. Kerry, and M. A. Morris. 2018. Food packaging: Surface engineering and commercialization. In *Nanomaterials for food packaging*, eds. M. A. P. R. Cerqueira, J. M. Lagaron, L. M. P. Castro, and A. A. M. O.S. Vicente, 301–328. Elsevier.

Park, C. H., and J. Lee. 2009. Electrosprayed polymer particles: Effect of the solvent properties. *J. Appl. Polym. Sci.* 114:430–437.

Patel, D. K., S. D. Dutta, J. Hexiu, K. Ganguly, and K. T. Lim. 2020. Bioactive electrospun nanocomposite scaffolds of poly(lactic acid)/cellulose nanocrystals for bone tissue engineering. *Int. J. Biol. Macromol.* 162:1429–1441.

Rojas, A., E. Velásquez, L. Garrido, M. J. Galotto, and C. L. de Dicastillo. 2020. Design of active electrospun mats with single and core-shell structures to achieve different curcumin release kinetics. *J. Food Eng.* 273:109900.

Santoro, M., S. R. Shah, J. L. Walker, and A. G. Mikos. 2016. Poly(lactic acid) nanofibrous scaffolds for tissue engineering. *Adv. Drug Deliv. Rev.* 107:206–212.

Scaffaro, R., L. Botta, F. Lopresti, A. Maio, and F. Sutera. 2017. Polysaccharide nanocrystals as fillers for PLA based nanocomposites. *Cellulose.* 24:447–478.

Shah T. V., and D. V. Vasava. 2019. A glimpse of biodegradable polymers and their biomedical applications. *e-Polymers.* 19:385–410.

Siakeng, R., M. Jawaid, H. Ariffin, S. M. Sapuan, M. Assim, and N. Saba. 2018. Natural fiber reinforced polylactic acid composites: A review. *Polym. Compos.* 40:446–463.

Somord, K., K. Somord, O. Suwantong, C. Thanomsilp, T. Peijs, and N. Soykeabkaew. 2018. Self-reinforced poly(lactic acid) nanocomposites with integrated bacterial cellulose and its surface modification. *Nanocomposites.* 4:102–111.

Songsurang, K., N. Praphairaksit, K. Siraleartmukul, and N. Muangsin. 2011. Electrospray fabrication of doxorubicin-chitosan-tripolyphosphate nanoparticles for delivery of doxorubicin. *Arch Pharm Res.* 34:583–592.

Stoyanova, N., D. Paneva, R. Mincheva et al. 2014. Poly(L-lactide) and poly(butylene succinate) immiscible blends: From electrospinning to biologically active materials. *Mater. Sci. Eng. C.* 41:119–126.

Tiimob, B. J., V. K. Rangari, G. Mwinyelle et al. 2018. Tough aliphatic-aromatic copolyester and chicken egg white flexible biopolymer blend with bacteriostatic effects. *Food Packag. Shelf Life.* 15:9–16.

Vatansever, E., D. Arslan, and M. Nofar. 2019. Polylactide cellulose-based nanocomposites. *Int. J. Biol. Macromol.* 137:912–938.

Vorawongsagul, S., P. Pratumpong, and C. Pechyen. 2020. Preparation and foaming behavior of poly(lactic acid)/poly(butylene succinate)/cellulose fiber composite for hot cups packaging application. *Food Packag. Shelf Life.* 27:100608.

Walther, A., and A. H. E. Muller. 2013. Janus particles: Synthesis, self-assembly, physical properties, and applications. *Chem. Rev.* 113:5194–5261.

Wang, K., X. K. Liu, X. H. Chen, D. G. Yu, Y. Y. Yang, and P. Liu. 2018. Electrospun hydrophilic Janus nanocomposites for the rapid onset of therapeutic action of Helicid. *ACS Appl. Mater. Interfaces.* 10:2859–2867.

Wang, F., M. Zhou, and Q. Jia. 2021. Evaluation of dynamic mechanical and cytotoxic properties of electrospun poly(lactic acid)/cellulose nanocrystalline composite membranes. *J. Phys. Conf.* 1759:012031.

Xiao, Q., and L. T. Lim. 2018. Pullulan-alginate fibers produced using free surface electrospinning. *Int. J. Biol. Macromol.* 112:809–817.

Xue, Y., Z. Mou, and H. Xiao. 2017. Nanocellulose as a sustainable biomass material: structure, properties, present status and future prospects in biomedical applications. *Nanoscale.* 9:14758–14781.

Yan, D., Q. Yao, F. Yu et al. 2020. Surface modified electrospun poly(lactic acid) fibrous scaffold with cellulose nanofibrils and Ag nanoparticles for ocular cell proliferation and antimicrobial application. *Mater. Sci. Eng. C.* 111:110767.

Yang, Z., J. Si, and Z. Cui et al. 2017. Biomimetic composite scaffolds based on surface modification of polydopamine on electrospun poly(lactic acid)/cellulose nanofibrils. *Carbohydr. Polym.* 174:750–759.

Yin, X., Y. Li, and P. Weng et al. 2018. Simultaneous enhancement of toughness, strength and superhydrophilicity of solvent-free microcrystalline cellulose fluids/poly(lactic acid) fibers fabricated via electrospinning approach. *Compos. Sci. Technol.* 167:190–198.

Yu, D. G., J. J. Li, M. Zhang, and G. R. Williams. 2017a. High-quality Janus nanofibers prepared using three-fluid electrospinning. *Chem. Commun.* 53:4542–4545.

Yu, H. Y., C. Wang, and S. Y. H. Abdalkarim. 2017b. Cellulose nanocrystals/polyethylene glycol as bifunctional reinforcing/compatibilizing agents in poly(lactic acid) nanofibers for controlling long-term in vitro drug release. *Cellulose*. 24:4461–4477.

Zaaba, N. F., and M. Jaafar. 2020. A review on degradation mechanisms of polylactic acid: Hydrolytic, photodegradative, microbial, and enzymatic degradation. *Polym. Eng. Sci.* 60:2061–2075.

Zare, Y., and K. Y. Rhee. 2019. Following the morphological and thermal properties of PLA/PEO blends containing carbon nanotubes (CNTs) during hydrolytic degradation. *Compos. B. Eng.* 175:107132.

Zawawi, E. Z. E., A. H. N. Hafizah, A. Z. Romli, N. Y. Yuliana, and N. N. Bonnia. 2020. Effect of nanoclay on mechanical and morphological properties of poly(lactide) acid (PLA) and polypropylene (PP) blends. *Mater. Today-Proc.* 46:1778–1782.

Zhang, X., W. Megone, T. Peijs, and J. E. Gaut. 2020. Functionalization of electrospun PLA fibers using amphiphilic block copolymers for use in carboxy-methyl-cellulose hydrogel composites. *Nanocomposites*. 6:85–98.

Zong, X., K. Kim, D. Fang, S. Ran, B. S. Hsiao, and B. Chu. 2002. Structure and process relationship of electrospun bioabsorbable nanofiber membranes. *Polymer*. 43:4403–4412.

15 Applications of PLA/ Cellulose Composites

Shiji Mathew
Mahatma Gandhi University
Kottyam, India

CONTENTS

15.1 INTRODUCTION

The massive use of petrochemical-based plastic products has resulted in the over-accumulation of their non-degradable waste in landfills and marine and environmental systems worldwide. This in turn poses extreme risk and challenges to the health and life of biotic components. These growing environmental concerns have tilted

DOI: 10.1201/9781003160458-15

researchers' minds to mainly focus on exploring and producing novel eco-friendly, renewable, and biodegradable resources (Mishra et al., 2020).

Polylactic acid (PLA), being renewable, biodegradable, biocompatible, possesses good thermomechanical, optical, and barrier properties and is extensively used as an alternative for non-biodegradable plastics. At the same time, PLA also possesses some drawbacks such as inherent brittleness and low toughness, poor processability, lower heat distortion temperature, lower service temperature, lower thermal resistance and formability, and foamability because of slow crystallization rate and melt strength (Gupta et al., 2007; Nofar et al., 2019; Xu, 2020). These limitations of PLA can be improved by the addition of nanofillers. Among the varied bio-based polymers, cellulose-based nanofillers, due to their non-toxic nature and abundance in nature, as well as their higher aspect ratio and larger surface area, are one of the best bio-based nanofillers that can be used as a reinforcement in PLA composites with improved properties. Cellulose particles having one dimension in the nanoscale are considered nanocellulose.

Nanocellulose can exist in different dimensions and is named cellulose nanocrystals (CNC), cellulose nanofibers (CNF), and bacterial celluloses (BC). But still, nanocellulose also presents certain demerits such as low specific density, high strength, and modulus (Miao and Hamad, 2013), which can in turn be improved after blending with PLA. Recently, a huge amount of attention has been paid to the development of hybrid PLA/cellulose composites, enabling commercialization and potential applications in medical, engineering and electrical, food packaging, and automotive industries. The first part of this chapter mainly details the individual applications of PLA and cellulose composites in different fields with more emphasis on the medical field. This is then followed by the combined applications of PLA/cellulose composites in various sectors.

15.2   APPLICATIONS OF PLA-BASED COMPOSITES

Owing to the many advantages of PLA over other petroleum-based products such as biodegradability, biocompatibility, renewability, thermo-plasticity, thermomechanical properties, and ease in availability, numerous daily use commercial consumer products such as bottles, nonwovens, fabrics, bags, and bottles are manufactured using PLA around the world. Besides this, PLA finds immense wider applications in areas such as the biomedical field, paper-coating applications, food packaging. thermoforming, injection molding and 3D printing, and many other industrial applications (Battù, 2018; Gotro, 2012).

15.2.1   BIOMEDICAL APPLICATIONS OF PLA-BASED COMPOSITES

One of the most common applications of PLA-based composites is in the biomedical field. PLA-based composites are extensively used for drug delivery purposes, in tissue engineering, for developing surgical sutures and medical implants, as scaffolds in the cosmetics and dermatology department, for tissue regeneration, and for producing medical packaging systems (Saba et al., 2017). Figure 15.1 depicts some of the important applications of PLA-based composites in the biomedical sector.

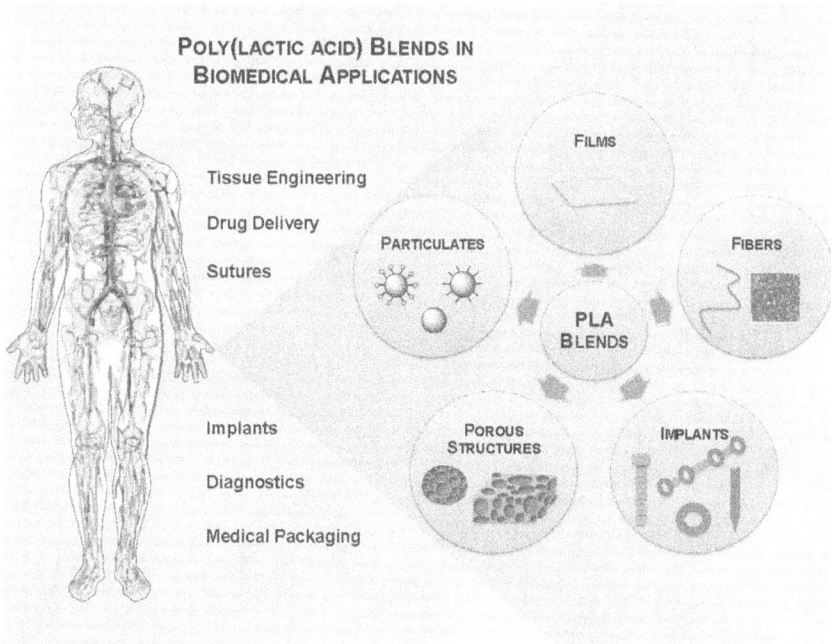

FIGURE 15.1 Biomedical applications of PLA. (Reprinted with permission from Saini et al. (2016). Copyrights © 2016 Elsevier B.V.)

15.2.2 MEDICAL 3D PRINTING TECHNOLOGY

PLA is a promising candidate in the medical and pharmaceutical industry, which is greatly exploited for fabricating personalized materials using medical 3D or 3D printing (3DP) technology. 3DP technology has emerged as an important and innovative tool for the development of therapeutic approaches for personalized medicine. 3DP has been extensively used to develop innovative, multifunctional, versatile, and smart medical and pharmaceutical products (dos Santos et al., 2021). Such smart materials can be produced using additive manufacturing (AM) technology, which performs many processes in a layer-by-layer pattern to fabricate a solid object. The additive 3D printing technologies commonly used in the medical and pharmaceutical industry include extrusion-based fused deposition modeling (FDM), stereolithography (SLA), digital light processing (DLP), selective laser sintering (SLS), selective laser melting (SLM), electron beam melting (EBM), and inkjet-based 3DP (Mardis, 2018; Mathew et al., 2020). Some examples of medical care products developed using PLA-based composites are given in the following section.

15.2.2.1 3D-Printed Implants

PLA has gained immense attention as a suitable material for the fabrication of medical implants and sutures using 3DP technology. Tappa et al. developed patient-specific orthopedic fixation gears such as screws, pins, and bone plates composed of PLA by 3D printing and loaded them with gentamicin (GS) and methotrexate (MTX) for

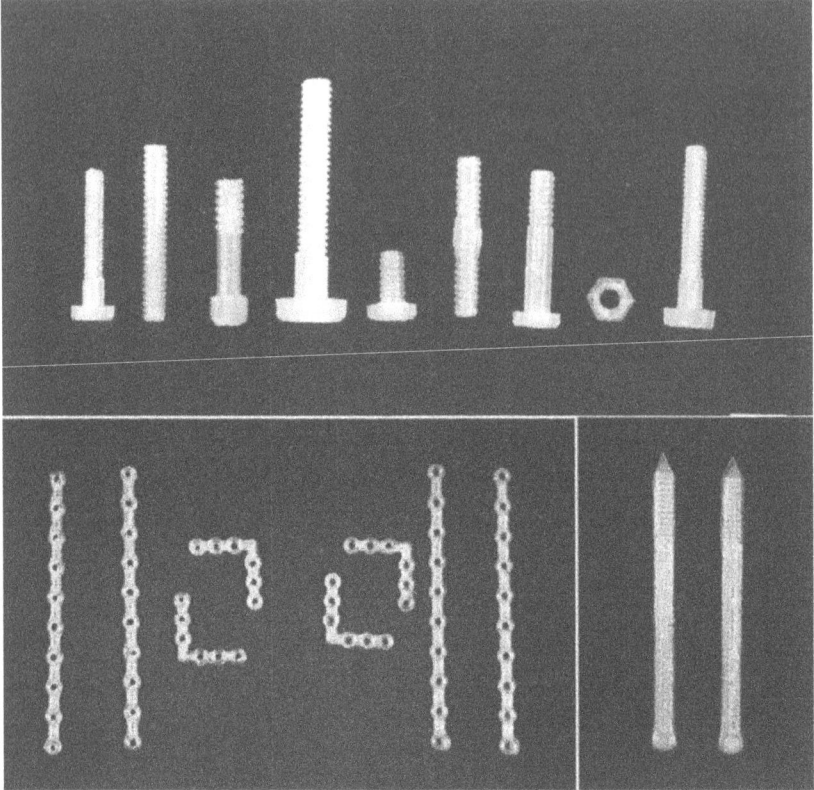

FIGURE 15.2 3D-Printed PLA orthopedic screws, pins, and plates. (Reprinted with permission from Tappa et al. (2019) under the Creative Commons Attribution (CC BY) license. Copyright © 2019 by the authors. Licensee MDPI, Basel, Switzerland.)

localized drug delivery application. Figure 15.2 shows the images of screws, pins, and bone plates made of PLA by using 3DP technology. These drug-impregnated implants were found to possess antibacterial and chemotherapeutic effects at the expense of decreased flexural and compressive strengths when compared to control implants.

15.2.2.2 3D-Printed Sutures

In another study, (Gayer et al., 2019) developed a PLA/calcium carbonate composite powder using the SLS method. Using a modified EOS Formiga P110 additive manufacturing system, the composite powder was utilized to fabricate a patient-specific implant demonstrator with interconnected pore structure, which paved potential promises as medical sutures to promote healing of critical size bone defects in craniomaxillofacial surgery (Figure 15.3).

15.2.2.3 3D-Printed Face Masks

The outburst of the COVID-19 pandemic has caused a significant lack of personal protective equipment such as face masks. In this context, the manufacturing of face

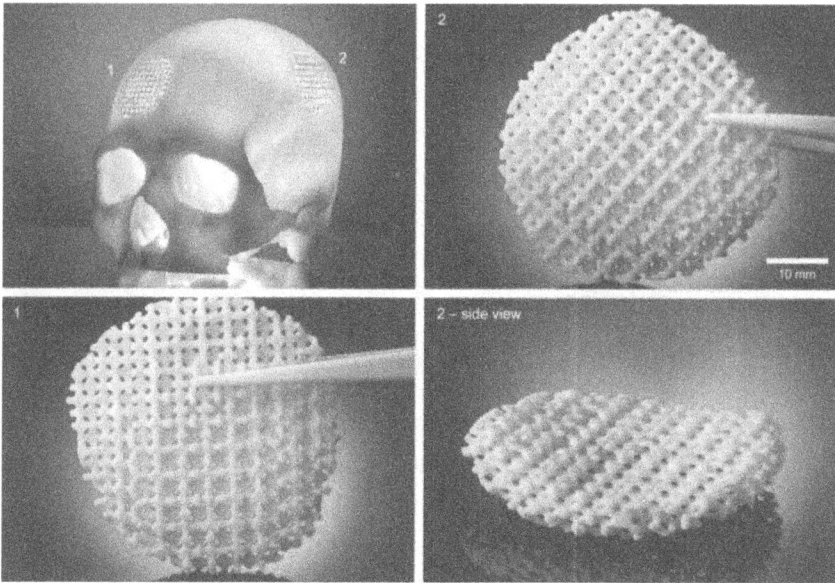

FIGURE 15.3 Patient-specific cranial implant demonstrators manufactured from the PLA-1.0/calcium carbonate (77/23) composite using a modified Formiga P 110 laser sintering machine. (Reprinted with permission from Gayer et al. (2019) under the Creative Commons Attribution (CC BY) license. Copyright © 1969, Elsevier.)

protective masks using 3D printing is a promising technology. Interestingly, due to its low printing temperatures of 200°C–210°C, favorable mechanical properties, smooth appearance, and low toxicity, PLA has proven to be one of the most attractive materials for 3D printing (Vicente et al., 2019).

In one such attempt, 3D-printed PLA-based masks were fabricated using the FDM technique (Figure 15.4), which was found to be suitable for protection against various microorganisms, thereby proving to be a promising material for application in the current pandemic crisis (Vaňková et al., 2020). It was also investigated that the PLA masks contaminated with various bacteria, fungi, and SARS-CoV-2 can be successfully disinfected with common chemical disinfectants such as ethyl alcohol, isopropyl alcohol, or sodium hypochlorite without causing any changes to PLA structural properties.

15.2.2.4   3D-Printed PLA-Based Scaffolds

Another application of PLA-based composites is for developing scaffolds for tissue engineering purposes. In a recent study conducted by Rages-Martinez et al., keratin and chitosan were used as a reinforcement in PLA to develop 3D printable polymer composite scaffolds for tissue engineering application. Keratin was isolated from chicken feathers and hair (Rojas-Martínez et al., 2020). Chitosan and keratin were found to increase cell growth in the PLA matrix. Figure 15.5 shows the schematic representation of the fabrication of PLA/keratin/chitosan scaffolds for tissue engineering application.

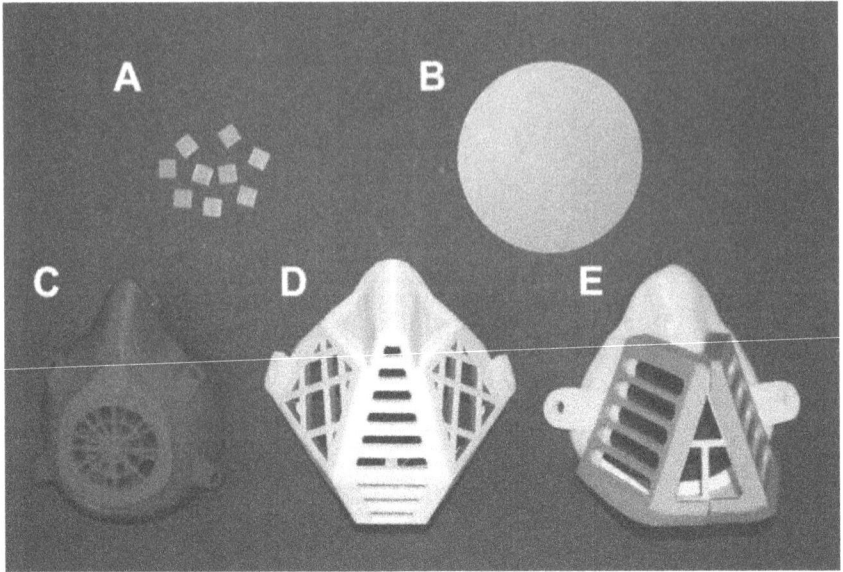

FIGURE 15.4 Different types of masks made from PLA filaments using 3D printing by FDM technology. (A) PLA carriers (1 × 1 cm). (B) Circular plate with a diameter of 10 cm (printed vertically). (C–E) Different types of PLA masks. (Reprinted with permission from Vaňková et al. (2020) under the Creative Commons Attribution (CC BY) license. Copyright © 2020, Vaňková et al.)

15.3 APPLICATIONS OF NANOCELLULOSE-BASED COMPOSITES

Cellulose is considered the most abundant, sustainable, and inexhaustible polymer available in nature. Nanostructured cellulose is termed as nanocellulose. It can exist in many forms such as cellulose nanocrystal (CNC or NCC), cellulose nanofibers or

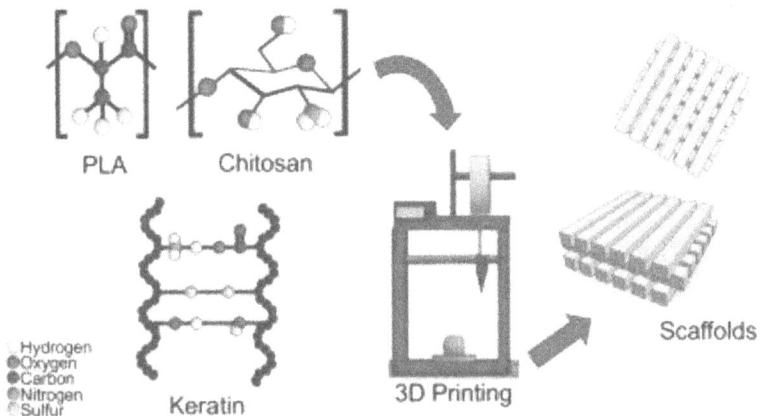

FIGURE 15.5 Schematic showing the development of PLA-based scaffolds using 3DP technology. (Reprinted with permission from Rojas-Martínez et al. (2020). Copyright © 2020 Elsevier Ltd.)

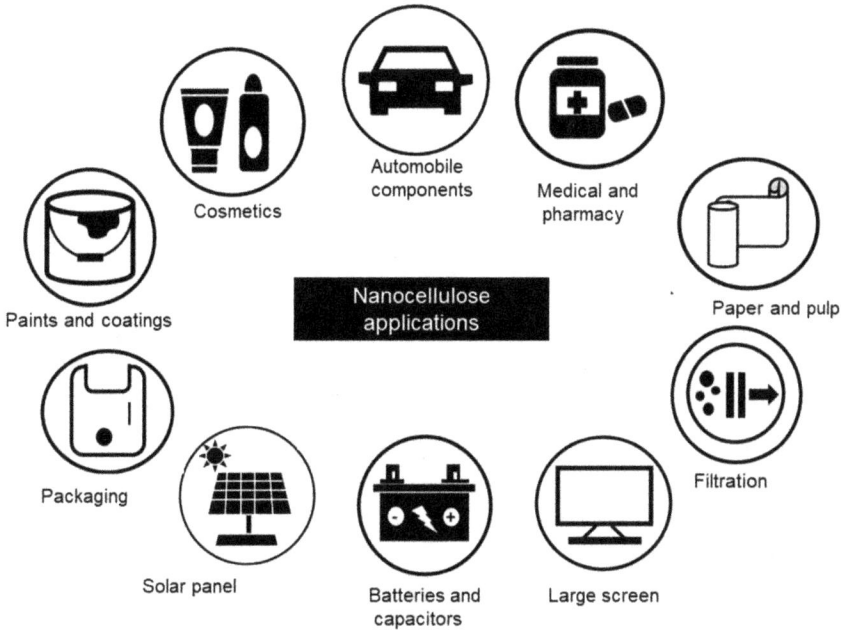

FIGURE 15.6 Varied applications of nanocellulose-based composites.

nanofibrillated cellulose (CNF), or nanostructured cellulose by bacteria, known as bacterial nanocellulose (Phanthong et al., 2018). The versatile characteristics of nanocellulose enable its further processing into bionanocomposites with potential applications. One of the most important applications of nanocellulose is in the paper and packaging industry. Owing to its nanoscale dimension, high strength, and stiffness, it is widely accepted for use in varied industries such as construction companies, automobile industry, electronics, pharmacy, and cosmetics industry. Other applications of nanocellulose include the development of paints and coatings, as 3D printing bioink. Figure 15.6 summarizes the main application areas of nanocellulose. Nanocellulose can be fabricated into paper and hydrogel for versatile applications. Some of the major applications of nanocellulose papers and hydrogels are detailed below.

15.3.1 APPLICATIONS OF NANOCELLULOSE-BASED HYDROGELS

Nanocellulose-based hydrogels have got many properties that enable them to be used as stimuli-responsive, self-healing, and shape-memory materials. The examples of each are discussed in the following section.

15.3.1.1 Stimuli-Responsive Nanocellulose Hydrogels

Stimuli-responsive hydrogels are composite materials that respond to variations in pH, temperature, chemicals, ionic, and magnetic strength. The main application of such stimuli-responsive hydrogels is applied for controlled drug delivery approaches. A smart, biocompatible, dual-responsive hydrogel was prepared by grafting cellulose

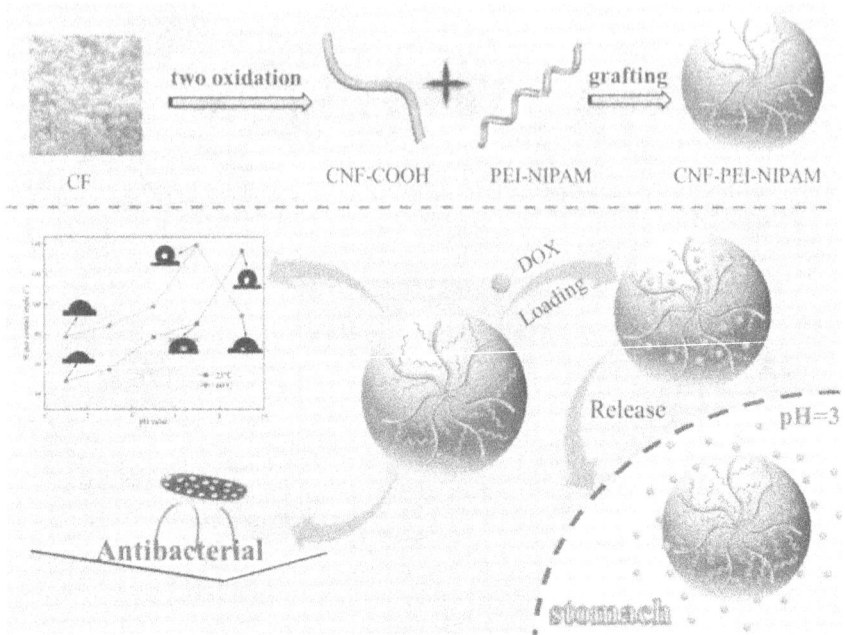

FIGURE 15.7 Schematic representation of CNF-PEI-NIPAM hydrogel preparation and the mechanism of controlled release of doxorubicin controlled by temperature and pH. (Reprinted with permission from Liang et al. (2020). Copyrights © 2020 Elsevier Ltd.)

nanofibers into dual-responsive temperature- and pH-sensitive polyethyleneimine-N-isopropylacrylamide (CNF-PEI-NIPAM) (Liang et al., 2020). The developed nano-fibers gels were used for sustained release of doxorubicin. The results showed that CNF-PEI-NIPAM hydrogels exhibited an excellent dual-response performance to pH and temperature, affording the CNF-PEI-NIPAM aerogel a high loading capacity (330.12 mg/g) for doxorubicin and a high cumulative release rate (59.45%) at 37°C and pH 3. Figure 15.7 depicts the development and application potential of CNF-PEI-NIPAM hydrogels for pH- and temperature-controlled delivery of doxorubicin into the stomach.

15.3.1.2 Shape-Memory Hydrogels

Hybrid nanocellulose hydrogel systems can even be applied for shape-memory applications. In such materials, the systems are initially locked kinetically into a temporary state by the first stimulus and then released from the kinetic trap by a second stimulus to regain equilibrium shape (Löwenberg et al., 2017). Recently, Sain and colleagues developed biocomposites using nanocellulose networks (CNC and CNF) extracted from onion skin and bioresin derived from vegetable oil (tung oil) and furan derivative, which exhibited significant shape-memory behavior (Sain et al., 2020). Figure 15.8 shows the schematic illustration for the development of the biocomposites film. A significant degree of shape recovery (~80%–96%) was observed in the biocomposites, which was comparable to the shape recovery properties of polyurethane.

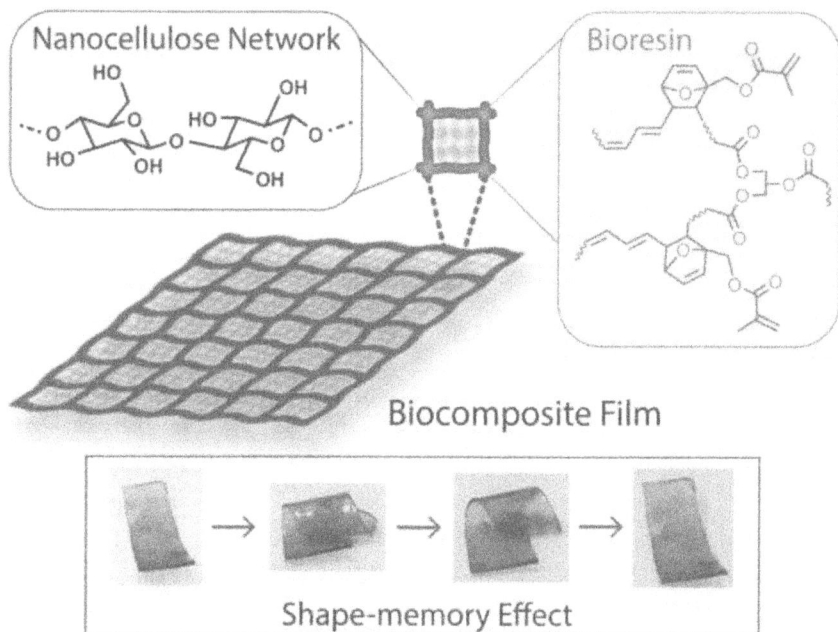

FIGURE 15.8 Schematic representation of nanocellulose biocomposites and their shape-memory effect. (Reprinted with permission from Sain et al. (2020) under common creative license. Copyright © 2020 by the authors. Licensee MDPI, Basel, Switzerland.)

15.3.1.3 Self-Healing Hydrogels

Conductive and self-healing nanocellulose-based hydrogels can mimic human skin and find immense applications in soft robots and wearable electronics. In a study, a physiochemically dual-crosslinked chemically modified cellulose nanofibers-carbon nanotubes/polyacrylic acid (TOCNF-CNTs-PAA) hydrogel was developed, which could serve the function as a self-healing material with the ability to monitor human activity for multipurpose applications such as use in wearable strain sensors, health monitoring systems, and smart robots (Jiao et al., 2021). Figure 15.9 illustrates the formation, application, and properties of TOCNF-CNTs-PAA hydrogel as a self-healing material.

15.3.2 Applications of Nanocellulose-Based Papers

Nanocellulose can be fabricated into thin sheets of paper, which can perform important functions in the medical as well as engineering and electronics field. Some of the major applications of nanocellulose-based papers are discussed below.

15.3.2.1 Paper and Packaging Industry

The papermaking industry mainly relies on the availability of cellulose from lignocellulosic materials, primarily wood. Nanocellulose, being a flexible, cheap, biodegradable, eco-friendly, and lightweight material, has attracted the paper industry as

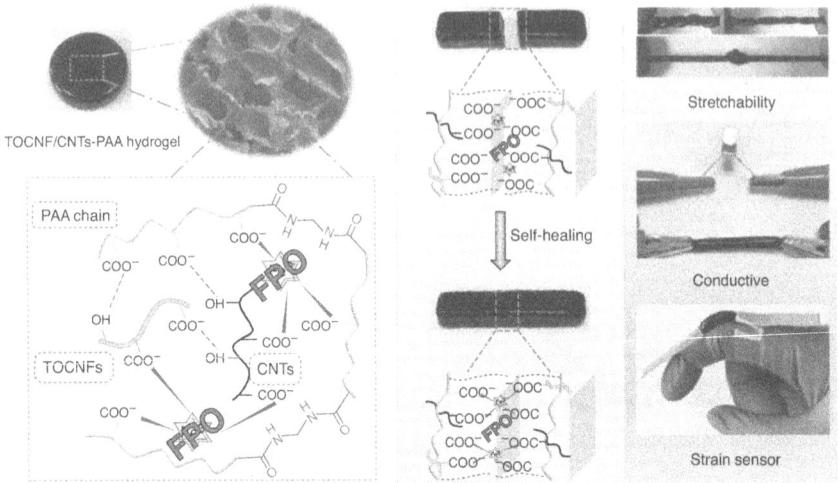

FIGURE 15.9 Schematic showing fabrication, properties, and applications of TOCNF-CNTs-PAA hydrogel. (Reprinted with permission from Jiao et al. (2021). Copyright © 2021, The Author(s), under exclusive license to Springer Nature B.V.)

an excellent reinforcement for improving the quality of paper and cardboard. The use of petroleum products has been replaced with the advent of nanocellulose-based composites in the packaging industry. The favorable properties of nanocellulose such as its availability, cost-effectiveness, and low toxicity, together with the improved mechanical, gas, and barrier properties, make it a suitable candidate in the packaging industry.

Based on this, a recent study developed a cellulose nanofiber/cellulose nanofiber oxidized coating (CNF/CNF-OX), which was used to coat the surface of fluting and core board papers (Ozcan et al., 2021). Figure 15.10 demonstrates the synthesis of CNC-based coating and the coating and printing of CNC-coated papers. The results showed that the CNF/CNF-OX coated papers had smoother surfaces and gave better results in terms of both gloss properties and printability.

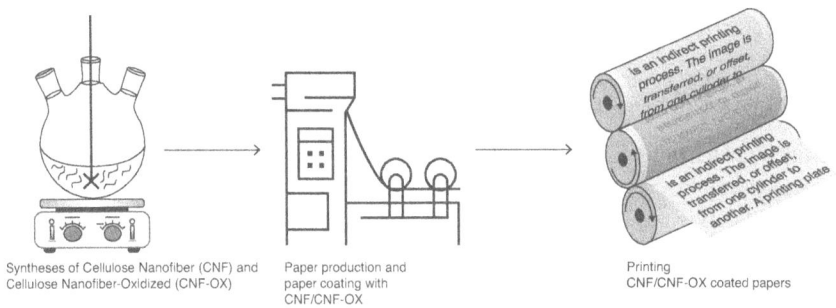

FIGURE 15.10 Schematic showing synthesis, coating, and printing of CNF/CNF-OX-coated papers. (Reprinted with permission from Ozcan et al. (2021).)

15.3.2.2 Cosmetics

Exposure to UV radiation from the sun can cause photoaging, photo-carcinogenesis, and photo-immunosuppression that can result in skin damage and facial aging. Early aging is one of the commonest problems people face with symptoms such as the appearance of wrinkles, fine lines, and skin pigmentation (Uitto, 1997). With the evolution of cosmetics, the lives of people suffering from early aging got transformed. Anti-aging creams are widely used by people to combat skin aging and other related problems. Most cosmetic formulations contain metal oxide nanoparticles to provide immediate anti-aging effects (Raj et al., 2012). But these creams are often linked with the risk of toxicity concerns. Nanocellulose is one of the promising biomaterials that can be included in cosmetic products as an alternative to the use of metal oxide nanoparticles.

In a recent study by Fonseca et al., 2021, bacterial nanocellulose (BC) was used as a back layer for developing innovative patches based on dissolvable hyaluronic acid microneedles (HA MNs), loaded with rutin for dermo-cosmetic applications. The in vivo and in vitro efficiency of the patches was investigated. The results showed that the presence of BC provided a molecular support for the incorporation of active ingredients and improved the mechanical resistance of HA MNs. The in vivo safety measures of the patches were evaluated on human volunteers and no cytotoxicity was noted. Figure 15.11 depicts the schematic for the development of rutin-loaded BC-HA microneedle patches and their skin applications.

15.3.2.3 Engineering and Electronics Field

The increased flexibility and conductivity of nanocellulose-based paper offers it greater opportunities to be used in the electronics industry for developing conductive and luminescent materials such as solar panels/cells, batteries and capacitors, large flexible display screens, and many promising applications in printed electronics, for instance, in supercapacitors, transparent conductive electrodes, electroluminescent devices, flexible circuits, organic light-emitting diodes, organic solar cells, touch screens, transistors, and conductive lines.

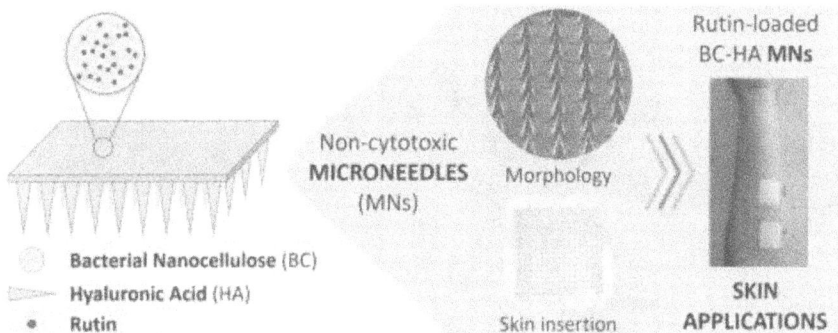

FIGURE 15.11 Schematic illustration of rutin-loaded BC-HA microneedle patches and their skin applications. (Reprinted with permission from Fonseca et al. (2021). Copyright © 2020 Elsevier B.V.)

15.3.2.3.1   Use in Solar Panels/Cells

Nanocellulose-based paper exhibits unique features such as excellent mechanical, thermal, and chemical stability; high transparency; and low surface roughness suitable for the development of optoelectronic devices (Zhang et al., 2016). In a recent study, Gao et al. (2019a) used nanocellulose derived from cotton for the development of nanocellulose paper-based flexible perovskite solar cells (PSCs). The nanocellulose paper (NCP) was made from the viscous solution of nanocellulose. Then the so-formed NCP was coated with acrylic resin to form a waterproof layer and further used for the fabrication of PSCs, which could be easily disposed of by flame burning (Figure 15.12).

15.3.2.3.2   Nanocellulose as Supercapacitors

Flexible, lightweight, high-performance, eco-friendly supercapacitors are always in great demand in appliances such as wearable electronics, hybrid electric vehicles,

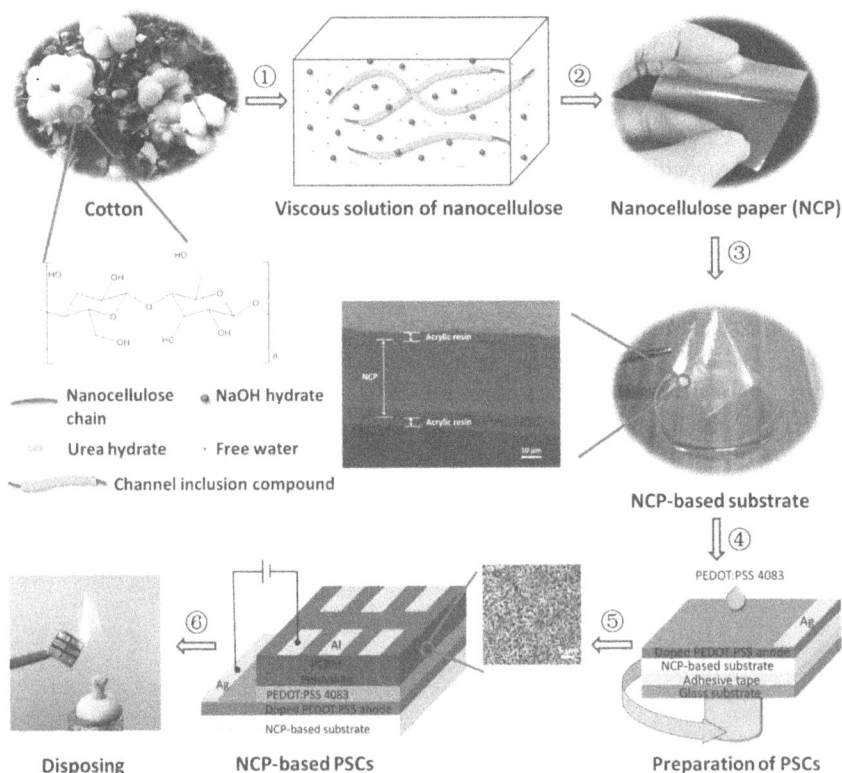

FIGURE 15.12 Schematic showing preparation process of NCP-based substrate and NCP-based PSCs. (1) Extraction of nanocellulose from cotton to form viscous solution. (2) From this viscous solution of nanocellulose, NCP was made. (3) NCP was coated by acrylic resin to form a waterproof layer on its surface. (4, 5) The development of NCP-based PSCs. (6) Disposal of flexible PSCs by flame burning. (Reprinted under Creative Common license from Gao et al. (2019a).)

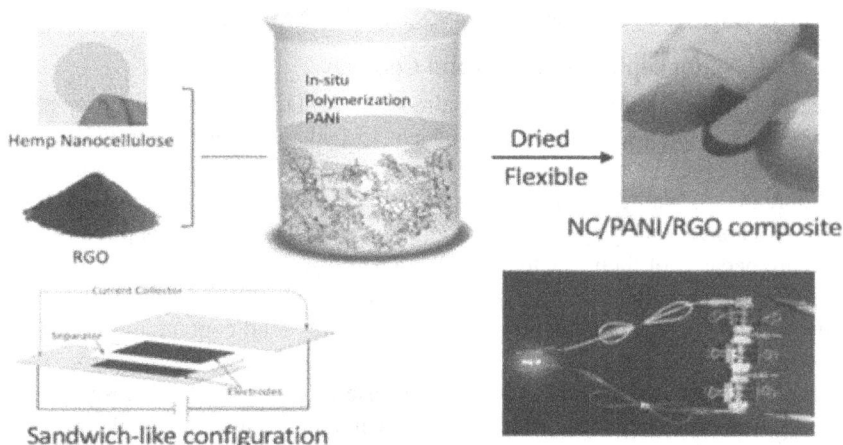

FIGURE 15.13 Development of NC/PANI/RGO composite-based supercapacitors. (Reprinted with permission from Hsu et al. (2019). Copyright © 2019, American Chemical Society.)

and industrial grid storage. One of the most potential optoelectronic applications of nanocellulose is in the development of supercapacitors.

In a study, nanocellulose was isolated from clean hemp fibers. As seen in Figure 15.13, a nanocellulose-based polyaniline/reduced graphene oxide (NC/PANI/RGO) electrode was developed by filtrations driven by a vacuum and assembled into a sandwich like supercapacitors (Hsu et al., 2019). The developed composite electrode exhibited good mechanical properties, large active materials mass loading ratio of 16.5 mg/cm^2, and the assembled supercapacitor gave small impedance at 3.90 Ω, suggesting an excellent conductivity. The high content of nanocellulose in the electrode composite resulted in higher mechanical properties and low electrochemical performance.

15.3.2.3.3 Nanocellulose-Based Sensors and Actuators

The inclusion of nanocellulose in sensing applications has been a promising one as nanocellulose is highly versatile and biocompatible. Because of this, nanocellulose can be combined with other stimuli-responsive compounds for fabricating rapid detection sensors such as gas (Listyarini et al., 2018; Zhang et al., 2017), chemical (Chen et al., 2019; Wu et al., 2020), protein (Fontenot et al., 2017; Naghdi et al., 2019), ion (Liu et al., 2020; Silva et al., 2020), and glucose (Prapaporn et al., 2020; Uddin et al., 2019) detection and also in light sensors (Wu et al., 2019; Xiong et al., 2020). Other applications include use in self-healing strain sensors (Zheng et al., 2019; Zhou and Hsieh, 2018), skin sensors (Han et al., 2019; Zheng et al., 2020), and touch sensors and pressure sensors (Gao et al., 2019b; Yang et al., 2021).

Actuators are components of a machine or device that promotes its movement. Cellulose-based paper can be effectively converted into actuators. Cellulose-based electroactive paper (EAPap) is emerging as a promising actuator as it is lightweight and has large bending deformation and low actuation voltage and power consumption. Based on the functions they perform, EAPaps are of three types: piezoelectric, ionic, and hybrid EAPap (Kim, 2021).

15.3.3 Other Applications of Nanocellulose-Based Composites

15.3.3.1 Nanocellulose in Paints and Coatings

Coatings and paints are applied over almost all surfaces such as wood or plastic furniture, automobiles, construction products, and others not only to provide decorative effect for the materials but also to protect these surfaces from the environment, microorganisms, and man-made damage. Paints composed of waterborne polyurethane (PU) is an emerging promising option and has been widely applied in wood furniture because of their low temperature flexibility, acid-alkali resistance, solvent resistance, and excellent weather resistance (Kong et al., 2019). But one of its major drawbacks is low mechanical properties due to lower solid content and weak intermolecular force.

In their recent study, Kong and colleagues used 0.1 wt% nanocellulose as a reinforcing agent for preparing waterborne polyurethane emulsion by the chemical grafting method. As shown in Figure 15.14, the developed coating was found to be green and sustainable, with less release of volatile organic compounds (VOCs), and showed excellent improvement in the comprehensive properties of the PU coating (Kong et al., 2019). The results showed that the tensile strength, elongation at break, hardness, and abrasion resistance of the waterborne PU paint increased by up to 58.7%, ~55%, 6.9%, and 3.45%, respectively, compared to the control PU, while the glossiness and surface drying time were hardly affected.

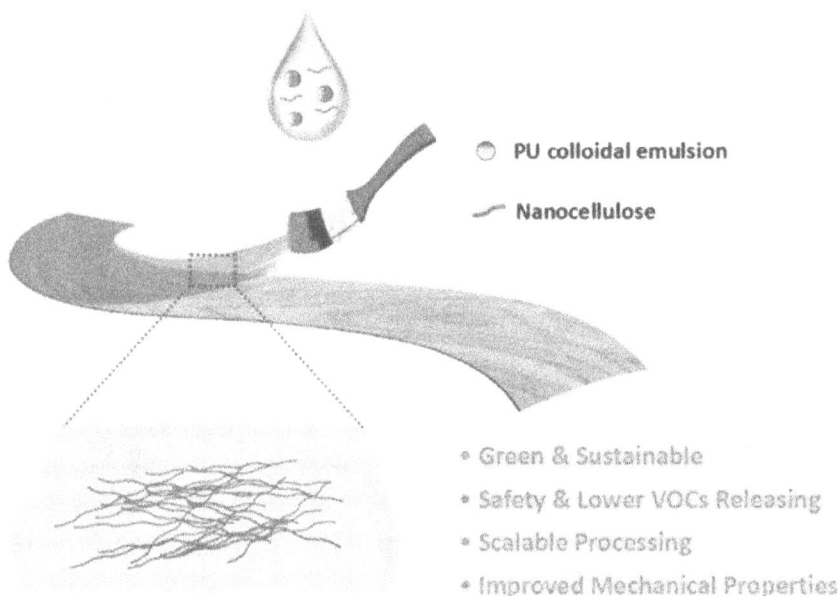

FIGURE 15.14 Schematic of nanocellulose-reinforced waterborne polyurethane wood coating material. (Reprinted with permission from Kong et al. (2019) under Common Creative license. Copyright © 2019 by the authors.)

15.3.3.2 Nanocellulose-Based Bioink for 3D Printing

Low cytotoxicity and the structural similarity of nanocellulose to extracellular matrices enable it to stand out as a platform material for bioink formation, which can be used for 3D printing purposes. Recently, low concentration inks formulated with nanocellulose assisted with gelatin methacrylate (GelMA) were used for the preparation of 3D-printed scaffolds for wound healing applications (Xu et al., 2019). Figure 15.15 demonstrates the fabrication of nanocellulose-GelMA bioink formulation and its application in extrusion-based 3D printing.

15.3.3.3 Nanocellulose in Medical Applications

The favorable mechanical features, eco-friendly nature, and nanofibrous structure of nanoforms of cellulose enable its extensive use in the medical field for varied applications. The primary applications include a wound dressing scaffold, a controlled drug delivery system, a medical sensor, and tissue and bone engineering. Table 15.1 provided below summarizes the recent works related to the application of nanocellulose in the medical and pharmaceutical industry.

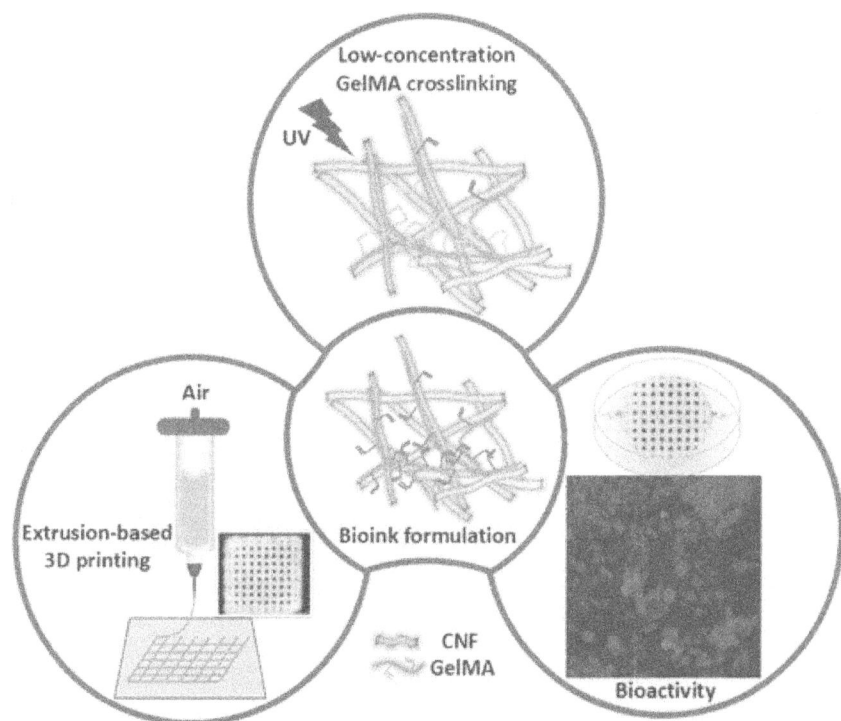

FIGURE 15.15 Illustration depicting the preparation of nanocellulose-GelMA bioink formulation, its application in 3D printing, and bioactivity with 3T3 fibroblast cell culture. (Reprinted with permission from Xu et al. (2019) under common creative license. Copyright © 2019 American Chemical Society.)

TABLE 15.1

Recent Studies on Nanocellulose-Based Composites for Medical Applications

Cellulose Source	Method of Development	Applications	References
Nanocellulose from wood	Mechanical defibrillation	Wound dressing scaffold	Claro et al. (2020)
Nanocellulose from *Gluconacetobacter xylinus*	Microbial	Printed circuit boards for medical sensing	Yuen et al. (2020)
Sponge with Janus character made of cellulose nanofibers	Heterogeneous mixing and freeze drying	Treatment of hemorrhagic wounds	Cheng et al. (2020)
Nanocellulose from oil palm empty fruit bunches	Wet spinning	Biocompatible fiber	Fahma et al. (2020)
Bacterial nanocellulose-based paper	Laser printing technology	Artificial sensor tongue for chemical discrimination	Abbasi-Moayed et al. (2018)
Bacterial nanocellulose-based scaffold from *Gluconacetobacter xylinus*	3D printing	Preparation of kidney model for tissue engineering	Sämfors et al. (2019)
Cellulose nanofiber from cotton	Acid-based hydrolysis and sonication	As drug delivery system for honey as an antimicrobial wound dressing	Md Abu et al. (2020)

15.3.3.4 Nanocellulose for Water Purification and Filtration

Membrane separation technology is one of the most successful methods used for removing contaminants from water (van Reis and Zydney, 2001). In this technology, ultrafiltration (UF) membranes are considered important for wastewater treatment (Cheryan, 1998). The preparation of ultrafiltration membranes requires relatively expensive synthetic materials and hence is often unsustainable (Sharma et al., 2020). Composite membranes made up of nanocellulose have emerged as an excellent alternative for ultrafiltration membranes.

Figure 15.16 shows the preparation of ultrafiltration membranes composed of filter paper (FP) as a support membrane and nanocellulose (NC) as a surface barrier layer (Wang et al., 2019). Two types of nanocelluloses with different sizes were chosen for this study: (1) CNCs prepared from microcrystalline cellulose and (2) CNFs prepared from hardwood bleached kraft pulp (HBKP). The results showed that the NC/FP composite membranes prepared by vacuum drying at 60°C with 0.1% CNFs exhibited excellent ultrafiltration properties with retention rates as high as 97.14% and an acceptable flux of 46,279 L m^{-2} h^{-1}.

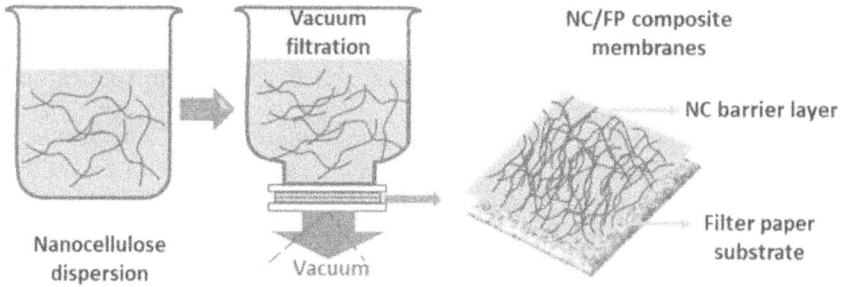

FIGURE 15.16 Schematic representation showing the fabrication of NC/FP composite membranes for filtration performance. (Reprinted with permission from Wang et al. (2019). Copyright © 2018, Springer Nature B.V.)

15.4 APPLICATIONS OF PLA/CELLULOSE COMPOSITES

In this chapter, we have seen the important individual applications of PLA and nanocellulose-based composites. Likewise, the PLA composites reinforced with nanocellulose fillers have been demonstrated to have efficient and improved properties with promising applications. Eventually, the development of PLA/cellulose composites is still under progress and is limited only to lab-scale experiments (Vatansever et al., 2019). The following table (Table 15.2) gives a list of recent studies based on various applications of PLA/cellulose composites.

TABLE 15.2

Various Studies Based on PLA/Cellulose Composites and Their Applications

Sl. No	Aim of Study	Method of Preparation	Applications	References
1	To prepare cellulose nanofibrils from microcrystalline cellulose and its reinforcement in PLA/starch nanocomposite film	High-pressure homogenization and hot water pre-treatment	Food packaging	Mao et al. (2019)
2	To develop three-phase multilayered materials by complexing a dry with CNF/CNC film with two layers of PLA sheets	Heat pressing process	High barrier and antioxidant Multilayer food packaging	Le Gars et al. (2020)
3	To develop PLA-based nanocomposites using CNF as reinforcing agent and maleated PLA (PLA-g-MA) as compatibilizer	Melt mixing and twin-screw extrusion	Nanocomposites with improved mechanical and physical properties	Ghasemi et al. (2018)

(Continued)

TABLE 15.2 (*Continued*)
Various Studies Based on PLA/Cellulose Composites and Their Applications

Sl. No	Aim of Study	Method of Preparation	Applications	References
4	Continuous processing of nanocellulose and PLA	Slot die coating and extrusion coating	Multilayer barrier coatings	Koppolu et al. (2019)
5	To develop an emulsion of PLA/*Apocynum venetum* cellulose nanofibers with controlled sea buckthorn extract	Electrospinning	Drug delivery system	Wang et al. (2021)
6	To develop PLA/cellulose/ ZnO nanocomposite film for the electrostimulated release of poorly water-soluble curcumin	Solvent casting	Drug delivery	Gunathilake et al. (2020)
7	To develop superabsorbent hydrogel PLA/cellulose composite	–	As water and fertilizer reservoir for agricultural applications	Calcagnile et al. (2019)
8	To develop PLA/CNF composites	Melt spinning	Composites with high stiffness	Clarkson et al. (2019)

15.5 CONCLUSIONS

Consumer demands for biobased polymers have increased recently. In this context, PLA is considered one of the popular and promising biopolymers, which can be used for several applications such as in food packaging, medical, automotive, electrical, and engineered equipment. However, the application of PLA is limited at times due to some drawbacks such as inherent brittleness and low toughness, poor processability, lower heat distortion temperature, lower service temperature, lower thermal resistance, formability and foamability because of slow crystallization rate and melt strength. Research has shown that the reinforcement with nanostructured cellulose can have a great impact on improving these limitations of PLA. This chapter gives vast information on the individual application potentials of PLA and nanocellulose-based composites with suitable examples. As already mentioned, the development of PLA/cellulose composites is still in progress and limited to laboratory trials. Still, there is a long way to develop PLA/cellulose composites, and hence some studies reporting the preparation and applications of PLA/cellulose composites are also discussed herewith.

REFERENCES

Abbasi-Moayed, S., Golmohammadi, H., Hormozi-Nezhad, M.R., 2018. A nanopaper-based artificial tongue: a ratiometric fluorescent sensor array on bacterial nanocellulose for chemical discrimination applications. *Nanoscale*. 10, 2492–2502. https://doi.org/10.1039/C7NR05801B

Battù, A., 2018. PLA | Sulzer [WWW Document]. Sulzer Technical Review. https://www. sulzer.com/en/shared/about-us/leading-technology-for-biobased-pla-plastics (accessed 6.8.21).

Calcagnile, P., Sibillano, T., Giannini, C., Sannino, A., Demitri, C., 2019. Biodegradable poly(lactic acid)/cellulose-based superabsorbent hydrogel composite material as water and fertilizer reservoir in agricultural applications. *J. Appl. Polym. Sci.* 136, 47546. https://doi.org/10.1002/app.47546

Chen, J., Huang, M., Kong, L., Lin, M., 2019. Jellylike flexible nanocellulose SERS substrate for rapid in-situ non-invasive pesticide detection in fruits/vegetables. *Carbohydr. Polym.* 205, 596–600. https://doi.org/10.1016/j.carbpol.2018.10.059

Cheng, H., Xiao, D., Tang, Y., Wang, B., Feng, X., Lu, M., Vancso, G.J., Sui, X., 2020. Sponges with Janus character from nanocellulose: Preparation and applications in the treatment of hemorrhagic wounds. *Adv. Healthc. Mater.* 9, 1901796. https://doi. org/10.1002/adhm.201901796

Cheryan, M., 1998. *Ultrafiltration and Microfiltration Handbook.* CRC Press. https://doi. org/10.1201/9781482278743

Clarkson, C.M., El Awad Azrak, S.M., Chowdhury, R., Shuvo, S.N., Snyder, J., Schueneman, G., Ortalan, V., Youngblood, J.P., 2019. Melt spinning of cellulose nanofibril/polylactic acid (CNF/PLA) composite fibers for high stiffness. *ACS Appl. Polym. Mater.* 1, 160–168. https://doi.org/10.1021/acsapm.8b00030

Claro, F.C., Jordão, C., de Viveiros, B.M., Isaka, L.J.E., Villanova Junior, J.A., Magalhães, W.L.E., 2020. Low cost membrane of wood nanocellulose obtained by mechanical defibrillation for potential applications as wound dressing. *Cellulose* 27, 10765–10779. https://doi.org/10.1007/s10570-020-03129-2

dos Santos, J., Oliveira, R.S., Oliveira, T.V., Velho, M.C., Konrad, M.V., da Silva, G.S., Deon, M., Beck, R.C.R., 2021. 3D printing and nanotechnology: A multiscale alliance in personalized medicine. *Adv. Funct. Mater.* 31, 2009691. https://doi.org/10.1002/ adfm.202009691

Fahma, F., Lisdayana, N., Abidin, Z., Noviana, D., Sari, Y.W., Mukti, R.R., Yunus, M., Kusumaatmaja, A., Kadja, G.T.M., 2020. Nanocellulose-based fibres derived from palm oil by-products and their *in vitro* biocompatibility analysis. *J. Text. Inst.* 111, 1354–1363. https://doi.org/10.1080/00405000.2019.1694353

Fonseca, D.F.S., Vilela, C., Pinto, R.J.B., Bastos, V., Oliveira, H., Catarino, J., Faísca, P., Rosado, C., Silvestre, A.J.D., Freire, C.S.R., 2021. Bacterial nanocellulose-hyaluronic acid microneedle patches for skin applications: In vitro and in vivo evaluation. *Mater. Sci. Eng. C* 118, 111350. https://doi.org/10.1016/j.msec.2020.111350

Fontenot, K.R., Edwards, J.V., Haldane, D., Pircher, N., Liebner, F., Condon, B.D., Qureshi, H., Yager, D., 2017. Designing cellulosic and nanocellulosic sensors for interface with a protease sequestrant wound-dressing prototype: Implications of material selection for dressing and protease sensor design. *J. Biomater. Appl.* 32, 622–637. https://doi. org/10.1177/0885328217735049

Gao, L., Chao, L., Hou, M., Liang, J., Chen, Y., Yu, H.-D., Huang, W., 2019a. Flexible, transparent nanocellulose paper-based perovskite solar cells. *NPJ Flex. Electron.* 3, 4. https://doi.org/10.1038/s41528-019-0048-2

Gao, L., Zhu, C., Li, L., Zhang, C., Liu, J., Yu, H.-D., Huang, W., 2019b. All paper-based flexible and wearable piezoresistive pressure sensor. *ACS Appl. Mater. Interfaces* 11, 25034–25042. https://doi.org/10.1021/acsami.9b07465

Gayer, C., Ritter, J., Bullemer, M., Grom, S., Jauer, L., Meiners, W., Pfister, A., Reinauer, F., Vučak, M., Wissenbach, K., Fischer, H., Poprawe, R., Schleifenbaum, J.H., 2019. Development of a solvent-free polylactide/calcium carbonate composite for selective laser sintering of bone tissue engineering scaffolds. *Mater. Sci. Eng. C* 101, 660–673. https://doi.org/10.1016/j.msec.2019.03.101

Ghasemi, S., Behrooz, R., Ghasemi, I., Yassar, R.S., Long, F., 2018. Development of nano-cellulose-reinforced PLA nanocomposite by using maleated PLA (PLA-g-MA). *J. Thermoplast. Compos. Mater.* 31, 1090–1101. https://doi.org/10.1177/0892705717734600

Gotro, J., 2012. Poly lactic acid (PLA) is gaining traction in the Market. Polymer Innovation Blog. https://polymerinnovationblog.com/poly-lactic-acid-pla-is-gaining-traction-in-the-market/ (accessed 6.1.21).

Gunathilake, T.M.S.U., Ching, Y.C., Chuah, C.H., Hai, N.D., Nai-Shang, L., 2020. Electro-stimulated release of poorly water-soluble drug from poly(lactic acid)/carboxymethyl cellulose/ZnO nanocomposite film. *Pharm. Res.* 37, 178. https://doi.org/10.1007/s11095-020-02910-z

Gupta, B., Revagade, N., Hilborn, J., 2007. Poly(lactic acid) fiber: An overview. *Prog. Polym. Sci.* 32, 455–482. https://doi.org/10.1016/j.progpolymsci.2007.01.005

Han, L., Cui, S., Yu, H.-Y., Song, M., Zhang, H., Grishkewich, N., Huang, C., Kim, D., Tam, K.M.C., 2019. Self-healable conductive nanocellulose nanocomposites for biocompat-ible electronic skin sensor systems. *ACS Appl. Mater. Interfaces* 11, 44642–44651. https://doi.org/10.1021/acsami.9b17030

Hsu, H.H., Khosrozadeh, A., Li, B., Luo, G., Xing, M., Zhong, W., 2019. An eco-friendly, nanocellulose/RGO/in situ formed polyaniline for flexible and free-standing supercapacitors. *ACS Sustain. Chem. Eng.* 7, 4766–4776. https://doi.org/10.1021/acssuschemeng.8b04947

Jiao, Y., Lu, K., Lu, Y., Yue, Y., Xu, X., Xiao, H., Li, J., Han, J., 2021. Highly viscoelastic, stretchable, conductive, and self-healing strain sensors based on cellulose nanofiber-reinforced poly-acrylic acid hydrogel. *Cellulose* 28, 4295–4311. https://doi.org/10.1007/s10570-021-03782-1

Kim, J., 2021. Nanocellulose-based paper actuators, in: *Nanocellulose Based Composites for Electronics.* Elsevier, pp. 163–183. https://doi.org/10.1016/B978-0-12-822350-5.00007-2

Kong, L., Xu, D., He, Z., Wang, F., Gui, S., Fan, J., Pan, X., Dai, X., Dong, X., Liu, B., Li, Y., 2019. Nanocellulose-reinforced polyurethane for waterborne wood coating. *Molecules* 24, 3151. https://doi.org/10.3390/molecules24173151

Koppolu, R., Lahti, J., Abitbol, T., Swerin, A., Kuusipalo, J., Toivakka, M., 2019. Continuous processing of nanocellulose and polylactic acid into multilayer barrier coatings. *ACS Appl. Mater. Interfaces* 11, 11920–11927. https://doi.org/10.1021/acsami.9b00922

Le Gars, M., Dhuiège, B., Delvart, A., Belgacem, M.N., Missoum, K., Bras, J., 2020. High-barrier and antioxidant poly(lactic acid)/nanocellulose multilayered materials for pack-aging. *ACS Omega* 5, 22816–22826. https://doi.org/10.1021/acsomega.0c01955

Liang, Y., Zhu, H., Wang, L., He, H., Wang, S., 2020. Biocompatible smart cellulose nano-fibres for sustained drug release via pH and temperature dual-responsive mechanism. *Carbohydr. Polym.* 249, 116876. https://doi.org/10.1016/j.carbpol.2020.116876

Listyarini, A., Imawan, C., Amalia, B., Fauzia, V., 2018. Chitosan/nanocellulose with natural dye as a new developed colorimetric film for ammonia detection. Presented at the Proceedings of The 3rd International Symposium on Current Progress in Mathematics and Sciences 2017 (ISCPMS2017), Bali, Indonesia, p. 020028. https://doi.org/10.1063/1.5064025

Liu, Z., Chen, M., Guo, Y., Zhou, J., Shi, Q., Sun, R., 2020. Oxidized nanocellulose facilitates preparing photoluminescent nitrogen-doped fluorescent carbon dots for Fe3+ ions detec-tion and bioimaging. *Chem. Eng. J.* 384, 123260. https://doi.org/10.1016/j.cej.2019.123260

Löwenberg, C., Balk, M., Wischke, C., Behl, M., Lendlein, A., 2017. Shape-memory hydro-gels: evolution of structural principles to enable shape switching of hydrophilic polymer networks. *Acc. Chem. Res.* 50, 723–732. https://doi.org/10.1021/acs.accounts.6b00584

Mao, J., Tang, Y., Zhao, R., Zhou, Y., Wang, Z., 2019. Preparation of nanofibrillated cellulose and application in reinforced PLA/starch nanocomposite film. *J. Polym. Environ.* 27, 728–738. https://doi.org/10.1007/s10924-019-01382-6

Mardis, N.J., 2018. Emerging technology and applications of 3D printing in the medical field. *Mo. Med.* 115, 368–373.

Mathew, E., Pitzanti, G., Larrañeta, E., Lamprou, D.A., 2020. 3D Printing of pharmaceuticals and drug delivery devices. *Pharmaceutics.* 12, 266. https://doi.org/10.3390/pharmaceutics12030266

Md Abu, T., Zahan, K.A., Rajaie, M.A., Leong, C.R., Ab Rashid, S., Mohd Nor Hamin, N.S., Tan, W.N., Tong, W.Y., 2020. Nanocellulose as drug delivery system for honey as antimicrobial wound dressing. *Mater. Today Proc.* 31, 14–17. https://doi.org/10.1016/j.matpr.2020.01.076

Miao, C., Hamad, W.Y., 2013. Cellulose reinforced polymer composites and nanocomposites: a critical review. *Cellulose* 20, 2221–2262. https://doi.org/10.1007/s10570-013-0007-3

Mishra, D., Shanker, K., Khare, P., 2020. Nanocellulose-mediated fabrication of sustainable future materials, in: *Sustainable Nanocellulose and Nanohydrogels from Natural Sources.* Elsevier, pp. 217–236. https://doi.org/10.1016/B978-0-12-816789-2.00010-9

Naghdi, T., Golmohammadi, H., Vosough, M., Atashi, M., Saeedi, I., Maghsoudi, M.T., 2019. Lab-on-nanopaper: An optical sensing bioplatform based on curcumin embedded in bacterial nanocellulose as an albumin assay kit. *Anal. Chim. Acta* 1070, 104–111. https://doi.org/10.1016/j.aca.2019.04.037

Nofar, M., Sacligil, D., Carreau, P.J., Kamal, M.R., Heuzey, M.-C., 2019. Poly (lactic acid) blends: Processing, properties and applications. *Int. J. Biol. Macromol.* 125, 307–360. https://doi.org/10.1016/j.ijbiomac.2018.12.002

Ozcan, A., Tozluoglu, A., Arman Kandirmaz, E., Tutus, A., Fidan, H., 2021. Printability of variative nanocellulose derived papers. *Cellulose* 28, 5019–5031. https://doi.org/10.1007/s10570-021-03861-3

Phanthong, P., Reubroycharoen, P., Hao, X., Xu, G., Abudula, A., Guan, G., 2018. Nanocellulose: Extraction and application. *Carbon Resour. Convers.* 1, 32–43. https://doi.org/10.1016/j.crcon.2018.05.004

Prapaporn, S., Arisara, S., Wunpen, C., Wijitar, D., 2020. Nanocellulose films to improve the performance of distance-based glucose detection in paper-based microfluidic devices. *Anal. Sci.* 36, 1447–1452. https://doi.org/10.2116/analsci.20P168

Raj, S., Sumod, U., Jose, S., Sabitha, M., 2012. Nanotechnology in cosmetics: Opportunities and challenges. *J. Pharm. Bioallied Sci.* 4, 186. https://doi.org/10.4103/0975-7406.99016

Rojas-Martínez, L.E., Flores-Hernandez, C.G., López-Marín, L.M., Martinez-Hernandez, A.L., Thorat, S.B., Reyes Vasquez, C.D., Del Rio-Castillo, A.E., Velasco-Santos, C., 2020. 3D printing of PLA composites scaffolds reinforced with keratin and chitosan: Effect of geometry and structure. *Eur. Polym. J.* 141, 110088. https://doi.org/10.1016/j.eurpolymj.2020.110088

Saba, N., Jawaid, M., Al-Othman, O., 2017. An overview on polylactic acid, its cellulosic composites and applications. *Curr. Org. Synth.* 14, 156–170. https://doi.org/10.2174/1570179413666160921115245

Sain, S., Åkesson, D., Skrifvars, M., Roy, S., 2020. Hydrophobic shape-memory biocomposites from tung-oil-based bioresin and onion-skin-derived nanocellulose networks. *Polymers* 12, 2470. https://doi.org/10.3390/polym12112470

Saini, P., Arora, M., Kumar, M.N.V.R., 2016. Poly(lactic acid) blends in biomedical applications. *Adv. Drug Deliv. Rev.* 107, 47–59. https://doi.org/10.1016/j.addr.2016.06.014

Sämfors, S., Karlsson, K., Sundberg, J., Markstedt, K., Gatenholm, P., 2019. Biofabrication of bacterial nanocellulose scaffolds with complex vascular structure. *Biofabrication* 11, 045010. https://doi.org/10.1088/1758-5090/ab2b4f

Sharma, P.R., Sharma, S.K., Lindström, T., Hsiao, B.S., 2020. Nanocellulose-enabled membranes for water purification: Perspectives. *Adv. Sustain. Syst.* 4, 1900114. https://doi.org/10.1002/adsu.201900114

Silva, R.R., Raymundo-Pereira, P.A., Campos, A.M., Wilson, D., Otoni, C.G., Barud, H.S., Costa, C.A.R., Domeneguetti, R.R., Balogh, D.T., Ribeiro, S.J.L., Oliveira Jr., O.N., 2020. Microbial nanocellulose adherent to human skin used in electrochemical sensors to detect metal ions and biomarkers in sweat. *Talanta* 218, 121153. https://doi.org/10.1016/j.talanta.2020.121153

Tappa, K., Jammalamadaka, U., Weisman, J., Ballard, D., Wolford, D., Pascual-Garrido, C., Wolford, L., Woodard, P., Mills, D., 2019. 3D printing custom bioactive and absorbable surgical screws, pins, and bone plates for localized drug delivery. *J. Funct. Biomater.* 10, 17. https://doi.org/10.3390/jfb10020017

Uddin, K.M.A., Jokinen, V., Jahangiri, F., Franssila, S., Rojas, O.J., Tuukkanen, S., 2019. Disposable microfluidic sensor based on nanocellulose for glucose detection. *Glob. Chall.* 3, 1800079. https://doi.org/10.1002/gch2.201800079

Uitto, J., 1997. Understanding premature skin aging. *N. Engl. J. Med.* 337, 1463–1465. https://doi.org/10.1056/NEJM199711133372011

van Reis, R., Zydney, A., 2001. Membrane separations in biotechnology. *Curr. Opin. Biotechnol.* 12, 208–211. https://doi.org/10.1016/S0958-1669(00)00201-9

Vaňková, E., Kašparová, P., Khun, J., Machková, A., Julák, J., Sláma, M., Hodek, J., Ulrychová, L., Weber, J., Obrová, K., Kosulin, K., Lion, T., Scholtz, V., 2020. Polylactic acid as a suitable material for 3D printing of protective masks in times of COVID-19 pandemic. *PeerJ* 8, e10259. https://doi.org/10.7717/peerj.10259

Vatansever, E., Arslan, D., Nofar, M., 2019. Polylactide cellulose-based nanocomposites. *Int. J. Biol. Macromol.* 137, 912–938. https://doi.org/10.1016/j.ijbiomac.2019.06.205

Vicente, C., Fernandes, J., Deus, A., Vaz, M., Leite, M., Reis, L., 2019. Effect of protective coatings on the water absorption and mechanical properties of 3D printed PLA. *Frat. Ed Integrità Strutt.* 13, 748–756. https://doi.org/10.3221/IGF-ESIS.48.68

Wang, Lu, Wang, C., Wang, Ling, Zhang, Q., Wang, Y., Xia, X., 2021. Emulsion electrospun polylactic acid/*Apocynum venetum* nanocellulose nanofiber membranes with controlled sea buckthorn extract release as a drug delivery system. *Text. Res. J.* 91, 1046–1055. https://doi.org/10.1177/0040517520970171

Wang, Z., Zhang, W., Yu, J., Zhang, L., Liu, L., Zhou, X., Huang, C., Fan, Y., 2019. Preparation of nanocellulose/filter paper (NC/FP) composite membranes for high-performance filtration. *Cellulose* 26, 1183–1194. https://doi.org/10.1007/s10570-018-2121-8

Wu, J., Feng, Y., Zhang, L., Wu, W., 2020. Nanocellulose-based surface-enhanced Raman spectroscopy sensor for highly sensitive detection of TNT. *Carbohydr. Polym.* 248, 116766. https://doi.org/10.1016/j.carbpol.2020.116766

Wu, B., Zhu, G., Dufresne, A., Lin, N., 2019. Fluorescent aerogels based on chemical cross-linking between nanocellulose and carbon dots for optical sensor. *ACS Appl. Mater. Interfaces* 11, 16048–16058. https://doi.org/10.1021/acsami.9b02754

Xiong, R., Yu, S., Kang, S., Adstedt, K.M., Nepal, D., Bunning, T.J., Tsukruk, V.V., 2020. Integration of optical surface structures with chiral nanocellulose for enhanced chiroptical properties. *Adv. Mater.* 32, 1905600. https://doi.org/10.1002/adma.201905600

Xu, Z., 2020. Recently progress on polylactide/nanocellulose nanocomposites. *IOP Conf. Ser. Mater. Sci. Eng.* 772, 012006. https://doi.org/10.1088/1757-899X/772/1/012006

Xu, W., Molino, B.Z., Cheng, F., Molino, P.J., Yue, Z., Su, D., Wang, X., Willför, S., Xu, C., Wallace, G.G., 2019. On low-concentration inks formulated by nanocellulose assisted with Gelatin Methacrylate (GelMA) for 3D printing toward wound healing application. *ACS Appl. Mater. Interfaces* 11, 8838–8848. https://doi.org/10.1021/acsami.8b21268

Yang, J., Li, H., Cheng, J., He, T., Li, J., Wang, B., 2021. Nanocellulose intercalation to boost the performance of MXene pressure sensor for human interactive monitoring. *J. Mater. Sci.* 56, 13859–13873. https://doi.org/10.1007/s10853-021-05909-y

Yuen, J.D., Shriver-Lake, L.C., Walper, S.A., Zabetakis, D., Breger, J.C., Stenger, D.A., 2020. Microbial nanocellulose printed circuit boards for medical sensing. *Sensors* 20, 2047. https://doi.org/10.3390/s20072047

Zhang, J., Jiang, G., Goledzinowski, M., Comeau, F.J.E., Li, K., Cumberland, T., Lenos, J., Xu, P., Li, M., Yu, A., Chen, Z., 2017. Green solid electrolyte with cofunctionalized nanocellulose/Graphene Oxide interpenetrating network for electrochemical gas sensors. *Small Methods* 1, 1700237. https://doi.org/10.1002/smtd.201700237

Zhang, J., Luo, N., Zhang, X., Xu, L., Wu, J., Yu, J., He, J., Zhang, J., 2016. All-cellulose nanocomposites reinforced with *in Situ* retained cellulose nanocrystals during selective dissolution of cellulose in an ionic liquid. *ACS Sustain. Chem. Eng.* 4, 4417–4423. https://doi.org/10.1021/acssuschemeng.6b01034

Zheng, C., Lu, K., Lu, Y., Zhu, S., Yue, Y., Xu, X., Mei, C., Xiao, H., Wu, Q., Han, J., 2020. A stretchable, self-healing conductive hydrogels based on nanocellulose supported graphene towards wearable monitoring of human motion. *Carbohydr. Polym.* 250, 116905. https://doi.org/10.1016/j.carbpol.2020.116905

Zheng, C., Yue, Y., Gan, L., Xu, X., Mei, C., Han, J., 2019. Highly stretchable and self-healing strain sensors based on nanocellulose-supported Graphene dispersed in electro-conductive hydrogels. *Nanomaterials* 9, 937. https://doi.org/10.3390/nano9070937

Zhou, J., Hsieh, Y.-L., 2018. Conductive polymer protonated nanocellulose aerogels for tunable and linearly responsive strain sensors. *ACS Appl. Mater. Interfaces* 10, 27902–27910. https://doi.org/10.1021/acsami.8b10239

Index

Note: Page numbers in *italics* represent figures and **bold** indicate tables in the text.

For Product Safety Concerns and Information please contact our EU
representative GPSR@taylorandfrancis.com
Taylor & Francis Verlag GmbH, Kaufingerstraße 24, 80331 München, Germany